Introduction to Non-equilibrium Physical Chemistry

Towards Complexity and Non-linear Science

Introduction to Non-equilibrium Physical Chemistry

Towards Complexity and Non-linear Science

R.P. Rastogi

Gorakhpur University, Gorakhpur, India

ELSEVIER

Amsterdam • Boston • Heidelberg • London • New York • Oxford
Paris • San Diego • San Francisco • Singapore • Sydney • Tokyo

Elsevier
Radarweg 29, PO Box 211, 1000 AE Amsterdam, The Netherlands
Linacre House, Jordan Hill, Oxford OX2 8DP, UK

First edition 2008

Library of Congress Cataloging-in-Publication Data
A catalog record for this book is available from the Library of Congress

British Library Cataloguing in Publication Data
A catalogue record for this book is available from the British Library

ISBN: 978-0-444-52188-0

For information on all Elsevier publications
visit our website at books.elsevier.com

Printed and bound by CPI Group (UK) Ltd, Croydon, CR0 4YY

Transferred to Digital Print 2011

Dedicated to

My Life-long Companion

Mrs Kamla Rastogi

TABLE OF CONTENTS

PREFACE

The traditional Physical Chemistry has largely been concerned with equilibrium phenomena and ideal situations. In the early part of the twentieth century, interest developed in investigating non-equilibrium phenomena and real situations involving social dynamics and living state. The obvious tools had been non-equilibrium thermodynamics, statistical mechanics, molecular dynamics and computer simulation. Significant advances were made in understanding non-equilibrium processes and discovering new laws and paradigms for which Onsager was awarded Nobel Prize in 1968 and Prigogine in 1977. Their researches gave further impetus to the study of more and more complex phenomena. Complex systems are multi-variable systems involving several processes, which cannot be understood by using the traditional reductionist approach by dissecting the phenomenon into parts.

Non-equilibrium phenomena have wide applications in real systems. Living state is such a typical non-equilibrium state. These have relevance in Physiology, Geology, Physical Science and Biological Science, Economics and Social Dynamics.

One can have the following type of situations as we move away from equilibrium.

Equilibrium state ⟶ Linear steady state close to equilibrium ⟶ Steady state ⟶ Non-linear steady state ⟶ Bifurcation phenomena ⟶ Multi-stability ⟶ Temporal and spatio-temporal oscillations ⟶ More complex situations (chaos, turbulence, pattern formation, fractal growth). All these stages have been discussed in different chapters of the book.

During the latter half of the last century and to date, important developments have taken place and new paradigms have been developed in the field of non-equilibrium phenomena, giving rise to a new discipline of non-equilibrium Physical Chemistry. Its importance is further enhanced by the fact that physico-chemical experiments provide instructive and useful models for Biological and Social Sciences, etc. It may be noted that several monographs have appeared in the last few decades dealing with these developments. Lots of experimental studies in these areas have also been reported during this period which provide a clearer picture. However, so far conventional text books have laid emphasis on equilibrium phenomena. Undoubtedly there is detailed reference to chemical kinetics and ion migration, but the above important developments involving entirely new concepts in the non-equilibrium region have escaped attention. The purpose of this book is to fill this gap so that it can be used as supplementary material for teaching as well as for further research. The idea is to present the material in a sequential, coherent and comprehensive manner with greater emphasis on concepts so that it may be useful from pedagogical angle. Additional purpose has been to present an elementary account to provide an insight into non-linear science and complexity for scientists in other disciplines such as Economics, Physiology and Biological Sciences.

I am extremely grateful to Professors A.C. Chatterji, B.N. Srivastava, M.N. Saha, K.G. Denbigh, Karl Popper and Nobel laureate Prof. Ilya Prigogine for stimulating my interest in the study of non-equilibrium phenomena from various angles.

I am indebted to Lucknow University, Punjab University, Gorakhpur University and Banaras Hindu University and Central Drug Research Institute where most of the experimental work in different areas could be accomplished. The financial support of funding agencies like University Grants Commission, Council of Scientific and Industrial Research and Department of Science and Technology (Government of India) and Indian National Science Academy for carrying out various projects related to the theme of the present text are acknowledged. The secretarial assistance provided by Prof. K.D.S. Yadav, Head, Chemistry Department, Gorakhpur University, is also gratefully acknowledged.

ACKNOWLEDGEMENTS

I pay my tribute to Prof. B.N. Srivastava, who was primarily responsible for the development of my interest in Irreversible Thermodynamics. I am equally grateful to my former Ph.D. students and other collaborators who had been involved in theoretical and experimental studies in the areas covered in the book, whose references appear in the book.

I am particularly thankful to Profs. R.C. Srivastava (Chapters 1–7, 11), Kehar Singh (3, 4), Ishwar Das (9, 10, 13), Kalanand Prasad (10) and Dr Pankaj Mathur (8, 12, 14) for collaborating in writing a couple of chapters.

I am also happy to acknowledge the involvement of Dr Ashtabhuja Prasad Mishra, Sharwan Kumar, Prof. A.K. Jain and Dr Mukul Das for their involvement in the write-up for specific chapters

I gratefully acknowledge the assistance rendered by Profs. A.K. Dutt and A.A. Bhalekar related to Appendices II and III.

It is a pleasure to acknowledge the stimulating discussions which I had with Prof. Raghuveer Singh, Prof. Hem Chandra Joshi, S.C. Mishra, Dr Gopishyam and Dr Ghanshyam Das in connection with socio-political and financial dynamics.

Further, the help rendered by Prof. N.B. Singh, Dr Vishnu Ji Ram, Dr S.S. Das, Dr Pankaj Mathur and Mr. Ramendra Pratap in connection with the preparation of the manuscript is gratefully acknowledged.

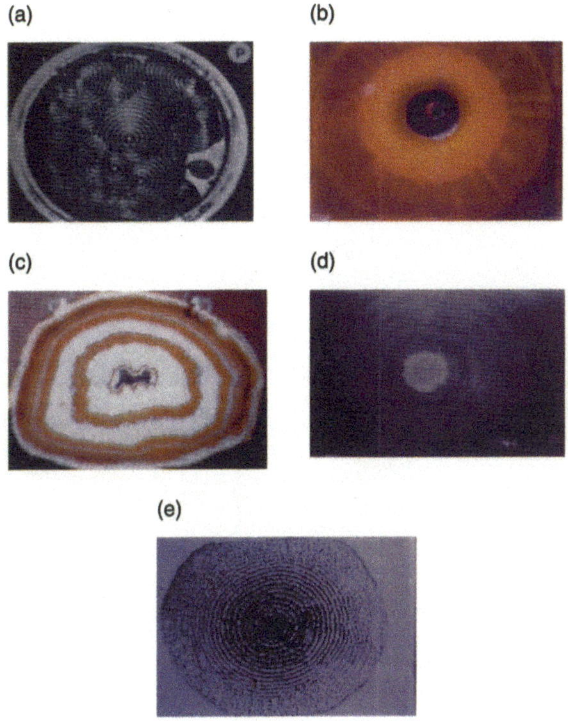

Chapter 10, Figure 10.3, p. 177.

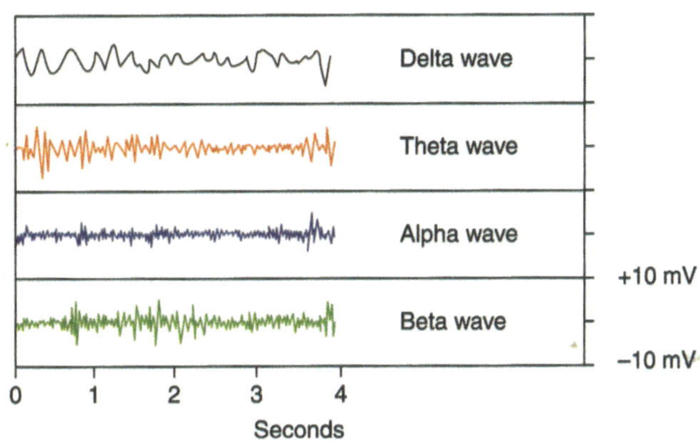

Chapter 12, Figure 12.2, p. 218.

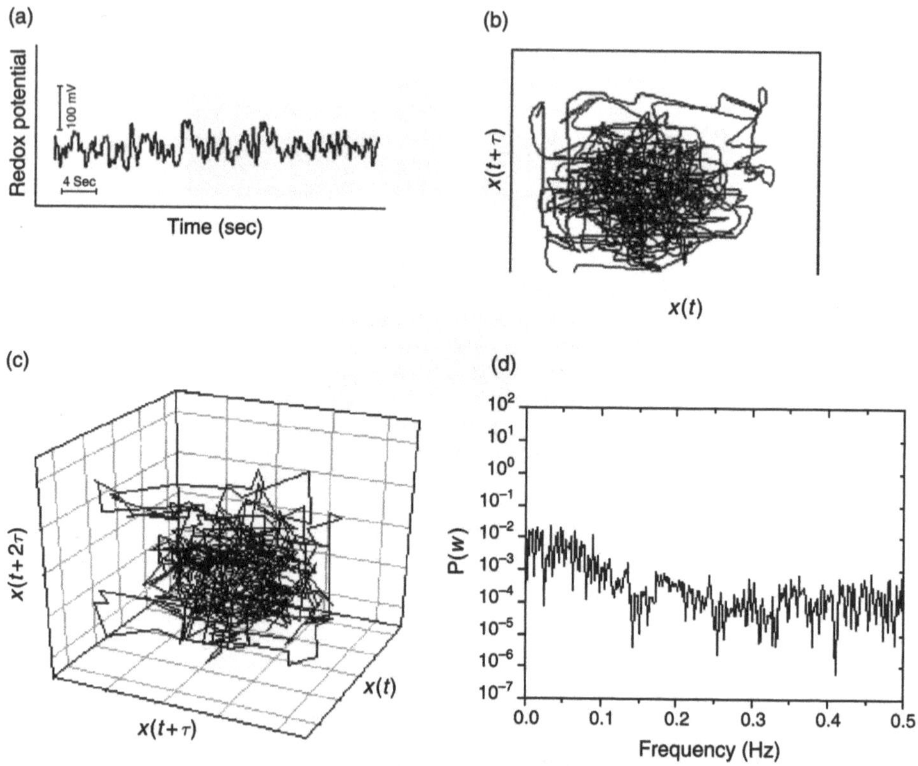

Chapter 12, Figure 12.9, p. 225.

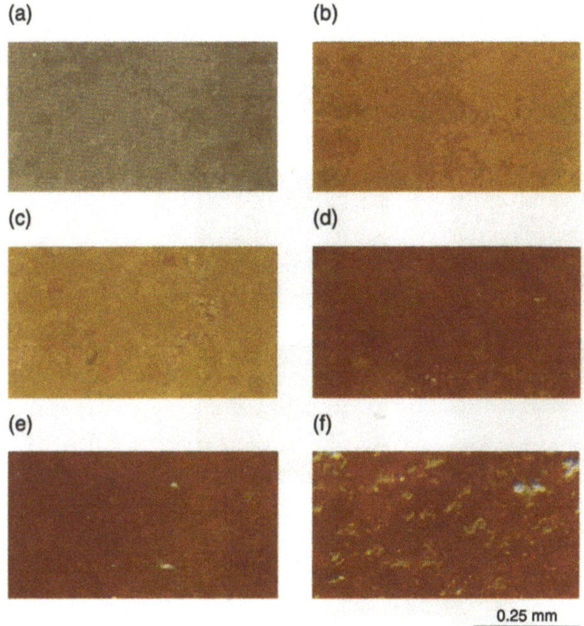

Chapter 13, Figure 13.17, p. 254.

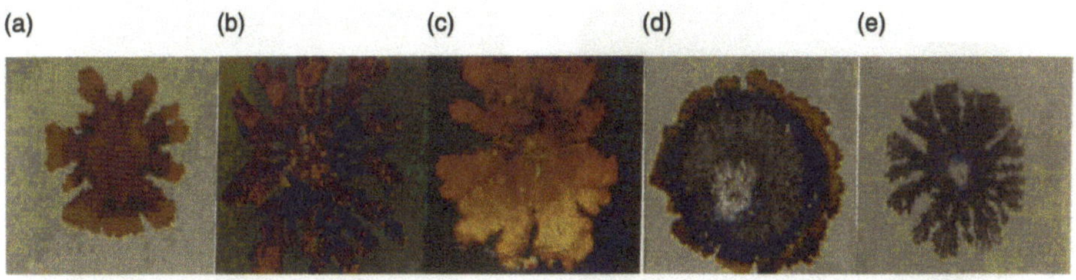

Chapter 13, Figure 13.21, p. 257.

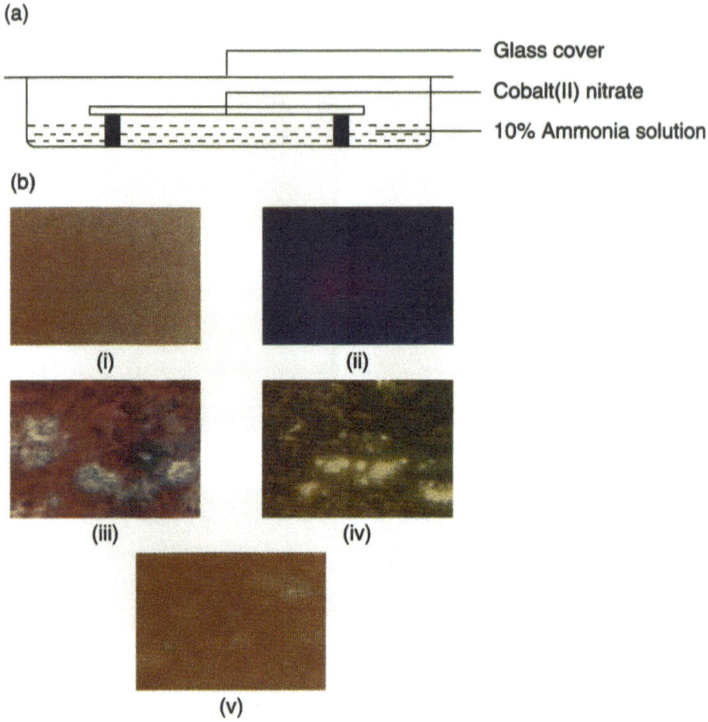

Chapter 13, Figure 13.18, p. 255.

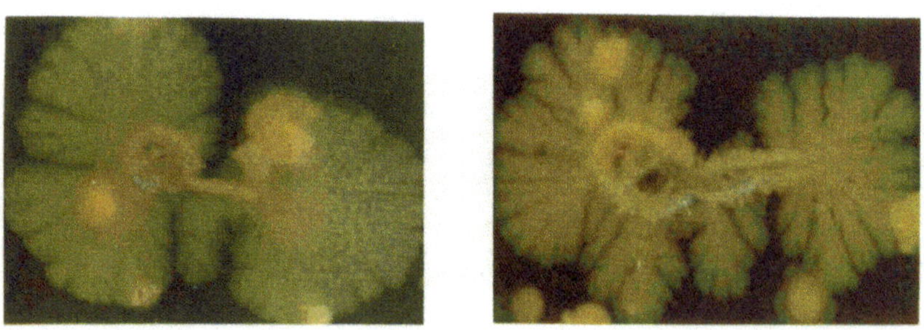

Chapter 13, Figure 13.28, p. 266.

Chapter 1

INTRODUCTION

1.1. Real systems

Traditional Physical Chemistry has largely been concerned with state of matter in equilibrium. Non-equilibrium aspects come up in the case of kinetic theory, chemical kinetics and ion transport. Real systems like living state and socio-economic systems are always in non-equilibrium and invoke a number of complicated phenomena. Science is now getting oriented to the study of such complex systems. From the philosophical angle the question whether science deals with real world [1a] has been raised. The concept of evolution in social systems in contrast to biology is moving towards an interdisciplinary theory of change of state [1b]. This is provoking interest in experimental and theoretical studies of analogous non-equilibrium phenomena in physico-chemical systems which can serve as a model for economists and biologists. Joint effort in understanding complex phenomena with the help of various disciplines is leading to the growth of Synergetics meaning, a new discipline.

1.2. Equilibrium and non-equilibrium states

The important feature of the equilibrium state is that variables such as temperature T, pressure P, chemical potential μ and electric potential ϕ everywhere are the same in the system. There can be two types of equilibrium states, viz. (a) dynamic equilibrium and (b) static equilibrium. Vapour–liquid equilibrium and chemical equilibrium are typical examples of dynamic equilibrium, where in the first case, the rate of condensation and rate of vaporization are equal while in the second case, the rates of forward and backward reactions are equal. Simple crystals belong to the class of static equilibrium.

In non-equilibrium state, the thermodynamic variables are not the same everywhere in the system, on account of which gradients of variables (e.g. grad P, grad T, etc.) develop which act as the cause (force) for generating effects (flows) such as volume flow or heat flow (fluxes). Non-equilibrium thermodynamics has the advantage of being used for identifying cause and effect, i.e. forces and fluxes, and also coupling between the fluxes without a detailed knowledge of the systems. However, for real systems problem arises in identifying variables, fluxes and forces, involved in processes and cross-phenomena. Further in many systems, additional difficulty arises on account of lack of knowledge about the nature and magnitude of the coupling coefficients between fluxes and forces [1].

The systems can be of three types:

Isolated systems in which there is no exchange of matter or energy with the surroundings;
closed systems in which there is no exchange of matter with the surroundings but exchange of energy can occur; and
open systems, which exchange both matter and energy with the surroundings.

1.3. Open systems

Open system is always in non-equilibrium. A closed system can be in non-equilibrium depending on the circumstances. It may have subsystems between which exchange of matter and energy can take place or in the system itself, thermodynamic variables may not be constant in space. A typical example of the former type is thermo-osmosis, which is discussed in Chapter 3, where the two subsystems are separated by a membrane. Example of the latter type is thermal diffusion, which has been discussed in Chapter 5. When the flows and counter-flows in opposite directions are generated by corresponding gradients, steady state is obtained. Both equilibrium and non-equilibrium steady states are time-invariant states, but in the latter case both flows and gradients are present.

Real systems are open systems and may consist of numerous subsystems; global system, human society and human body are typical examples. The nature of subsystems, variables, fluxes and forces, their coupling leading to cross-phenomena, temporal and spatio-temporal changes, pattern formation and self-organization would be discussed in the subsequent chapters.

1.4. Approach to equilibrium

For taking a comprehensive view, it is also desirable to keep in mind the process of approach to equilibrium. Chemical kinetics and kinetic theory of gases have been the traditional tools. Simple reactions have been studied by Monte-Carlo technique or stochastic approach by monitoring random picks of molecules represented by digits on the computer and employing a criterion that accepts or discards potential conversions. The methodology adopted for the study of simple set of simultaneous reactions has been received by Gupta, which involves comparison of experimental results with postulated mechanism [2]. For complex reactions, numeric integration techniques are employed to abstract concentration profiles. The methodology involved is essentially linear kinetics. Chemical kinetics is now moving towards the study of more and more complex reaction network which may simultaneously involve (i) electron transfer reaction, (ii) free-radical reaction, (iii) organic reaction, (iv) inorganic reaction and (v) reaction between organic and inorganic species. For such type of systems, a new methodology called non-linear kinetics involving non-linear differential equations is emerging.

There is considerable error in thermodynamic prediction if true equilibrium is not maintained, a condition never maintained in industry due to time factor. Rastogi and Denbigh [3] investigated this aspect theoretically. As an illustration, they examined the reaction

$$2HI \rightleftharpoons H_2 + I_2$$

for which the equilibrium constant K is of the order of 2.18×10^{-2} at 763.8 K. The energy of activation of the forward reaction and ΔH, the enthalpy change of the reaction, has values of 44 000 cal mol^{-1} and 3000 cal mol^{-1}, respectively. Let r denote the ratio of the true temperature coefficient of the yield to the temperature coefficient of the yield, which would be predicted on the assumption that the system is at equilibrium. For f = 0.9, the true temperature coefficient of the yield is 10.6 times the value predicted thermodynamically on the supposition of equilibrium. Further when $f = 0.99$, the corresponding factor is as much as 2.8.

In a similar manner, cooling rate at the rocket nozzle throat used to be computed by assuming isentropic flow [4]. However, it has been shown that the cooling rate at the throat is likely to increase when departure from equilibrium becomes significant [5].

1.5. Non-equilibrium states

There was tremendous interest in mid-twentieth century in exploring general principles for understanding non-equilibrium phenomena along with the development of non-equilibrium thermodynamics and non-equilibrium statistical mechanics. Pioneering work of Professor Prigogine [6] and his school in Brussels stimulated a good deal of interest in the field of non-equilibrium statistical mechanics. Formal solutions of Liouville equation [7] in terms of a Greenian and complete internal propagator leads to a theoretical expression for electrical conductivity tensor, which easily leads to classical formula for electrical conductivity of metals based on the free-electron model [8].

Kinetic theory, non-equilibrium statistical mechanics and non-equilibrium molecular dynamics (NEMD) have proved to be useful in estimating both straight and cross-coefficients such as thermal conductivity, viscosity and electrical conductivity. In a typical case, cross-coefficient in case of electro-osmosis has also been estimated by NEMD. Experimental data on thermo-electric power has been analysed in terms of free electron gas theory and non-equilibrium thermodynamic theory [9]. It is found that phenomenological coefficients are temperature dependent. Free electron gas theory has been used for estimating the coefficients in homogeneous conductors and thermo-couples.

Onsager relations are satisfied, showing that free electron gas theory is consistent with thermodynamic theory. The free electron theory correctly predicts the temperature dependence of thermo-electric power. Similarly, the interpretation of the phenomenon of thermo-osmosis of gases on the basis of non-equilibrium thermodynamics and kinetic theory of gases is mutually consistent.

1.6. Complex non-equilibrium phenomena

The departure from equilibrium occurs primarily on account of appearance of gradients such as temperature and concentration leading to flow of heat or of some species and subsequently leading to a specific non-equilibrium state. Earlier in the first instance, uncoupled flows, e.g. heat conduction, Poisseuille flow and electrical conduction, were the subject of investigation. Discussion of such processes has been given due attention in conventional Physical Chemistry texts. However, complex and exotic phenomena in the non-equilibrium thermodynamics provide a good tool for understanding such phenomena.

Far from equilibrium, one comes across exotic phenomena as pointed out earlier. The study of these involved novel theoretical approaches and novel experimental studies on a variety of phenomena to check the validity of phenomenological relations, Onsager reciprocity relations and thermodynamic predictions regarding steady state along with analysis of data in the non-equilibrium region on the basis of non-equilibrium thermodynamics. These developments led to the study of non-linear steady states, non-linear flux equations along with the attempts to understand the origin of non-linear terms. Further away, bifurcation phenomena are encountered involving multi-stability and oscillatory behaviour, followed by spatio-temporal oscillations, chaos and noise. All these phenomena attracted good deal of attention from theoretical and experimental angle in view of practical interest in Physiology and other disciplines.

1.7. Scope

The great importance of thermodynamics and hydrodynamic methods lies in the fact that these provide us with a reduced description in simplified language to describe macroscopic systems as stated by Glansdroff and Prigogine. The present contribution is intended to present a coherent account of developments, both theoretical and experimental, in the advancing field of knowledge related to complex phenomena from equilibrium to far from equilibrium region. From this angle, it is reasonable to expect that the concepts and thought methodology would be useful for taking a synergetic view of real systems in nature and social surroundings. Recent developments in non-equilibrium Physical Chemistry have also been examined and discussed in this context.

The book is divided into four parts. Part One, which consists of six chapters, deals with basic principles and concepts of non-equilibrium thermodynamics along with discussion of experimental studies related to test and limitation of formalism. Chapter 2 deals with theoretical foundations involving theoretical estimation of entropy production for open system, identification of fluxes and forces and development of steady-state relations using Onsager reciprocity relation. Steady state in the linear range is characterized by minimum entropy production. Under these circumstances, fluctuations regress exactly as in thermodynamics equilibrium.

Chapter 3 deals with theoretical and experimental studies of thermo-osmosis of liquids and gases along with thermo-osmotic concentration differences. Correlation with kinetic theory has also been attempted. Chapter 4 is concerned with experimental and theoretical studies of electro-kinetic phenomena, e.g. electro-osmosis and streaming in a system containing two subsystems separated by a membrane. Relationship with Helmholtz double layer theory has been examined with a view to provide physical interpretation of phenomenological coefficients. Theory and experiment have been compared to assess the range of validity of thermodynamic theory. Chapters 3 and 4 are concerned with discontinuous systems involving a membrane as barrier.

Chapters 5 and 6 deal with systems where interaction between temperature gradient, concentration gradient and potential gradients without any barrier are involved. In these chapters, theoretical and experimental studies relating to thermal diffusion, Dufour effect, Soret effect, thermal diffusion potential, thermo-cells, precipitation and dissolution potential have been described. Physical implications of the experimental results have also been described.

Appendices follow Chapter 6. In Chapter 2, it has been pointed that local entropy may be expressed in terms of same independent variables as if the system were at equilibrium (local equilibrium). The limitations of Gibbs equation have been discussed in Appendix I. At no moment, molecular distribution function of velocities or of relative positions may deviate strongly from their equilibrium form. This is a sufficient condition for the application of thermodynamics method. Some new developments related to alternative theoretical formalism such as *extended irreversible* are discussed in Appendices II and III.

In Part One, steady state corresponding to the situation when linear phenomenological relations hold have been discussed. As we move further away from equilibrium, we come across non-linear stable steady states, wherein non-linear flux equation hold. Experimental studies for volume flux and streaming current have been reported. Theoretical interpretation of second-order coefficients in terms of double layer theory has also been discussed. These aspects have been discussed in Chapter 7 (Part Two).

Equilibrium structures are formed and maintained through reversible transformation implying no appreciable deviation from equilibrium. Non-equilibrium dissipative structures, on the other hand, have quite different character. They are formed and maintained through the effect of exchange of energy and matter in non-equilibrium condition.

Bifurcation from steady states to different types of dissipative structures takes place depending on the bifurcation parameter. The transition takes place at the specific bifurcation point. Phenomenon of bistability has been discussed in Chapter 8 which deals with experimental and mechanistic studies for different types of situation.

Moving still further away from equilibrium, one gets different types of dissipative structures involving time order and space order. Chapter 9 is concerned with time order involving chemical oscillations. Around 1968, the credibility of Belousov–Zhabotinskii reaction involving the reactants BrO^{3-}, Ce^{4+}, malonic acid and H_2SO_4 was well established after the discovery of Brusselator model [10]. This triggered detailed investigations on similar oscillatory reactions in order to have a reliable picture of mechanism [11]. This initiated use of non-linear kinetics and computer modelling with the result that insight into the mechanism could be obtained. The mechanism involved the concept of positive and negative feedback through autocatalysis and inhibitory reactions. The field of oscillatory reactions has been covered in Chapter 9.

When phenanthroline is added to the B–Z system, in a test tube, travelling red and blue bands are observed displaying spatio-temporal oscillations. This feature denotes a specific type of time order and space order. In the system, export of entropy through diffusion occurs and a specific type of dissipative structure is obtained. Both travelling waves and stationary structures are formed depending on the experimental conditions. Experimental studies, mechanism and mathematical modelling of such structures have been discussed in Chapter 10. Turning patterns, mosaic structure, precipitation patterns (Liesegang rings) and their mechanism have also been discussed in detail. Essentially non-linear kinetics involving reaction–diffusion is the basis of mathematical modelling.

Part Three deals with complex non-equilibrium phenomena, which occur very far from equilibrium (Chapters 11–13).

Chapter 11 is concerned with oscillatory phenomena at (i) solid–liquid interface, (ii) liquid–liquid interface, (iii) solid–liquid and liquid–liquid interface together, (iv) liquid–vapour interface and solid–vapour interface. Both experimental and theoretical studies along with applications have been discussed in this chapter.

In Chapter 12, special attention has been given to non-periodic oscillations of various types, including deterministic chaos and random motion (noise). Mathematical formalism, characterization and control of chaos have also been discussed.

Complex pattern formation and fractal growth are the theme of Chapter 13. Experimental studies relating to fractal growth during crystallization, electro-deposition, bacterial growth and polymerization along with the interpretation in the context of growth models based on fractal geometry have been discussed in this chapter.

In Part Four, attempt has been made to point out in what way the concepts developed in previous chapters could be utilized to analyse and get an insight into the behaviour of real systems, including socio-economic, socio-political and biological systems. This is the need of the hour, which could promote Synergetics (constructed from Greek words indicating "joint efforts"). Synergetics denotes a new discipline, which aims at multi-disciplinary and interdisciplinary approach involving comparison and cooperative effort in different fields. It may be noted that experimental studies of particular physico-chemical systems from regions close to equilibrium to regions very far from equilibrium can provide appropriate models and mechanisms for the analysis of dynamics of real systems. Chapter 14 has been divided into three subsections dealing with Social Sciences including Sociology, Psychology and Economics. Membrane transport and biochemical oscillators have been discussed in Chapter 15 dealing with living state.

References

1. (a) R. Tigg, Interdiscipl. Sci. Rev., 27 (2002) 94. (b) L. Gabora and D. Aerts, Interdiscipl. Sci. Rev., 30 (2005) 69.
2. Y.K. Gupta, J. Indian Chem. Soc., 80 (2003) 218.
3. R.P. Rastogi and K.G. Denbigh, Chem. Engg. Sci., 7 (1958) 261.
4. S.S. Penner, Chemical Reactions in Flow Systems, Butterworths Scientific Publications, London, 1955, Ch. 3.
5. R.P. Rastogi and T.P. Pandya, J. Chem. Phys., 25 (1956) 1009.
6. I. Prigogine, Non-equilibrium Statistical Mechanics, Interscience, New York, 1975.
7. R. Baleseu, Equilibrium and Non-equilibrium Statistical Mechanics, Wiley, New York, 1975.
8. R.P. Rastogi and M.C. Gupta, J. Chem. Educ., 59 (1982) 822.
9. R.P. Rastogi and B.P. Mishra, Physica, 36 (1967) 315.
10. G. Nicolis and I. Prigogine, Self-Organization in Non-equilibrium Systems, Wiley, New York, 1977.
11. R.J. Field and M. Berger (eds), Oscillations and Chemical Waves in Chemical Systems, Wiley, New York, 1985.

Part One

NON-EQUILIBRIUM STEADY STATES CLOSE TO EQUILIBRIUM

Chapter 2

BASIC PRINCIPLES OF NON-EQUILIBRIUM THERMODYNAMICS

2.1. Introduction

Systems in equilibrium are simple so far as analysis is concerned. The moment we move away from equilibrium or far away from equilibrium, we immediately enter the region of complexity. For analysis of such situations, new procedures have to be involved. Non-equilibrium thermodynamics is amongst such powerful tools for which a number of monographs are available [1–11]. We shall give a brief and concise account of the subject in this chapter with special reference to steady states close to equilibrium.

Isolated systems, which do not exchange both matter and energy with the surroundings, ultimately attain equilibrium. In case of closed system, like a conductor in contact with the surroundings at a lower temperature, there would be transfer of heat from conductor to the surroundings and ultimately equilibrium would be attained. Open systems may consist of various sub-systems, through which exchange of matter and energy can take place between one another. We may call such systems as discontinuous systems, where the intensive thermodynamics variable such as temperature T, pressure P, chemical potential μ and electrical potential ϕ may differ in different systems, but may be the same in each system. On the other hand, we can have continuous systems, where such variables can vary from point to point.

Living systems and social systems are quite complex open systems. They consist of many sub-systems between which interaction can take place. In the world of today, we have continents and nation states between which continuous interaction takes place. Even such nations can have various sub-units as we have in the United States, which can act for separate sub-systems for a limited purpose. The interaction between such sub-systems may be quite complex depending on the geographical location, ethnic and political considerations.

Non-equilibrium thermodynamics (NET) as it developed turned out to be quite useful in basic understanding of non-equilibrium steady states close to equilibrium and also in the region far from equilibrium. The theoretical and experimental studies can serve as a model for studies of similar type of real systems including biological, social and economic systems.

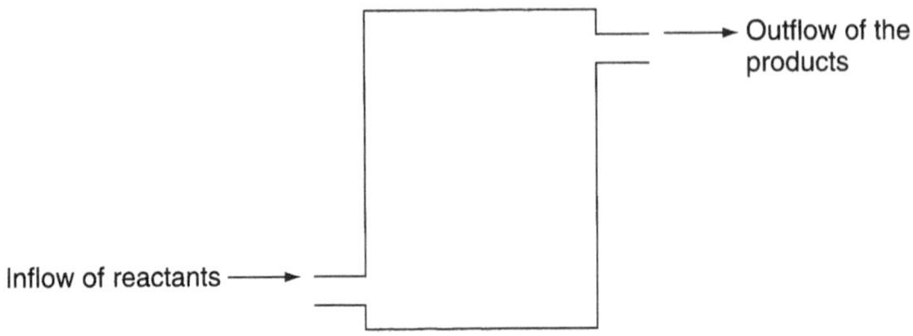

Figure 2.1. Continuous stirred-tank reactor.

In this chapter, the main objective is to present the basic principles of NET in a concise form. Applications to specific systems and phenomena along with correlation of theory and experiment have been discussed in subsequent chapters.

Open systems are typical non-equilibrium systems, which display a variety of exotic phenomena. Typical examples are Continuous Stirred Tank Reactor (CSTR) (Fig. 2.1) and the living state.

Living organisms are never at equilibrium with the surroundings. Molecular composition reflects a dynamic steady state. This involves exchange of matter and energy with the surroundings through the following types of metabolism as an example,

(a) Ingestion → Glucose (in blood) → Utilization for energy
(b) Carbohydrates utilization → (i) waste CO_2, (ii) storage of fats and (iii) other products

Similar examples can be encountered in Social Dynamics, Economics (Finance), Geophysics and Biological Sciences.

There can be two types of open systems.

Discontinuous systems (*membrane phenomena*), where two or more homogeneous systems are separated by a boundary. In such a system, the intrinsic properties are discontinuous at the boundary of homogenous systems. Typical example of a discontinuous system is the one, which has two sub-systems I and II containing a single gas separated by a membrane displaying the non-equilibrium phenomenon of thermo-osmosis (Fig. 2.2). In this case, $P_2 > P_1$ and $T_2 > T_1$. Further volume flux (thermo-osmotic flow)

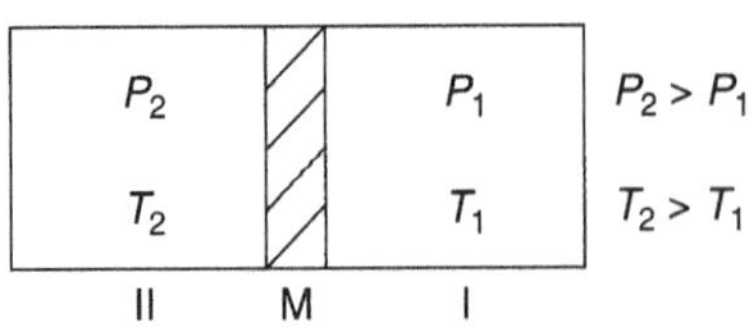

Figure 2.2. Two sub-systems separated by a membrane; M, membrane; I, chamber I; II, chamber II, P_1 and T_1, pressure and temperature in chamber I; P_2 and T_2, pressure and temperature in chamber II.

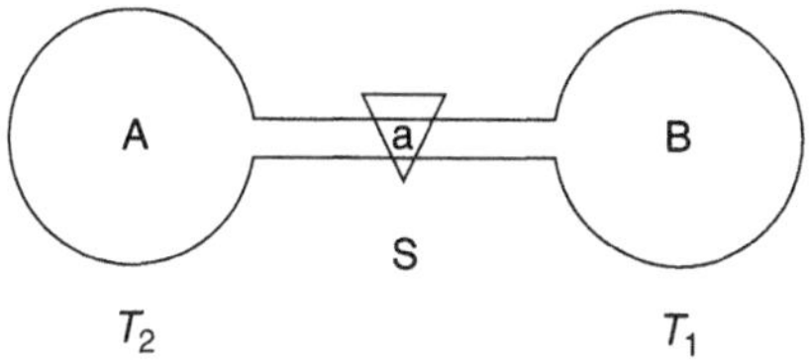

Figure 2.3. Continuous system with no barrier; A, B, chambers containing same solution or gaseous mixture; S, stopcock.

occurs on account of temperature difference $\Delta T = T_2 - T_1$. On the other hand, back flow occurs from chamber II to chamber I due to the pressure difference $\Delta P = P_2 - P_1$. The two flows continue to proceed till a steady state is reached.

Continuous systems (*with no barrier*): In this case, the intensive variables are not only functions of time but also continuous functions of space coordinates. Such a typical system is illustrated in Fig. 2.3, where there is interaction of flows due to temperature gradient and concentration gradient.

The following effects are observed in such a case.

Thermal diffusion/Soret effect: Establishment of steady concentration gradient due to fixed temperature gradient.

Dufour effect: Establishment of steady temperature gradient due to fixed concentration gradient. There can be two types of time-invariant states. One such is equilibrium state. In the equilibrium state, thermodynamic variables such as temperature T, pressure P and chemical potentials μ_i are adjusted in a way so that there is no (i) flow of matter, (ii) flow of energy and current and (iii) occurring in the system. Typical examples are vapour–liquid, liquid–liquid, solid–liquid and chemical equilibria. However, time-invariant non-equilibrium steady states are also possible when opposite flows are balanced and gradients are maintained constant.

2.2. Second law of thermodynamics for open systems

In an open system, entropy production $\mathrm{d}S$ is made up of $\mathrm{d_e}S$, entropy exchanged with the surroundings, and $\mathrm{d_i}S$, the internal production of entropy within the system itself. Thus,

$$\mathrm{d}S = \mathrm{d_e}S + \mathrm{d_i}S \tag{2.1}$$

According to second law of non-equilibrium, $\mathrm{d}S > 0$ as well as $\mathrm{d_i}S > 0$. However, for open systems, it can be positive, negative or zero. For a closed system at equilibrium $\mathrm{d}S = 0$ and entropy is maximum.

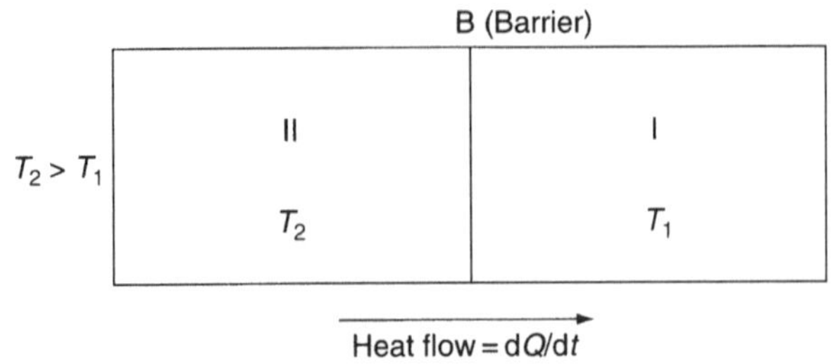

Figure 2.4. Thermo-osmosis through a barrier.

Entropy production for a non-equilibrium close to equilibrium is estimated with the help of Gibbs equation with the objective to estimate internal entropy production $\sigma = \mathrm{d_i}S/\mathrm{d}t$ which is needed for characterization of fluxes J and forces X since as we shall later that σ can be expressed as sum of product of fluxes and forces. To illustrate this point, we consider a discontinuous system involving two chambers separated by a barrier but maintained at different temperatures T_1 and T_2. In the present case, heat flow only occurs on account of force generated due to temperature difference (Fig. 2.4).

$$\text{Heat flow} = \mathrm{d}Q/\mathrm{d}t$$

We note that

$$\mathrm{d}S = (\mathrm{d}S)_{\mathrm{I}} + (\mathrm{d}S)_{\mathrm{II}} \tag{2.2}$$

and

$$\mathrm{d}S = (\mathrm{d}Q/T_{\mathrm{I}}) + (\mathrm{d}Q/T_{\mathrm{II}}), \tag{2.3}$$

Further,

$$(\mathrm{d}Q/T_{\mathrm{I}}) = (\mathrm{d_i}Q/T_{\mathrm{I}}) + (\mathrm{d_e}Q/T_{\mathrm{I}}) \tag{2.4}$$

$$(\mathrm{d}Q/T_{\mathrm{II}}) = (\mathrm{d_i}Q/T_{\mathrm{II}}) + (\mathrm{d_e}Q/T_{\mathrm{II}}) \tag{2.5}$$

It should be noted that since $T_2 > T_1$, heat will flow from chamber II to chamber I and $(\mathrm{d_i}Q)_{\mathrm{I}} = -(\mathrm{d_e}Q)_{\mathrm{II}}$, and hence,

$$\mathrm{d}S = (\mathrm{d_e}Q/T_{\mathrm{I}}) + (\mathrm{d_e}Q/T_{\mathrm{II}}) + \mathrm{d_i}Q\{(1/T_{\mathrm{I}}) - (1/T_{\mathrm{II}})\} > 0 \tag{2.6}$$

The sum of the first two terms on the right-hand side of the above equation would be equal to $\mathrm{d_e}S$, while the third term would be equal to $\mathrm{d_i}S$. Thus, the rate of entropy production σ would be given by

$$\sigma = \mathrm{d_i}S/\mathrm{d}t = (\mathrm{d}Q/\mathrm{d}t)\{(1/T)_{\mathrm{I}} - (1/T)_{\mathrm{II}}\} \tag{2.7}$$

where $(\mathrm{d_i}Q/\mathrm{d}t)_{\mathrm{I}}$ = heat flux and $\Delta T/(T_{\mathrm{I}} \cdot T_{\mathrm{II}})$ = force where $\Delta T = T_{\mathrm{II}} - T_{\mathrm{I}}$. If flux is denoted by $\boldsymbol{J}$ and force by X,

$$\sigma = LX \tag{2.8}$$

For a general case X = grad T/T (thermal force). For a more general case where other thermodynamic variables are involved, we have to make use of Gibbs equation along with laws of conservation of mass, energy and electric charge.

2.3. Law of conservation of mass, charge and energy

In any process, there has to be conservation of mass, charge and energy, and hence it has to be taken into consideration in all phenomena. We shall consider both (a) discontinuous systems and (b) continuous systems for fluxes and forces with the help of Gibbs equation; one has to use law of conservation of mass, charge and density from this angle to deduce explicit relations.

Let us consider a system as indicated in the figure. Let us suppose the internal transfer of mass energy and charge from one sub-system to other. The law of conservation of mass would take the form,

$$(\mathrm{d_i}M_k)_{\mathrm{I}} = -(\mathrm{d_i}M_k)_{\mathrm{II}} \tag{2.9}$$

and when r chemical reactions involve k chemical components,

$$\Sigma(\mathrm{d}_r M_k)_{\mathrm{I}} = 0; \quad \Sigma(\mathrm{d}_r M_k)_{\mathrm{II}} = 0 \tag{2.10}$$

Conservation of charge and energy would be expressed by Eqs. (2.11) (2.12), respectively,

$$\Sigma(e_k\,\mathrm{d}M_k)_{\mathrm{I}} + \Sigma(e_k\,\mathrm{d}M_k)_{\mathrm{II}} = 0 \tag{2.11}$$

and

$$\mathrm{d}U_{\mathrm{I}} + \mathrm{d}U_{\mathrm{II}} = \mathrm{d}U \tag{2.12}$$

where "U" is internal energy.

For continuous systems, the expression for mass and energy are little complex, since local variation of properties has to be taken into account. We will consider each separately.

(a) **Conservation of mass**

If ρ_k is the density (mass per unit volume) of k and v_k velocity of species/component k, then the local variation of ρ_k is given by

$$\text{Local time derivative} = \delta\rho_k/\mathrm{d}t = -\mathrm{div}\rho_k \cdot v_k + \Sigma v_{kj} J_c \qquad (2.13)$$

Equation (2.13) has the form of balance equation. The local charge on the left-hand side is equal to the sum of negative divergence of a flow term and a source term giving the production and destruction of the species k. The divergence of the flow has the simple physical meaning of giving per unit volume, the excess of the flow, which leaves a small volume to the flow, which penetrates into it.

(b) **Conservation of energy**

For continuous systems, the force equation can be written in the following form:

$$\rho \cdot \mathrm{d}V/\mathrm{d}T = -\ \mathrm{grad}\ P + \Sigma F_k \rho_k \qquad (2.14)$$

V = centre of mass movement
F_k = external force per unit mass of substance k

The energy equation for u, the energy per unit mass worth the exclusion of bar centric kinetic energy, would be given by

$$\rho \cdot d\{(1/2)v^2 + u\}/\mathrm{d}t = -\mathrm{div}\,(Pv + \mathbf{J}_q) + \Sigma F_k v_k \rho_k \qquad (2.15)$$

$\mathbf{J}_q$ = flow of heat as part of energy flow
u = thermal/heat energy
v = centre of mass velocity

2.4. Gibbs equation

2.4.1. For discontinuous systems

Gibbs equation for more complex situations is used for calculating the rate of internal production of entropy in contrast with the case discussed in Section 2.2, with the objective of identifying fluxes and forces for typical phenomena (systems).

Using second law of thermodynamics, Gibbs equation for discontinuous systems can be written as

$$T\,\mathrm{d}S = \mathrm{d}U + P\,\mathrm{d}V - \Sigma\mu_k \mathrm{d}_\mathrm{i} N_k \qquad (2.16)$$

where T is the temperature, P the pressure, S the entropy, U the energy, V the volume and N_k the number of molecules.

It may be noted that volume and entropy are extensive properties, while temperature, pressure and chemical potential are intensive properties.

2.4.2. For continuous systems

In most situations, we may assume that equilibrium thermodynamic relations are valid for the thermodynamic variables assigned to an elemental volume.

When this is done, all thermodynamic variables become functions of position x and time t, so that

$$T = T(x, t), \quad P = P(x, t), \quad \mu = \mu(x, t) \tag{2.17}$$

The intensive properties are replaced by densities s, u and n_k defined as

$s(x, t)$ = entropy per unit volume,
$u(x, t)$ = energy per unit volume,
n_k = mole number per unit volume of reactant k.

Thus Gibbs equation is assumed to be valid for small elements with

$$U = Vu, \quad S = sV \text{ and } N_k = V_k n_k$$

Gibbs equation is used for calculating the rate of internal production of entropy. It may be noted that Gibbs equation provides a simple route for identifying fluxes and forces. Gibbs equation strictly holds for equilibrium. However, it is found to be valid even in non-equilibrium close to equilibrium.

2.5. Phenomenological equations for single flows

Some heat flows in connection with entropy production are associated with other thermodynamic variables. Typical single fluxes and forces are summarized above. It may be noted that steady fluxes are considered. Kinetic theory provides theoretical justification of some of these flux force relations ($J = LX$). Here, "L" is called phenomenological coefficient. But kinetic theory has limitation in the sense that first approximation to distribution function corresponds to local equilibrium hypothesis. It may be noted that non-equilibrium molecular dynamics (model and simulation) provides justification of these laws for a wide range. Nevertheless, justification has to be provided by experiments (Table 2.1).

2.6. Phenomenological equations for coupled flows

Depending on the number of fluxes in the system, one can have coupling of flows with some restrictions. For example, in case of thermo-osmosis, coupling of heat flow and mass flow takes place, since a membrane separates two compartments

Table 2.1. Fluxes and forces in non-equilibrium systems.

Force	Flux	Phenomenological coefficients	Law
Grad T	Heat flux J_q	Heat conduction coefficient	Fourier's Law
Concentration gradient	Mass flux J_M	Diffusion coefficient	Fick's Law
Pressure difference	Volume flux J_V	Related with viscosity	Poiseuille's Law
Potential gradient	Current I	Reciprocal of resistance R	Ohms Law
Affinity "A" of chemical reaction	Chemical reaction rate J_C	Related with rate constant	–

having different temperatures and pressures. Both temperature difference and pressure difference can be the cause of heat flux and volume flux. In such cases, the linear phenomenological relations represent the fluxes $\mathbf{J}_i$,

$$J_i = \Sigma L_{ik} X_k, \quad i, k = 1, 2, \ldots\ldots \tag{2.18}$$

close to equilibrium. L_{ik} are called "cross-phenomenological coefficients". It is assumed that,

$$J_i = L_{ij} X_j, \quad i, j = 1, 2, \ldots\ldots \tag{2.19}$$

also holds good. Table 2.2 gives some more examples of cross-phenomena where such coupling occurs.

Coupling can take place (i) amongst vectorial forces or (ii) amongst scalar forces, but no coupling can take place between vectorial and scalar forces (Curie–Prigogine principle).

Table 2.2. Coupling in cross-phenomena.

Phenomena	Coupling between flows	Coupling between forces
Thermo-osmosis	Heat flux and volume flux	Temperature and pressure gradient
Thermal diffusion (Soret effect and Dufour effect)	Heat flux and mass flux	Temperature and pressure gradient
Electro-osmosis and streaming potential	Volume and current flux	Potential and pressure gradient
Thermo-electricity and Peltier heat	Current and heat flux	Potential and temperature difference

2.7. Onsager reciprocity relation

Onsager reciprocity relation is based on (i) principle of microscopic reversibility, (ii) fluctuation theory and (iii) the assumption that decay of fluctuations follows ordinary macroscopic laws. We give below a brief account of its derivation.

We consider the fluctuations in two variables $a_i(t)$ and $a_j(t+\tau)$ where τ is the time interval. The fluctuation in the average value of the product of the two variables $[a_i(t) \cdot a_j(t+\tau)]$ and $[a_j(t)a_i(t+\tau)]$ would differ only by temporal order i.e. by the substitution of $t \rightarrow -t$. Now using this concept of microscopic reversibility, we obtain following relation

$$\overline{[a_i(t)a_j(t+\tau)]} = \overline{[a_j(t)a_i(t+\tau)]} \tag{2.20}$$

On subtracting $a_j(t)$ and $a_i(t)$ from both the sides and on dividing by τ, we get

$$\overline{\{a_i(t) \cdot [a_j(t+\tau) - a_j(t)]/\tau\}} = \overline{\{a_j(t) \cdot [a_i(t+\tau) - a_i(t)]/\tau\}} \tag{2.21}$$

or

$$\overline{[a_i, (\delta a_j/\delta t)]} = \overline{[a_j(t) \cdot (\delta a_i/\delta t)]} \tag{2.22}$$

Equation (2.22) involves the concept of microscopic reversibility. If we assume that decay of fluctuations follows ordinary macroscopic laws, we can write Eq. (2.18) in a form where $J_i = \delta a_i/\delta t$ and X_k is the force. Substituting Eq. (2.18) into Eq. (2.22) we obtain

$$\overline{\Sigma L_{ik}[a_i \cdot X_k]} = \overline{\Sigma L_{jk}[a_j \cdot X_k]}$$

which after using the result from fluctuation theory yields

$$\boldsymbol{L}_{ij} = \boldsymbol{L}_{ji}$$

In principle, the results based on fluctuation theory and principle of microscopic reversibility would only be true for systems close to equilibrium. This will also be true for the assumption of linear relation between fluxes and forces. One has to understand also the serious limitation of phenomenological linear laws. The condition is that the magnitude of τ should satisfy the inequality,

$$\tau_0 << \tau << \tau_r$$

where τ_0 is the characteristic molecular time, e.g. the time between two collisions or the time required to establish a steady flow in hydrodynamics, and τ_r is the regression time of fluctuations, i.e. a time in which the deviation from equilibrium is appreciably reduced. The Brownian motion is a phenomena which is outside the range of validity of phenomenological laws since $\tau > \tau_r$.

The adiabatically insulated system and in the absence of external magnetic field can be characterized by a number of independent parameters. These parameters can be of two types:

α-type variables; which are even functions of particle velocities, i.e. mass energy.
β-type variables which are odd functions of particle velocities, i.e. momentum.

The Onsager reciprocity relation takes the following form in case of fluxes involving different types of variables.

(a) When coupling between α-type variables is involved

$$L_{ji}(\alpha, \alpha) = L_{ij}(\alpha, \alpha) \tag{2.23}$$

(b) When coupling between β-type variables is involved

$$L_{ji}(\beta, \beta) = L_{ij}((\beta, \beta) \tag{2.24}$$

(c) When coupling between α-type and β-type variables is involved

$$L_{ij}(\alpha, \beta) = -L_{ij}(\beta, \alpha) \tag{2.25}$$

Forms of Onsager reciprocal relations are given in Table 2.3 along with the tensorial rank of fluxes.

Table 2.3. The form of Onsager reciprocal relations and tensorial rank of corresponding fluxes.

Pair of fluxes	Functional relation of fluxes to particle velocity	Onsager relation
Mass flux and heat flux	Mass is independent of particle velocity (α-type), kinetic energy is a function of square of particle velocity (α-type)	$L_{ik} = L_{ki}$
Chemical reaction and bulk viscosity	Concentration is independent of particle velocity (α-type); momentum is the product of mass and velocity; function of first power of velocity (β-type)	$L_{ik} = -L_{kl}$
Electric current heat flux	Charge is independent of particle velocity (α-type); kinetic energy is a function of square of velocity (α-type)	$L_{ik} = L_{kl}$

2.8. Entropy production in multi-variable systems

In a multi-variable system, entropy production σ is given by[1]

$$\sigma = \Sigma \cdot J_i X_i, \quad i = 1, 2, \ldots . \tag{2.26}$$

where forces X_i and fluxes J_i can be identified by using Gibbs equation. We consider phenomenological relations for two flows J_1 and J_2 which are given by

$$J_1 = L_{11} X_1 + L_{12} X_2 \tag{2.27}$$

$$J_2 = L_{21} X_1 + L_{22} X_2 \tag{2.28}$$

According to Onsager reciprocity relation,

$$L_{12} = L_{21} \tag{2.29}$$

Two important inferences follow from the above equations.

1. Since entropy production σ is positive definite, it can be easily shown that

$$L_{11} > 0; L_{22} > 0 \text{ and } L_{12} \cdot L_{21} > L_{11} \cdot L_{22} \tag{2.30}$$

2. Entropy production would be given by

$$\sigma = L_{11}(X_1)^2 + (L_{12} + L_{21}) X_1 \cdot X_2 + L_{22}(X_2)^2 \tag{2.31}$$

It can be shown that entropy production would be minimum in the steady state when (a) X_2 is kept constant or (b) X_1 is kept constant as indicated in Fig. 2.5.

$(\delta\sigma/\delta X_1) = 0$ when X_2 is constant, which corresponds to the situation when $J_1 = 0$ (steady state). Similarly $(\delta\sigma/\delta X_2) = 0$ when X_1 = constant which corresponds to $J_2 = 0$ (steady state).

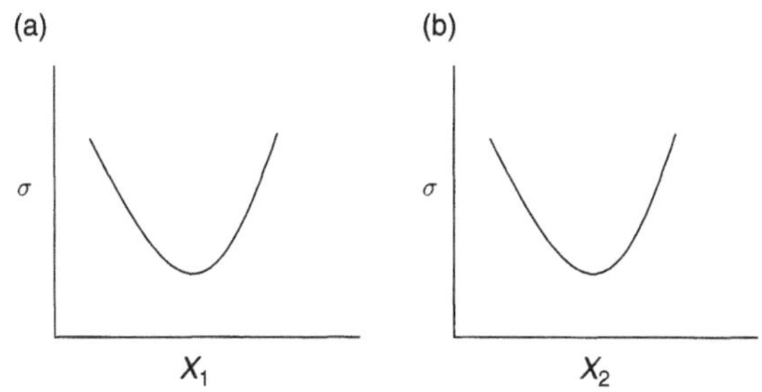

Figure 2.5. Minimum entropy production in steady state.

---

[1] Detailed derivation is given in Chapters 3–6

2.9. Basic postulates of non-equilibrium thermodynamics close to equilibrium

In the formalism for close to equilibrium, the basic steps postulated are as follows:

(a) Identification of appropriate fluxes and forces with the help of Gibbs equation assuming local equilibrium.
(b) Assumption of linear relation between single flux and single force.
(c) Assumption of linear phenomenological equation between fluxes and forces.
(d) Onsager reciprocity relation, which involves fluctuation theory and principle of microscopic reversibility.

While (b) and (c) are necessary conditions, (a) and (d) are necessary and sufficient conditions for deciding the domain of validity of linear non-equilibrium thermodynamics (LNT). In a sense, domain of validity of (c) is automatically covered once local equilibrium is maintained according to (a). However, justification and limitation of LNT can also be assessed experimentally.

2.10. Experimental test of LNT

Simplest case is when coupling between two forces is involved. In this case, one has to test

(a) Steady flows:

$$(J_1)_{X2=0} = L_{11}X_1;\ (J_2)_{X1=0} = L_{22}X_2$$
$$(J_1)_{X1=0} = L_{12}X_2;\ (J_2)_{X2=0} = L_{21}X_1$$

(b) Linear phenomenological equation

$$J_1 = [(J_1)_{X2=0} + (J_1)_{X1=0}]$$

(c) Steady state relations

$$(X_1/X_2)_{J1=0} = -L_{12}/L_{11} \text{ and } (X_2/X_1)_{J2=0} = -L_{21}/L_{22}$$

(d) Onsager reciprocity relation $L_{12} = L_{21}$

Although LNT is valid close to equilibrium, it is a good step for understanding the behaviour of systems beyond equilibrium as a first approximation. We will try to examine theory and experiment for typical steady states in order to assess to what extent theoretical predictions are supported by experimental observations in Chapters 3–6. Experimental results as well as theoretical analysis do show that the above formula is strictly valid close to equilibrium. During recent years, new tools have been developed, such as extended irreversible thermodynamics and non-equilibrium molecular dynamics, which are supposed to cover wider non-equilibrium region. These are discussed in Appendices II and III.

2.11. Application to other disciplines: sociology, economics and finance

The concept of forces and fluxes corresponding to cause and effect phenomena is quite relevant for the other disciplines. The only problem is the proper identification of the relevant fluxes and forces and the interrelationship. Processes involving positive and negative feedback have also to be identified. It has to be kept in mind that coupling of scalar and vectorial forces is not possible however, as a first approximation, the concept of coupling and linear phenomenological relations can be used. Although it is not easy to assess the magnitude of phenomenological coefficients in social systems, an approximate estimate can be made of relative magnitude depending on the nature of circumstances.

2.12. Concluding remarks

Thermodynamics of irreversible processes is a useful tool for the study of non-equilibrium phenomena close to equilibrium, the range of validity of Gibbs equation particularly identifying fluxes and forces (relevant cause and effect phenomena). Phenomena occurring beyond this region can be examined by extended irreversible thermodynamics, which has also its limitation as discussed in Appendix II. Beyond the range of validity of LNT, the concept of fluxes and forces is still valid, and the second law of thermodynamics for irreversible process is still a useful guide in the following form as pointed out by Prigogine in the following framework for understanding the emergence of ordered (dissipative) structure outside of equilibrium.

(a) For open system in non-equilibrium, the entropy production $\mathrm{d}S$ is made up of internal entropy production $\mathrm{d_i}S$ and $\mathrm{d_e}S$, the entropy exchanged with the surroundings, i.e.

$$\mathrm{d}S = \mathrm{d_i}S + \mathrm{d_e}S$$

(b) $\mathrm{d}S > 0$, since entropy increases in irreversible processes.
(c) However, although $\mathrm{d_i}S > 0$, $\mathrm{d_e}S$ can be greater, zero or negative [$\mathrm{d_e}S > 0$ or $\mathrm{d_e}S < 0$]

(d) Dissipative structure or ordered structure results depending on appropriately regulated flow of entropy to the surroundings and proper balance of forces and counter forces.
(e) The manner in which the entropy is exported to the surroundings determines the nature of time order and space order.

Under specific conditions, even deterministic chaos is possible during temporal oscillations.

If entropy export is random, noise and random behaviour in space and time would be observed.

Based on the above ideas, non-liner kinetics, non-linear dynamics and statistical techniques along with computer modelling, exotic phenomena up to far from equilibrium region have been investigated, and interesting results have been obtained which have been discussed in Part II and Part III of the book.

LNT helps in identifying (i) fluxes and forces and their interaction and (ii) fluxes and counter fluxes [positive and negative feedbacks]. The approach is useful for interpreting similar phenomena in the real systems.

In Chapters 3 and 4, it is intended to discuss and test the above formalism I, the context of experimental studies involving coupling of two or more forces and fluxes in simple systems.

In Chapter 3, coupling of heat flux and volume flux in a system containing two sub-systems separated by a membrane due to two forces, i.e. ΔP and ΔT, has been considered. Similarly, in Chapter 4, coupling of volume flux and electric current due to ΔP and $\Delta\phi$ involving transport phenomena through membrane has been considered. On the other hand, experimental and theoretical studies related to similar type of phenomena (e.g. thermal difussion and Dufour effect) involving two or three fluxes and forces but without a barrier have been discussed in Chapters 5 and 6.

It may be noted that in real systems, non-equilibrium phenomena are much more complex which involve multi-processes and complex coupling.

References

1. K.G. Denbigh, The Thermodynamics of Steady State, Methuen, London, 1951.
2. S.R. de Groot, Thermodynamics of Irreversible Processes, North-Holland, Amsterdam, 1952.
3. I. Prigogine, Introduction to Thermodynamics of Irreversible Processes, Wiley, New York, 1968.
4. D.D. Fitts, Non-equilibrium Thermodynamics, McGraw-Hill, New York, 1962.
5. S.R. de Groot and P. Mazur, Non-equilibrium Thermodynamics, North-Holland, Amsterdam, 1962.
6. A. Katchalsky and Peter F. Curran, Non-equilibrium thermodynamics in Biophysics, Howard University Press, Cambridge, MA, 1965.

7. H.J.M. Hanley, Transport Phenomena in Fluids, Chapters 3, 11, 12, Marcel Dekker, New York and London, 1969.
8. A.R. Peacocke, An Introduction to Physical Chemistry of Biological Organization, Clarendon Press, Oxford, 1983.
9. P. Glansdroff and I. Prigogine, Thermodynamic Theory of Structure, Stability and Fluctuations, Wiley-Interscience, London, New York, 1971.
10. G. Nicolis and I. Prigogine, Self-organization in Non-equilibrium Systems (from Dissipative Structures to order through Fluctuations), John Wiley and Sons, New York, 1977.
11. D. Kondepudi and I. Prigognic, Modern Thermodynamics (from Engines to Dissipative Structures), John Wiley and Sons, New York, Singapore, 1988.

Chapter 3

APPLICATIONS TO TYPICAL STEADY-STATES PHENOMENA

3.1. Introduction

In this chapter, we will consider simple non-equilibrium phenomena involving two fluxes and two forces. Such a situation arises in membrane transport phenomena (e.g. thermo-osmotic and electro-osmotic phenomena) involving two sub-systems separated by a membrane. In Chapters 5 and 6, transport phenomena in two flux–two force systems in continuous systems without a membrane would be discussed.

Thermo-osmosis is a phenomenon in which matter is driven through a membrane or an orifice from one chamber to another on account of the temperature difference between the two chambers. This can occur for a single fluid or a mixture of fluids.

In a system without a membrane or a barrier, temperature gradient can give rise to concentration gradient and this phenomenon is called thermal diffusion. Conversely, a concentration gradient can give rise to temperature gradient; this phenomenon is known as Dufour effect. This effect was discovered in gases by Waldmann [1], and in liquids by Rastogi and Madan [2].

Theories based on non-equilibrium thermodynamics [3–8] have been applied extensively to elucidate the phenomenon of thermo-osmosis. The methodology of non-equilibrium thermodynamics essentially involves the evaluation of entropy production by the application of the laws of conservation of mass and energy and Gibbs equation. Appropriate fluxes and forces are chosen by suitably splitting the expression for entropy production and subsequently, thermodynamic transport equations are written. The theory of thermo-osmosis based on non-equilibrium thermodynamics is discussed below.

3.2. Thermodynamic theory of thermo-osmosis

Let us consider an adiabatically insulated multicomponent system, divided into two subsystems I and II by a porous barrier or a membrane (Fig. 3.1).

Let us suppose that chemical reactions can occur in the two chambers. Chambers I and II having volumes V^{I} and V^{II} are maintained at temperatures T^{I} and T^{II} ($T^{\mathrm{II}} = T^{\mathrm{I}} + \Delta T$), pressures P^{I} and P^{II} ($P^{\mathrm{II}} = P^{\mathrm{I}} + \Delta P$), ΔT and ΔP represent the temperature

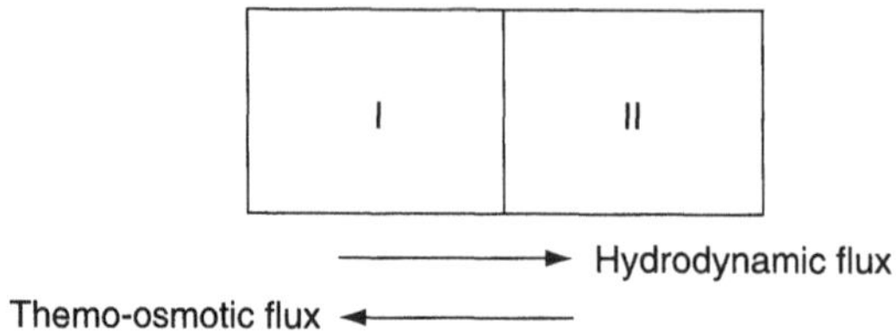

Figure 3.1. Opposing forces in thermo-osmosis.

and pressure difference, respectively in the two chambers. The system as a whole is a closed one, but the subsystems I and II are open systems, as matter can be transferred from one subsystem to the other. When a finite temperature difference is maintained in such a discontinuous system, a steady ΔP and a fixed concentration difference, ΔC_k, of any component, k, is established across the barrier in the steady state. For such a system, non-equilibrium thermodynamics can be successfully applied, assuming the concept of local equilibrium.

Now, dM_k^l, the change of mass M_k of component k and energy in reservoir I is given by

$$dM_k^{\mathrm{I}} = d_t M_k^l + d_r M_k^l \quad (k = 1, 2, \ldots, n) \tag{3.1}$$

Similarly, dU^{I} and the change of internal energy U in reservoir I, is given by

$$dU^{\mathrm{I}} = d_e U^{\mathrm{I}} + d\,U^{\mathrm{I}} \tag{3.2}$$

In the same manner, we have for reservoir II,

$$dM_k^{\mathrm{II}} = d_t M_k^{\mathrm{II}} + d_r M_k^{\mathrm{II}} \tag{3.3}$$

$$dU^{\mathrm{II}} = d_e U^{\mathrm{II}} + d\,U^{\mathrm{II}} \tag{3.4}$$

The superscripts I and II denote chambers I and II; subscript k, the component; $d_i M_i$, the transfer of mass from one reservoir to the other; and $d_t M_k$, the change in mass resulting from the chemical reaction in a particular chamber. The change of energy dU^{I} has been split into an external part $d_e U^{\mathrm{I}}$, the energy exchanged with the surroundings and $d_i U^{\mathrm{I}}$, the energy exchanged with another vessel internally. Similar splitting has been done for dU^{II} also.

Using the laws of conservation of mass and energy, we get

$$d_t M_k^{\mathrm{I}} + d_t M_k^{\mathrm{II}} = 0 \tag{3.5}$$

and

$$d_t U^{\mathrm{I}} + d_t U^{\mathrm{II}} = 0 \tag{3.6}$$

Further, for the chemical reaction in chamber I, we have

$$\sum_{k=1}^{n} \mathrm{d}_r M_k^{\mathrm{I}} = 0 \tag{3.7}$$

and similarly for chamber II, we have

$$\sum_{k=1}^{n} \mathrm{d}_r M_k^{\mathrm{II}} = 0 \tag{3.8}$$

The changes in the masses of k, caused by chemical reaction, are related by the relation

$$\frac{\mathrm{d}M_1}{\nu_1} = \frac{\mathrm{d}M_2}{\nu_2} = \cdots = \frac{\mathrm{d}M_n}{\nu_n} \tag{3.9}$$

where ν_k are the stoichiometric numbers. These are counted positive for the reactants and negative for the products.

Further, we can define a quantity ξ satisfying the relation

$$M \mathrm{d}\xi/_{\mathrm{d}t} = J_c \tag{3.10}$$

where $M = \sum_{k=1}^{n} M_k$ and J_c, the reaction rate, is defined by

$$J_c = \frac{1}{\nu_k}\left(\frac{\mathrm{d}_r M_k}{dt}\right) \tag{3.11}$$

The parameter ξ is called the degree of advancement of the chemical reaction and is a quantity of state which specifies the degree of freedom introduced by the chemical reaction. Combining Eqs. (3.10) and (3.11), we get for chamber I

$$\mathrm{d}_r M_k^{\mathrm{I}} = \nu_k M^{\mathrm{I}} \mathrm{d}\xi^{\mathrm{I}} \tag{3.12}$$

and for chamber II

$$\mathrm{d}_r M_k^{\mathrm{II}} = \nu_k M^{\mathrm{II}} \mathrm{d}\xi^{\mathrm{II}} \tag{3.13}$$

where

$$M^{\mathrm{I}} = \sum_{k=1}^{n} M_k^{\mathrm{I}} \tag{3.14}$$

and

$$M^{\mathrm{II}} = \sum_{k=1}^{n} M_k^{\mathrm{II}} \tag{3.15}$$

Our object now is to evaluate the entropy production d_iS due to irreversible processes inside the system. This is achieved by applying Gibbs equation to each chamber. Thus, for chamber I, we have

$$T^{\mathrm{I}}\mathrm{d}S^{\mathrm{I}} = \mathrm{d}U^{\mathrm{I}} + P^{\mathrm{I}}\mathrm{d}V^{\mathrm{I}} - \sum_{k=1}^{n} \mu_k^{\mathrm{I}} \mathrm{d}_t M_k^{I} - \sum_{k=1}^{n} \mu_k^{\mathrm{I}} \mathrm{d}_r M_k^{\mathrm{I}} \tag{3.16}$$

where T is the temperature; S, the entropy; U, the energy; P, the pressure; V, the volume; and μ_k, the chemical potential of substance k. Noting that the chemical affinity A is given by

$$A = -\sum_{k=1}^{n} \mu_k \nu_k \tag{3.17}$$

and using Eqs. (3.14)–(3.17), we get for subsystem I

$$T^{\mathrm{I}}\mathrm{d}S^{\mathrm{I}} = \mathrm{d}U^{\mathrm{I}} + P^{\mathrm{I}}\mathrm{d}V^{\mathrm{I}} - \sum_{k=1}^{n} \mu_k^{\mathrm{I}} \mathrm{d}_t M_k^{\mathrm{I}} + M^{\mathrm{I}} A^{\mathrm{I}} \mathrm{d}\xi^{\mathrm{I}} \tag{3.18}$$

Similarly, for subsystem II, we have

$$T^{\mathrm{II}}\mathrm{d}S^{\mathrm{II}} = \mathrm{d}U^{\mathrm{II}} + P^{\mathrm{II}}\mathrm{d}V^{\mathrm{II}} - \sum_{k=1}^{n} \mu_k^{\mathrm{II}} \mathrm{d}_t M_k^{\mathrm{II}} + M^{\mathrm{II}} A^{\mathrm{II}} \mathrm{d}\xi^{\mathrm{II}} \tag{3.19}$$

The change in entropy of the total system is given by

$$\begin{aligned} \mathrm{d}S = \mathrm{d}S^{\mathrm{I}} + \mathrm{d}S^{\mathrm{II}} = {} & \frac{(\mathrm{d_e}U^{\mathrm{I}} + P^{\mathrm{I}}\mathrm{d}V^{\mathrm{I}})}{T_{\mathrm{I}}} + \frac{(\mathrm{d_e}U^{\mathrm{II}} + P^{\mathrm{II}}\mathrm{d}V^{\mathrm{II}})}{T_{\mathrm{II}}} + \frac{\mathrm{d_i}U^{\mathrm{I}}}{T^{\mathrm{I}}} + \frac{\mathrm{d_i}U^{\mathrm{II}}}{T^{\mathrm{II}}} \\ & - \sum_{k=1}^{n} \frac{\mu_k^{\mathrm{I}} \mathrm{d}_t M_k^{\mathrm{I}}}{T^{\mathrm{I}}} - \sum_{k=1}^{n} \frac{\mu_k^{\mathrm{II}} \mathrm{d}_t M_k^{\mathrm{II}}}{T^{\mathrm{II}}} + \frac{M^{\mathrm{I}} A^{\mathrm{I}} \mathrm{d}\xi^{\mathrm{I}}}{T^{\mathrm{I}}} + \frac{M^{\mathrm{II}} A^{\mathrm{II}} \mathrm{d}\xi^{\mathrm{II}}}{T^{\mathrm{II}}} \end{aligned} \tag{3.20}$$

Further, on substituting relations (3.5) and (3.6) in Eq. (3.20), we have

$$\begin{aligned} \mathrm{d}S = {} & \frac{\mathrm{d_e}U^{\mathrm{I}} + P^{\mathrm{I}}\mathrm{d}V^{\mathrm{I}}}{T^{\mathrm{I}}} + \frac{\mathrm{d_e}U^{\mathrm{II}} + P^{\mathrm{II}}\mathrm{d}V^{\mathrm{II}}}{T^{\mathrm{II}}} + \left(\frac{\Delta T}{T^2}\right) \mathrm{d_i}U^{\mathrm{I}} \\ & + \sum_{k=1}^{n} \Delta(\mu_k/T) \mathrm{d}_t M_k^{\mathrm{I}} + \left(\frac{A^{\mathrm{I}}}{T^{\mathrm{I}}}\right) M^{\mathrm{I}} \mathrm{d}\xi^{\mathrm{I}} + \left(\frac{A^{\mathrm{II}}}{T^{\mathrm{II}}}\right) M^{\mathrm{II}} \mathrm{d}\xi^{\mathrm{II}} \end{aligned} \tag{3.21}$$

Here, the symbol Δ is used to denote the increase in magnitude of a quantity in subsystem II over that in I, and T is the mean temperature across the porous barrier defined by $T = \sqrt{T^{\mathrm{I}} T^{\mathrm{II}}}$. Expression (3.21) for the change in entropy can be split into two parts, one giving the entropy received from the surroundings, i.e.

$$\mathrm{d_e}S = \frac{\mathrm{d_e}U^{\mathrm{I}} + P^{\mathrm{I}}\mathrm{d}V^{\mathrm{I}}}{T^{\mathrm{I}}} + \frac{\mathrm{d_e}U^{\mathrm{II}} + P^{\mathrm{II}}\mathrm{d}V^{\mathrm{II}}}{T^{\mathrm{II}}} \tag{3.22}$$

and the other giving the internal production of entropy, which results from the action of irreversible processes inside the system, so that

$$\mathrm{d_i}S = (\Delta T/T^2)\mathrm{d_i}U^{\mathrm{I}} + \sum \Delta(\mu_k/T)\mathrm{d}_t M_k^{\mathrm{I}} + (A^{\mathrm{I}}/T^{\mathrm{I}})M^{\mathrm{I}}\mathrm{d}\xi^{\mathrm{I}} + (A^{\mathrm{II}}/T^{\mathrm{II}})M^{\mathrm{II}}\mathrm{d}\xi^{\mathrm{II}} \quad (3.23)$$

Consequently, the rate of entropy production, σ, can be written as

$$\sigma = \frac{\mathrm{d_i}S}{\mathrm{d}t} = \left(\frac{\Delta T}{T^2}\right)\frac{\mathrm{d_i}U^{\mathrm{I}}}{\mathrm{d}t} + \sum_{k=1}^{n}\Delta(\mu_k/T)\frac{\mathrm{d}_t M_k^{\mathrm{I}}}{\mathrm{d}t} + (A^{\mathrm{I}}/T^{\mathrm{I}})\frac{M^{\mathrm{I}}\mathrm{d}\xi^{\mathrm{I}}}{\mathrm{d}t} + (A^{\mathrm{II}}/T^{\mathrm{II}})\frac{M^{\mathrm{II}}\mathrm{d}\xi^{\mathrm{II}}}{\mathrm{d}t} \quad (3.24)$$

Equation (3.24) with the help of Eqs. (3.12) and (3.13) reduces to Eq. (3.25). The respective fluxes would then be given by

$$\sigma = \left(\frac{\Delta T}{T^2}\right)\frac{\mathrm{d_i}U^{\mathrm{I}}}{\mathrm{d}t} + \sum_{k=1}^{n}\Delta(\mu_k/T)\frac{\mathrm{d}_t M_k^{\mathrm{I}}}{\mathrm{d}t} + (A^{\mathrm{I}}/T^{\mathrm{I}})\frac{\mathrm{d}_r M_k^{\mathrm{I}}}{\nu_k}\frac{1}{\mathrm{d}t} + (A^{\mathrm{II}}/T^{\mathrm{II}})\frac{\mathrm{d}_r M_k^{\mathrm{II}}}{\nu_k}\frac{1}{\mathrm{d}t} \quad (3.25)$$

The respective fluxes would then be given by

$$J_u = -\mathrm{d_i}U^{\mathrm{I}}/\mathrm{d}t = \mathrm{d_i}U^{\mathrm{II}}/\mathrm{d}t \quad (3.26)$$

$$J_k = -\mathrm{d}_t M_k^{\mathrm{I}}/\mathrm{d}t = \mathrm{d}_t M_k^{\mathrm{II}}/\mathrm{d}t \quad (3.27)$$

$$J_{\mathrm{I}} = \left(\frac{1}{\nu_k}\right)\frac{\mathrm{d}_r M_k^{\mathrm{I}}}{\mathrm{d}t} \quad (3.28)$$

$$J_{\mathrm{II}} = \left(\frac{1}{\nu_k}\right)\frac{\mathrm{d}_r M_k^{\mathrm{II}}}{\mathrm{d}t} \quad (3.29)$$

The flows J_u and J_k are counted as positive for the flow from chambers I to II; J_{I} and J_{II} are positive when the chemical reaction proceeds from left to right according to the reaction equation.

Expressing the flows as linear functions and considering the vector and scalar characters of the various forces and flows, we have

$$J_i = -\sum_{k=1}^{n} L_{ik}\Delta(\mu_k/T) - L_{iu}(\Delta T/T^2) \quad (3.30)$$

$$J_u = -\sum_{k=1}^{n} L_{uk}\Delta(\mu_k/T) - L_{uu}(\Delta T/T^2) \quad (3.31)$$

$$J_{\mathrm{I}} = L_c M^{\mathrm{I}} A^{\mathrm{I}}/T^{\mathrm{I}} \quad (3.32)$$

$$J_{\mathrm{II}} = L_c M^{\mathrm{II}} A^{\mathrm{II}}/T^{\mathrm{II}} \quad (3.33)$$

In Eq. (3.32), the term $A^{\mathrm{II}}/T^{\mathrm{II}}$ and in Eq. (3.33), the term $A^{\mathrm{I}}/T^{\mathrm{I}}$ do not occur. This is due to the fact that such chemical cross-terms do not exist, as there is admittedly no interaction between the chemical reactions in the two chambers. Further, using Eqs. (3.30) and (3.31), we can rewrite Eq. (3.25) as

$$\sigma = \sum_{k=1}^{n} L_{ik}\Delta(\mu_i/T)\Delta(\mu_k/T) + \sum_{k=1}^{n}(L_{ku}+L_{uk})\Delta(\mu_k/T)(T/T^2) + L_{uu}(\Delta T/T^2) \\ + L_c M^{\mathrm{I}}(A^{\mathrm{I}}/T^{\mathrm{I}})^2 + L'_c M^{\mathrm{I}}(A^{\mathrm{I}}/T^{\mathrm{I}})^2 \tag{3.34}$$

where Onsager reciprocal relationship (ORR)

$$L_{ik} = L_{ki}, \quad L_{uk} = L_{ku} \quad (i, k = 1, 2, \ldots, n)$$

has been assumed to be valid for the phenomenological coefficients occurring in Eqs. (3.30) and (3.31).

deGroot [3] assumed the same coefficient for both the chambers, i.e. $L_c = L'_c$, stating that there was obviously no physical difference between the reaction velocities in chambers I and II. In general, however, these would be unequal. Srivastava *et al.* [9] examined this point from considerations based on chemical kinetics and showed that L_c was different for different temperatures, as the reaction velocities are different.

Let us consider a simple example of a mixture of two arbitrary isomers 1 and 2, capable of undergoing chemical reaction $1 \rightleftharpoons 2$. For the sake of simplicity, we assume that the sub-systems have equal masses ($M^{\mathrm{I}} = M^{\mathrm{II}}$). Further, we denote $L_{cc} = M^{\mathrm{I}}L_c$, $L'_{cc} = M^{\mathrm{II}}L'_c$.

Then the phenomenological equations for this particular case can be written as

$$J_1 = -L_{11}\Delta(\mu_1/T) - L_{12}\Delta(\mu_2/T) - L_{1u}(\Delta T/T^2) \tag{3.35}$$

$$J_1 = -L_{21}\Delta(\mu_1/T) - L_{22}\Delta(\mu_2/T) - L_{2u}(\Delta T/T^2) \tag{3.36}$$

$$J_1 = -L_{21}\Delta(\mu_1/T) - L_{22}\Delta(\mu_2/T) - L_{2u}(\Delta T/T^2) \tag{3.37}$$

$$J_u = -L_{u1}\Delta(\mu_1/T) - L_{u2}\Delta(\mu_2/T) - L_{uu}(\Delta T/T^2) \tag{3.38}$$

$$J_{\mathrm{I}} = -L'_{cc}(A^{\mathrm{I}}/T^{\mathrm{I}}) \tag{3.39}$$

$$J_{\mathrm{II}} = -L'_{cc}(A^{\mathrm{II}}/T^{\mathrm{II}}) \tag{3.40}$$

Between the coefficients we have three Onsager relations

$$L_{12} = L_{21}, L_{1u} = L_{u1} \text{ and } L_{2u} = L_{u2} \tag{3.41}$$

The equation for entropy production would be written as

$$\begin{aligned}\sigma = {} & L_{11}\Delta(\mu_1/T) + L_{22}\Delta(\mu_{12}/T^2) + L_{uu}(\Delta T/T^2) + 2L_{12}\Delta(\mu_1/T)\Delta(\mu_2/T) \\ & + 2L_{1u}\Delta(\mu_1/T)(\Delta T/T^2) + 2L_{2u}\Delta(\mu_2/T)(\Delta T/T^2) \\ & + L_{cc}(A^{\mathrm{I}}/T^{\mathrm{I}}) + L'_{cc}(A^{\mathrm{II}}/T^{\mathrm{II}})\end{aligned} \tag{3.42}$$

Stationary state—Let us consider the stationary state of the first order when ΔT is kept constant. Using Prigogine's theory of minimum entropy production [5], we have the following three conditions for the stationary state

$$\frac{\partial\sigma}{\partial\Delta(\mu_1/T)} = 0; \quad \frac{\partial\sigma}{\partial\Delta(\mu_1/T)} = 0; \quad \frac{\partial\sigma}{\partial\Delta(\mu_1/T)} = 0 \tag{3.43}$$

corresponding to the remaining three forces, $\Delta(\mu_1/T)$, $\Delta(\mu_2/T)$ and $A^{\mathrm{I}}/T^{\mathrm{I}}$. Using the following relation, which follows from relation (3.17), i.e.:

$$A^{\mathrm{II}}/T^{\mathrm{II}} = A^{\mathrm{I}}/T^{\mathrm{I}} + \Delta(\mu_1/T) - \Delta(\mu_2/T) \tag{3.44}$$

we can write Eq. (3.43) explicitly as follows:

$$L_{11}\Delta(\mu_1/T) + L_{12}\Delta(\mu_2/T) + L_{1u}(\Delta T/T^2) - L_{cc}(A^{\mathrm{I}}/T^{\mathrm{I}}) = 0 \tag{3.45}$$

$$L_{22}\Delta(\mu_2/T) + L_{12}\Delta(\mu_1/T) + L_{2u}(\Delta T/T^2) - L_{cc}(A^{\mathrm{I}}/T^{\mathrm{I}}) = 0 \tag{3.46}$$

$$L_{cc}(A^{\mathrm{I}}/T^{\mathrm{I}}) + L'_{cc}(A^{\mathrm{II}}/T^{\mathrm{II}}) = 0 \tag{3.47}$$

Writing the chemical potential in the form

$$\Delta(\mu_k/T) = -h_k(\Delta T/T^2) + v_k(\Delta P/T) + \left(\frac{\partial\mu_k}{\partial c_1}\right)_{T,P}(\Delta c_1/T) \quad (k = 1, 2) \tag{3.48}$$

where h_k and v_k are the partial specific enthalpy and partial specific volume, respectively of the component k and c_1, the concentration of species 1. Using the Gibbs–Duhem relation, one gets

$$c_1(\partial\mu_1/\partial c_1)_{T,P} + c_2(\partial\mu_2/\partial c_2)_{T,P} = 0 \tag{3.49}$$

Since out of the four variables ΔP, Δc_1, $A^{\mathrm{I}}/T^{\mathrm{I}}$ and ΔT, one, viz. ΔT is fixed, the other three variables can be calculated in terms of ΔT using Eqs. (3.45)–(3.47). Therefore,

$$(\Delta P/\Delta T) = \frac{h + c_1L_1 + c_2L_2}{vT} \tag{3.50}$$

where

$$L_1 = [(L_{cc} + L'_{cc})(L_{12}L_{2u} - L_{1u}L_{22}) - L_{cc}L'_{cc}(L_{1u} + L_{2u})]/Z \tag{3.51}$$

$$L_2 = [(L_{cc} + L'_{cc})(L_{12}L_{1u} - L_{2u}L_{11}) - L_{cc}L'_{cc}(L_{1u} + L_{2u})]/Z \tag{3.52}$$

with

$$Z = L_{cc}L'_{cc}(L_{11} + L_{22} + 2L_{12}) + (L_{cc} + L'_{cc})(L_{11}L_{22} - L_{12}^2) \tag{3.53}$$

The mean specific enthalpy h and volume v are given by

$$h = c_1 h_1 + c_2 h_2 \tag{3.54}$$

$$h = c_1 h_1 + c_2 h_2 \tag{3.55}$$

Further, the separation per unit temperature difference is given by

$$\frac{\Delta c_1}{\Delta T} = \frac{c_2 [v_2 (L_1 + h_1) - v_1 (L_2 + h_2)]}{vT\,(\partial \mu_1 / \partial c_1)_{T,P}} \tag{3.56}$$

and the chemical force is given by

$$A^{\mathrm{I}}/T^{\mathrm{I}}\Delta T = \frac{L'_{cc}}{L_{cc} + L'_{cc}} \left(\frac{L_2 - L_1}{T^2}\right) \tag{3.57}$$

$$A^{\mathrm{II}}/T^{\mathrm{II}}\Delta T = \frac{L'_{cc}}{L_{cc} + L'_{cc}} \left(\frac{L_2 - L_1}{T^2}\right) \tag{3.58}$$

The new results expressed by Eqs. (3.50), (3.56), (3.57) and (3.58) are of interest in the study of chemical reactions. The above treatment has been utilized successfully to develop the necessary formalism for thermal effusion and thermomolecular pressure for an isomeric reaction, particularly for liquid He II.

The treatment of de Groot [3] was extended by Rastogi and Srivastava [10, 11] to the phenomenon of thermal transpiration in the presence of dissociation. They examined a system containing a mixture of two species X_n and X capable of the reaction $X_n \rightleftharpoons nX$. Denoting X_n by the subscript 1 and X by the subscript 2, the following relations are obtained:

$$(\Delta P/\Delta T) = (h + c_1 L'_1 + c_2 L'_2)/vT \tag{3.59}$$

$$(\Delta c_1/\Delta T) = \frac{c_2}{vT}[v_2(h_1 + L'_1) - v_1(h_2 + L'_2)] \Big/ \left(\frac{\partial \mu 1}{\partial c_1}\right)_{T,P} \tag{3.60}$$

$$(A^{\mathrm{I}}/T^{\mathrm{I}}\Delta T) = \frac{L'_{cc}}{L_{cc} + L'_{cc}} \frac{1}{T^2}[nL'_2 - L'_1] \tag{3.61}$$

$$(A^{\mathrm{II}}/T^{\mathrm{II}}\Delta T) = \frac{L'_{cc}}{L_{cc} + L'_{cc}} \frac{1}{T^2}[nL'_2 - L'_1] \tag{3.62}$$

The above equations are interesting from the view point of estimating the sluft from thermodynamics equilibrium due to thermo-osmosis. Such an estimate has been made for the reaction $N_2O_4 \leftrightarrow 2NO_4$ using the available equilibrium data on thermal dissociation to estimate the magnitude of the shift from thermodynamic equilibrium [11]. The result reveals appreciable deviation from equilibrium depending on the temperature coefficient of reaction rate.

3.3. Thermodynamic theory of thermo-osmosis of gaseous non-reacting mixtures

(Thermo-osmotic pressure and concentration difference)

Using Eq. (3.25) for the general case, the expression for the rate of entropy production in the case of a mixture of non-reacting gases may be written as

$$\sigma = J_u X_u + \sum_{k=1}^{n} J_k X_k \tag{3.63}$$

The forces corresponding to J_u and J_k are given by

$$X_u = \Delta(l/T) = -(\Delta T/T^2) \tag{3.64}$$

$$X_k = \Delta(\mu_k/T) = -\Delta\mu_k/T + \mu_k(\Delta T/T^2) \tag{3.65}$$

The linear phenomenological equations may consequently be expressed as

$$J^i_{k=1} = \sum^{n} L_{ik} X_k + L_{iu} X_u \tag{3.66}$$

$$J_u = \sum_{k=1}^{n} L_{uk} X_k + L_{uu} X_u \tag{3.67}$$

On account of ORR, one gets

$$L_{ik} = L_{ki} \text{ and } L_{iu} = L_{ui} \quad (k = 1, \ldots, n) \tag{3.68}$$

The phenomenological coefficients L_{ik} and L_{iu} depend on temperature, pressure and composition of the gaseous mixture in addition to structure of the membrane. If we now introduce n quantities U_k^* defined by

$$L_{iu} = \sum_{k=1}^{n} L_{ik} U_k^* \quad (i = 1, 2, \ldots, n) \tag{3.69}$$

in Eq. (3.66) and combine it with Eq. (3.67) we get

$$J_u - \sum_{i}^{n} U_i^* J_i = \sum_{k=1}^{n} (L_{uk} - \sum L_{ik} U_i^*) X_k + \left(L_{uu} - \sum_{ik}^{n} L_{ik} U_i^* U_k^* \right) X_u \tag{3.70}$$

The first term on the right hand side of Eq. (3.70) vanishes on account of Onsager relations so that we have

$$J_u = \sum_{k=1}^{n} U_i^* J_i + \left(L_{uu} - \sum_{i,k}^{n} L_{ik} U_i^* U_k^* \right) X_u \tag{3.71}$$

From the above equation it is obvious that U_i^* is the energy carried per unit flow of component i at uniform temperature ($\Delta T = 0$ or $X_u = 0$). If we now combine Eq. (3.69) with Eq. (3.66) substituting the values of X_u and X_k from Eq. (3.65), we get

$$J_i = \sum_{k=1}^{n} L_{ik} \left[(-\Delta\mu_k/T) - (U_k^* - \mu_k)\Delta T/T^2 \right] \tag{3.72}$$

Substitution of values of μ_k from Eq. (3.48) in Eq (3.72) yields

$$J_i = \sum_{k=1}^{n} L_{ik}[-v_k(\Delta P/T) - \sum_{i=1}^{n} (\partial\mu_k/\partial C_i)_{T,P,C_j}(\Delta C_i/T) - (U_k^* - h_k)\Delta T/T^2] \tag{3.73}$$

Equation (3.73) is a general relationship for flow of mass of a multicomponent system without chemical reaction. In the steady state when all mass flows vanish but an energy flow persists, Eq. (3.73) becomes

$$v_k \,\Delta P + \sum_{i=1}^{n-1} (\partial\mu_k/\partial C_i)_{\underset{i=1}{T,P,C_j}} \Delta C_i^n + (U_k^* - h_k)\Delta T/T = 0 \tag{3.74}$$

The theory of non-equilibrium thermodynamics permits an alternative choice of fluxes and forces provided the sum of the product of fluxes and forces is equal to the entropy production. In the present case, let us define the new fluxes as

$$J_q = J_u - h \sum_{k=1}^{n} J_k \tag{3.75}$$

$$J_k' = J_k \tag{3.76}$$

The corresponding forces would be

$$X_q = -(\Delta T/T^2)$$

$$X_k' = -\Delta(\mu_k/T) - h(\Delta T/T^2) \tag{3.77}$$

The phenomenological equations consequently become

$$J_i = \sum_{k=1}^{n} L_{ik} X_k + L_{iq} X_q \tag{3.78}$$

$$J_q = \sum_{k=1}^{n} L'_{qk} + L'_{qq} X_q \tag{3.79}$$

We now introduce a quantity of heat transfer Q_k^* expressed as

$$L'_{iq} = \sum_{k=1}^{n} L'_{ik}\, Q_k^* \tag{3.80}$$

When temperature on both sides of the membrane is equal $\Delta T = 0$ in term X_q =0, Eqs. (3.78) and (3.79) yield

$$J_q = \sum_{k=1}^{n} Q_k^*\, J_k \tag{3.81}$$

Comparison of Eq. (3.81) with Eq. (3.71) for X_u = 0 shows that

$$Q_k^* = U_k^* - h \tag{3.82}$$

Further when we write

$$J_q = Q_k^* \sum_{k=1}^{n} J_k \tag{3.83}$$

and compare this equation with Eq. (3.75), we get

$$Q^* = U^* - h \tag{3.84}$$

where Q^* is the heat carried with one unit of mass of the permeant. U^* is the energy transported per unit mass of the permeant in the steady state when $\Delta T = 0$ and is defined as

$$J_u = U^* \sum_{k=1}^{n} J_k \tag{3.85}$$

Combining Eqs. (3.82) and (3.84) and writing U^* in terms of mass fraction i.e. $U^* = \Sigma c_i U_i^*$, we get

$$Q^* = \sum_{k=1}^{n} C_k\, Q_k^* \tag{3.86}$$

Further, Eq. (3.74) with the help of Eq. (3.84) can alternatively be written as

$$v_k\, \Delta P + \sum_{i=1}^{n-1} (\partial \mu_k / \partial c_i)_{T,P,C_j} \Delta c_i + Q_k^* (\Delta T/T) = 0 \quad (k = 1, 2, \ldots, n) \tag{3.87}$$

Equation (3.87) is a set of n relations which can be solved to yield thermo-osmotic pressure difference, ΔP and thermo-osmotic concentration difference $\Delta c_i (i = 1, 2, \ldots, n)$ as a function of temperature difference ΔT.

For the thermo-osmosis of a single gas which does not under go any chemical reaction, Eq. (3.87) yields

$$(\Delta P/\Delta T)J_{i=0} = -[Q^*]^0/vT \tag{3.88}$$

$[Q^*]^0$ is the heat of transport of the pure gas. For a two component gaseous mixture, Eq. (3.87) may be written as

$$v_1\Delta P + (\partial \mu_1/\partial c_1)_{T,P,C_2}\Delta c_1 + Q_1^*(\Delta T/T) = 0 \tag{3.89}$$

$$v_2\Delta P + (\partial \mu_2/\partial c_1)_{T,P,C_2}\Delta c_1 + Q_2^*(\Delta T/T) = 0 \tag{3.90}$$

Multiplication of Eqs. (3.89) and (3.90) by c_1 and c_2, respectively, and addition yields

$$\begin{aligned}&(c_1v_1 + c_2v_2)\Delta P + c_1(\partial \mu_1/\partial c_1)_{T,P,C_2}\Delta c_1 + c_2(\partial \mu_2/\partial c_1)_{T,P,C_2}\Delta c_1\\&\quad + [c_1Q_1^* + c_2Q_2^*]\Delta T/T = 0\end{aligned} \tag{3.91}$$

This equation in the light of Gibbs–Dubem relationship

$$c_1(\partial \mu_1/\partial c_1)_{T,P,C_2}\Delta c_1 + c_2(\partial \mu_2/\partial c_1)_{T,P,C_2}\Delta c_1 = 0 \tag{3.92}$$

leads to

$$(\Delta P/\Delta T)_{J_1=0,\ J_2=0} = -Q_m^*/vT \tag{3.93}$$

where

$$Q_m^* = c_1Q_1^* + c_2Q_2^* \tag{3.94}$$

and

$$v = c_1v_1 + c_2v_2 \tag{3.95}$$

Equation (3.93) is a useful relationship which gives magnitude of the thermo-osmotic pressure corresponding to a particular value of ΔT.

Further, from Eqs. (3.89) and (3.90), it follows that

$$(\Delta c_1/\Delta T) = c_2(v_1Q_2^* + v_2Q_1^*)(\partial \mu/\partial c_1)_{T,P,C_2}/vT \tag{3.96}$$

Equation (3.96) is another important relationship, which predicts that a concentration difference Δc_1, would be developed corresponding to ΔT maintained across the membrane.

Assuming ideal behaviour for components of the gaseous mixture, Eq. (3.93) may also be written as

$$(\Delta P/\Delta T)_{J_1=0,\ J_2=0} = -(c_1 Q_1^* + c_2 Q_2^*) \times P/RT^2/(c_1 M_1 + c_2 M_2) \qquad (3.97)$$

Using Eqs. (3.94) and (3.95) with similar considerations as applied to Eq. (3.96), the following equation can be written with the help of relation:

$$v_k = \mathrm{RT}/M_k P \quad (k = 1, 2, \ldots, n)$$

$$(\Delta c_1/\Delta T) = [c_1 c_2 (Q_2^*/M_1 - Q_1^*/M_2/c_1/M_1 + c_2/M_2](\overline{M}/\mathrm{RT}^2) \qquad (3.98)$$

where $\overline{M} = c_1 M_2 + c_2 M_1$. M_1 and M_2 represents molecular weights of the gas components 1 and 2, respectively. The above equation yields the relation between thermo-osmotic concentration difference Δc_1 and temperature difference ΔT.

Furthermore, using Eqs. (3.97) and (3.98) it is possible to obtain

$$Q_1^* = -RT^2/\overline{M}\Delta T[\Delta c/c_1 + \Delta P/P\overline{M}/M_1] \qquad (3.99)$$

and

$$Q_2^* = -RT^2/\overline{M}\Delta T[\Delta c_2/c_2 + \Delta P/P\overline{M}/M_2] \qquad (3.100)$$

It may be noted that Q_1^* and Q_2^* represent heats of transport of component gases in the binary gaseous mixtures and not heats of transport of the pure components. The above theoretical treatment can be easily extended to the case of ternary and quaternary mixtures [12, 13].

Thermodynamics of irreversible processes discussed above by itself cannot give an insight into the mechanism of gaseous transport through membranes. The molecular kinetic theory can be used with advantage for this purpose.

3.3.1. Kinetic theory of thermo-osmosis

A membrane is a heterogeneous barrier between otherwise two homogeneous systems. It consists of a complicated network of pores which may be connected to each other in a complex manner. The mechanism of transport depends on (i) the size and shape of the pores, and (ii) nature of the permeant. The phenomenon of thermo-osmosis can occur only when the diameter of pores is comparable to mean free path of the permeating species. For gaseous systems, the mean free path can be controlled by controlling the mean pressure. In view of the membrane being made up of a complex network of pores, some pores may have diameters considerably less than the mean free path while some others may have pore diameters considerably larger than the mean free path. In the latter case, viscous flow can occur and the net flow is a composite flow made up of

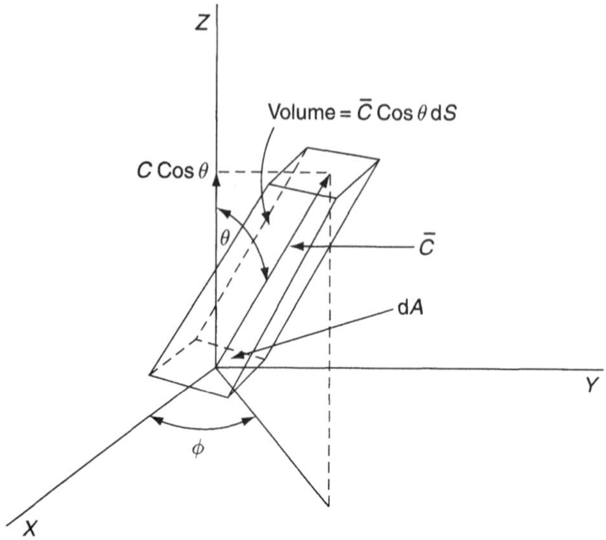

Figure 3.2. Cross-section of the orifice.

Knudsen flow as well as viscous flow. The kinetic theory of these different situations has been discussed by several workers [14–16].

For the sake of simplicity, we discuss below thermo-osmosis of a Knudsen gas.

We consider a gas contained in two chambers communicating through an orifice such that mean free path of the molecules is considerably larger than the orifice. For the evaluation of rate of transfer, let us consider a part dA of the cross-sectional area of the orifice as shown in Fig. 3.2.

The number of molecules hitting the area dA in a unit time are those contained in a cylinder of base dA and a slanting height $\overline{C}$, the distance from the origin which represents the magnitude of velocity. The angles θ and ϕ represent its direction. The volume of the cylinder is $\overline{C}\,\mathrm{Cos}\,\theta\,\mathrm{d}A$ and the number of molecules contained in it is $n\overline{C}\,\mathrm{Cos}\,\theta$. n denotes number of gas molecules per unit volume. Any particular direction from the origin is specified by the solid angle dω given by

$$\mathrm{d}\omega = \sin\theta\,\mathrm{d}\theta\,\mathrm{d}\phi \tag{3.101}$$

The fraction of total number of molecules having velocities with in this spread of direction is dω/4π. Since 4π is the total angle subtended by the surface of a sphere, the number of molecules hitting the surface dA in unit time from the given direction (θ, ϕ) becomes

$$1/4\pi n\,\overline{C}\cos\theta\,\sin\theta\,\mathrm{d}\theta\,\mathrm{d}\phi\,\mathrm{d}A$$

The number of molecules striking a unit surface is

$$1/4\pi n \overline{C} \cos\theta \sin\theta \, \mathrm{d}\theta \, \mathrm{d}\phi$$

Consequently the total number of molecules n' striking from all directions per unit time becomes

$$\mathrm{d}n/\mathrm{d}t = \int_{\theta=0}^{\theta=\pi/2} \int_{\phi=2}^{\phi=2\pi} 1/4\pi n \overline{C} \cos\theta \sin\theta \, \mathrm{d}\theta \, \mathrm{d}\phi = 1/4 n\overline{C} \tag{3.102}$$

n denotes number density.

The velocity is given by

$$\overline{C} = \sqrt{\frac{8RT}{\pi M}} \tag{3.103}$$

M denotes molecular weight of the gas. Eq. (3.102) may thus be written as

$$\mathrm{d}n'/\mathrm{d}t = \frac{1}{4} n \sqrt{\frac{8RT}{\pi M}} \tag{3.104}$$

The mass flowing per unit area per unit time, J_M is given by

$$J_M = \frac{1}{4} mn \sqrt{\frac{8RT}{\pi M}} \tag{3.105}$$

where m is the weight of one molecule of the gas. mn describes the density of the gas and is equal to M/V so that we may write

$$J_M = \frac{1}{4} \frac{M}{V} \sqrt{\frac{8RT}{\pi M}} \tag{3.106}$$

Replacing V by RT/P' we have

$$J_M = MP/\sqrt{2\pi MRT} \tag{3.107}$$

When the two compartments are maintained at temperatures T_1 and T_2; and pressures P_1 and P_2, respectively, the mass flux is given by

$$J_M = \frac{M}{\sqrt{2\pi RM}} \left[\frac{P_1}{\sqrt{T_1}} - \frac{P_2}{\sqrt{T_2}} \right] \tag{3.108}$$

The subscripts 1 and 2 refer to the chambers 1 and 2, respectively. In the steady state $J_M = 0$ so that

$$\left[\frac{P_1}{\sqrt{T_1}} = \frac{P_2}{\sqrt{T_2}} \right] \tag{3.109}$$

which is the well-known transpiration relation.

3.3.2. Energy flow of Knudsen gas

The energy flow J_u occurring on account of the gas flow is given by

$$J_u = 2KTn\sqrt{\frac{RT}{2\pi M}} \tag{3.110}$$

where Eq. (3.104) has been used. $2kT$ is the energy carried by a single molecule. $k = R/N$ where N is Avogadro's number. Eq. (3.110) can easily be transformed to

$$J_u = 2P\sqrt{\frac{RT}{2\pi M}} \tag{3.111}$$

It follows from first law of thermodynamics for open systems that the heat flow J_q is related to the energy flow J_u by the following relationship

$$J_q = J_u - h\,J_M \tag{3.112}$$

where h is the specific enthalpy. If we consider a monatomic gas and assume that the translational energy is exclusively transferred, we have

$$h = E + PV = (3/2RT) + RT = 5/2RT \tag{3.113}$$

so that

$$J_q = -P/2\sqrt{\frac{RT}{2\pi M}} \tag{3.114}$$

If the temperature and pressure on the two sides of the membrane are T_1, T_2 and P_1 and P_2, respectively, Eq. (3.114) leads to the following expression for net heat flow from one chamber to the other:

$$J_q = R/2\sqrt{2\pi RM}[P_2\sqrt{T_2} - \mathrm{P}_1\sqrt{T_1}] \tag{3.115}$$

where $T_1 > T_2$. Eq. (3.115) reduces to Eq. (3.114) when $P_2 = P$ and $T_2 = T_1 = T$; and $P_1 = 0$.

3.3.3. Kinetic molecular interpretation of heat of transport

The energy transferred per unit mass, using Eq. (3.103) is given by

$$U^* = 1/2m(\overline{C})^2 N = 2RT/M \tag{3.116}$$

N denotes the Avogadro number. By definition, the heat of transport Q^* is

$$Q^* = U^* - h$$

For monatomic gas we have,

$$Q^* = U^* - (5/2)RT \tag{3.117}$$

since $h = 5/2RT$. On substituting the value of U^* from Eq. (3.116) we get

$$Q^* = -(RT/2) \tag{3.118}$$

When pores of the membrane are much larger than the mean free path so that the gas moves through it as a whole, U^* includes a work term PV and

$$U^* = u + PV = h \tag{3.119}$$

where u is the specific energy. Hence

$$Q^* = U^* - h = 0 \tag{3.120}$$

Thus the heat of transport is zero for wide pores.
From Eq. (3.109) it follows that

$$P_1 T_1^{1/2} - P_2 T_2^{1/2} = 0 \tag{3.121}$$

We may define mean temperature T and mean pressure P as

$$T_1 = T - \Delta T/2; \quad T_2 = T + \Delta T/2$$
$$P_1 = P - \Delta P/2; \quad P_2 = P + \Delta P/2$$

Equation (3.121) thus may be written as,

$$PT^{-1/2}(1 - \Delta P/2P)(1 - \Delta T/2T)^{1/2} - (1 + \Delta P/2P)(1 + \Delta T/2T)^{1/2} = 0 \tag{3.122}$$

On expansion when $\Delta P/P < 1$ and $-\Delta T/T < 1$, and neglecting higher terms, we obtain

$$(1 - \Delta P/2P)(1 + \Delta T/4T) - (1 + \Delta P/2P)(1 - \Delta T/4T) = 0 \quad \text{or} \quad \Delta P/\Delta T = P/2T \tag{3.123}$$

On combining Eq. (3.123) with Eq. (3.122), Eq. (3.188) is obtained. It thus follows that the kinetic theory of thermo-osmosis is consistent with the thermodynamic theory of thermo-osmosis.

Let us now consider the case of a binary gaseous mixture in which number densities of species 1 and 2 are n_1 and n_2, respectively. The total number density of the mixture, n, is given by

$$n = n_1 + n_2 \tag{3.124}$$

Since the average energy per molecule of each species passing through the orifice is 2kT, the total energy per unit volume for the mixture would be $(n_1 + n_2)\, 2kT$. Therefore the energy U^*_{mix} for the mixture is given by

$$U^*_{\text{mix}} = 2n\, kT/\rho \tag{3.125}$$

where ρ is the density (mass/volume) of the mixture. Again, the energy of transfer for pure species U_i^* may be written as

$$U_i^* = 2RT/M_i \tag{3.126}$$

M_i denotes molecular weight of species i. If C_1 and C_2 are the mass fractions of the respective components in the binary gaseous mixture, it can be shown that

$$U^*_{\text{mix}} = C_1\, U_1^* + C_2\, U_2^* \tag{3.127}$$

since

$$C_1 = n_i M_i / \Sigma n_i\, M_i \tag{3.128}$$

Now it is easy to show that the heat of transport of the binary mixture is given by

$$Q^*{}_{\text{mix}} = \Sigma C_i Q_i^* \quad (i = 1, 2) \tag{3.129}$$

where Q_i^* is the heat of transport of the species i in the mixture. The result is consistent with the thermodynamic theory of thermo osmosis of a multicomponent system.

3.4. Experimental studies

Thermo-osmosis was the first non-equilibrium phenomenon which was extensively studied from experimental and theoretical angles, both for the case of liquids, gases and gaseous mixtures.

Haase [17] has reported some observations on thermo-osmosis of water through cellophane membrane with and without deposition of copper ferrocyanide in the pores. A well-authenticated instance of thermal migration of a liquid against a hydrostatic pressure through a permeable barrier is the fountain effect in liquid He II. Like thermo-osmosis, this process gives rise to a well-defined stationary pressure difference.

There are a number of factors, which govern the occurrences of the phenomenon of thermo-osmosis. It is essential to choose a system such that the dimensions of the pores of the membrane are comparable to the mean free path of the permeating molecules. The deficiency is all the more serious in case of liquids. In case of gases, appreciable difference of pressure has been observed as a result of therm-osmosis through natural rubber membranes [18–20]

Themo-osmosis of water through a cellophane membrane has been reported by Rastogi, Blokhra and Agrawal [21]. Experimental studies on thermo-osmosis of liquids (methanol and water) across Du pont 600 cellophane were reported for the first time by Rastogi and Singh [22].

Experimental set-up used for the measurement of hydrodynamic permeability, thermo-osmotic permeability and thermo-osmotic pressure difference is shown in Fig. 3.3. The membrane was fixed with araldite between two ground-glass joints. Thermocouples were inserted through the adjoining tubes so that the junctions touched the membrane.

The two parts of the cell were kept in two air thermostats maintained at different temperatures; the temperatures of one could be lowered below the room temperature by circulation of water through cooling coils. The movement of the liquid meniscus was observed with a cathetometer. Since pressure gradient is an important variable, the dependence of flow on the pressure gradient when $\Delta T = 0$ was also studied with the same apparatus.

Thermo-osmotic permeability and simple permeability were measured for water and methyl alcohol. Toluene and ethanol were also tried, but even simple permeation was not detectable for ethanol. Permeation occurred with ethanol only when the membrane was treated with zinc chloride solution.

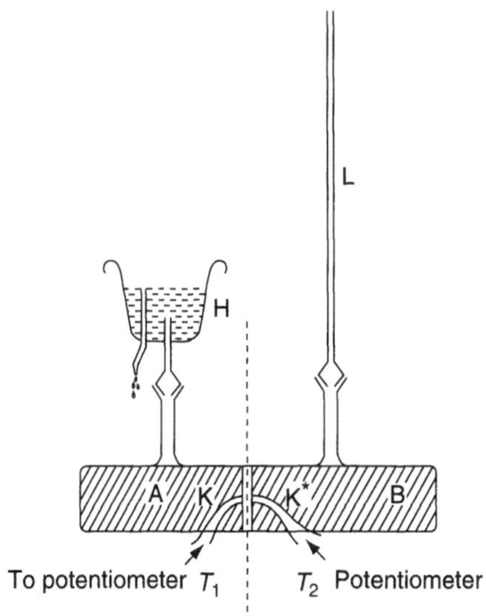

Figure 3.3. Cell for the measurement of hydrodynamic permeability, thermo-osmotic permeability and thermo-osmotic pressure, L, capillary tube, H, constant pressure head, K, thermo-couple, M, Membrane, T_1 = temperature of one air-thermostat, T_2 = temperature of second air thermostat.

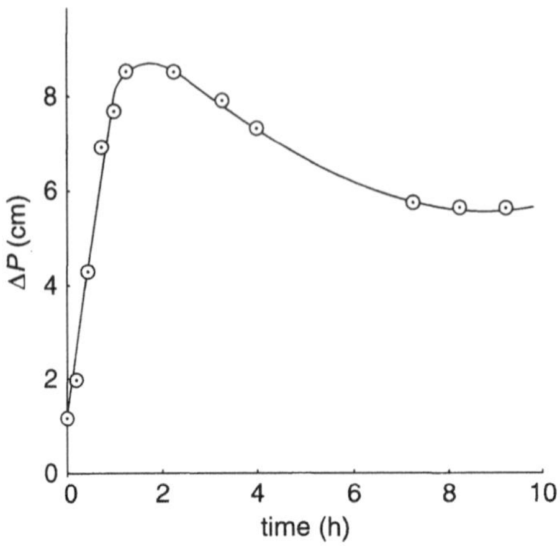

Figure 3.4. Approach to steady thermo-osmotic pressure.

Thermo-osmotic pressure was also measured with the experimental cell shown in Fig. 3.3. The steady state is attained in about 8 h. Figure 3.4. describes a typical run which shows how the steady-state thermo-osmotic pressure is developed in the course of time.

According to Eq. (3.78) based on non-equilibrium thermodynamic theory, hydrodynamic flux for a one-component system is given by

$$(J_M)_{\Delta T=0} = L_{11}(-v\Delta P/T) \tag{3.130}$$

Validity of the above equation was tested by plotting $(J_M)_{\Delta T=0}$ against measured values of ΔP for water and methanol. Straight lines are obtained in both the cases as indicated in Fig. 3.5 confirming the validity of linear flux equation.

Using Eq. (3.78), thermo-osmotic flux in a one-component system may be written as,

$$(J_M)_{\Delta P=0} = L_{12}(-\Delta T/T^2) \tag{3.131}$$

The above equation was tested by plotting experimental values of $(J_M)_{\Delta P=0}$ against ΔT as shown in Fig. 3.6 in the case of water and methanol.

Straight lines are obtained in conformity with the above equation. The cross-phenomenological coefficient L_{12} values estimated from the slopes are also given in the Table 3.1.

Results included in Figs. 3.5 and 3.6 do not afford a complete test of the linear following phenomenological equation based on Eq. (3.78) for a one-component system,

$$J_M = L_{11}(-v\Delta P/T) + L_{12}(-\Delta T/T^2) \tag{3.132}$$

The above equation can also be written as,

$$J_M = (J_M)_{\Delta T=0} + (J_M)_{\Delta P=0} \tag{3.133}$$

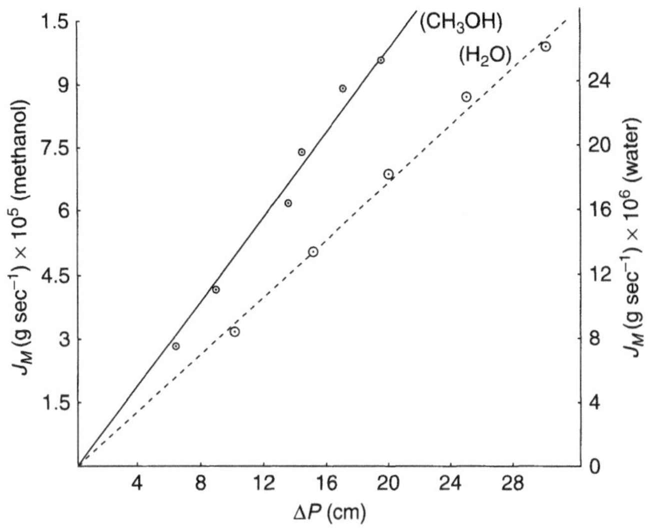

Figure 3.5. Estimation of hydrodynamic permeability of water and methanol.

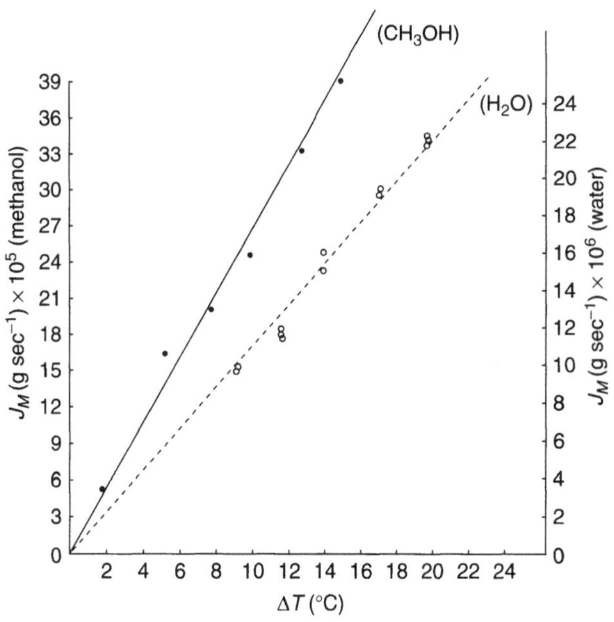

Figure 3.6. Estimation of thermo-osmotic permeability of water and methanol.

Table 3.1. Cross-phenomenological coefficient L_{12} (Mean temperature 50°C).

Quantities	First run	Second run	Third run
ΔP (cm H_2O)	20.35	29.35	23.75
ΔT (°C)	13.00	8.80	16.04
(J_M) g s^{-1}	2.14×10^{-6}	2.65×10^{-6}	2.54×10^{-6}
$(J_M)_{\Delta T=0}$ g s^{-1}	1.53×10^{-6}	2.20×10^{-6}	1.78×10^{-6}
$(J_M)_{\Delta P=0}$ g s^{-1}	0.61×10^{-6}	0.45×10^{-6}	0.76×10^{-6}
L_{12} (g s^{-1} deg)	4.88×10^{-3}	5.30×10^{-3}	4.93×10^{-3}

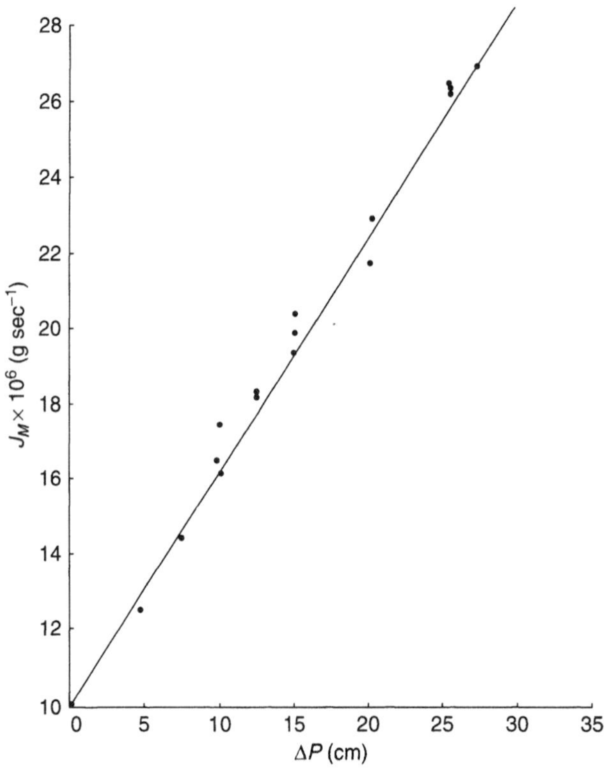

Figure 3.7. Test of linear phenomenological equation for mass transport – Experimental values and theoretical values.

It is thus essential to measure $(J_M)_{\Delta T=0}$, $(J_M)_{\Delta P=0}$ and J_M separately and show that sum of the first two is equal to the third one. Such measurements were made for several values of ΔT and ΔP. Values of J_M are plotted against ΔP at a fixed ΔT in Fig. 3.7.

These are compared with the values calculated from the above equation using measured values of $(J_M)_{\Delta T=0}$ and $(J_M)_{\Delta P=0}$. The agreement is satisfactory and shows that

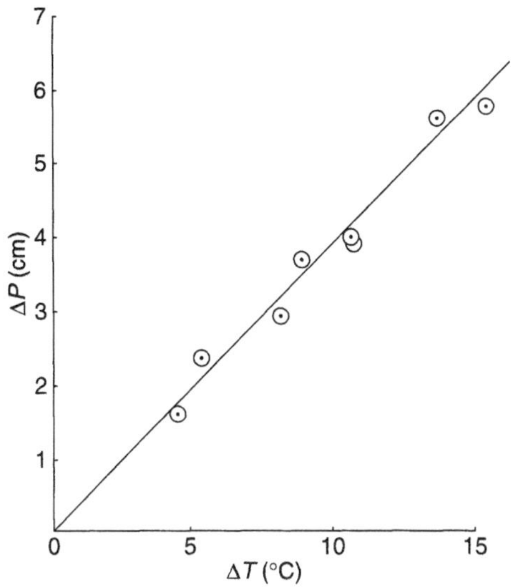

Figure 3.8. Dependence of thermo-osmotic pressure on temperature gradient.

the relation is valid within the range of ΔT and ΔP used. According to Eq. (3.132), when $J_M = 0$

$$[\Delta P/\Delta T] = [L_{11}/L_{12}]/vT \tag{3.134}$$

According to the above equation, thermo-osmotic pressure difference $(\Delta P/\Delta T)_{JM=0}$ should be a constant quantity. This is found to be so in Fig. 3.8 where ΔP has been plotted against ΔT.

Heat of transport values estimated from such data are also included in Table 3.1. The values are comparable for different membranes and are in reasonable agreement with those obtained by Alexander and Wirtz [23].

It may be noted that thermo-osmosis of water and methanol occurs through cellophane membrane but not for ethyl alcohol or toluene. Thermo-osmosis occurs if the radii of the pores are smaller than the Van der Waals radii, but not if the pore size is greater than the molecular radii but smaller than the mean free path. Since the mean free path of the molecules in the liquid is of molecular dimensions, the range of pore size for thermo-osmosis to occur is limited. The size of ethyl alcohol molecules appears to be comparable to the pore size, whereas for toluene the pore size is perhaps greater than the molecular radii and the mean free path. Probably for methyl alcohol and water the pore diameter of the membrane is between the mean free path and the Van der Waals radii. However, there is no independent evidence for the magnitude of mean free path of liquid water.

The thermo-osmotic permeability and the ordinary permeability of water is greater than that for methanol. For these reasons, thermo-osmosis of water could not be detected with sintered Pyrex where pore diameter ranged from 10^{-2} to 10^{-4} cm. These considerations suggest that mechanism of permeation of water and methanol through allophone membrane involves molecular flow and not Poiseuille flow. With molecular flow, either only one molecule may pass through the pore at a time or molecules may migrate along the surface of the pore [24].

Detailed studies on thermo-osmosis using highly selective cellulose acetate membrane in the presence and absence of osmotic pressure difference have also been carried out [25]. Using general description of thermo-osmosis based on irreversible thermodynamics, it was shown that coupling between the flow of heat and the flow of water is quite loose possibly on account of thermal leak between the compartments. Whatever the detailed structural interpretation, it was argued that in annealed, less-permeable membranes, the water–matrix interaction is increased relative to the water–water interaction and with only this type of interaction strong thermo-osmosis is expected.

Studies on thermo-osmosis of water through cellulose acetate membrane [26] support predictions by Rastogi and Singh [27] based on correlation between the existence of themo-osmosis and the membrane pore size.

Studies on hydrodynamic permeability, thermo-osmotic permeability for D_2O have also been carried out and heat of transport has been estimated. Similar measurements have been made for solutions of KCl in H_2O and D_2O using a copper ferrocyamide membrane. These results support the theory of non-equilibrium thermodynamics within a limited range of departure from equilibrium [28].

Extensive studies on thermo-osmosis have been carried out using various hydrophobic and hydrophilic membranes [29]. Using non-equilibrium thermodynamic principles, it was concluded that water is transferred through hydrophilic polymer membranes from the cold side to the hot side because the transported entropy of water in the membrane is smaller than molar entropy of water in the external free solutions. In contrast, water is transferred through hydrophobic polymer membranes from the hot side to the cold side because the transported entropy of water in the membrane is larger than the molar entropy of water in the external free solutions.

During the study of phenomenon of thermo-osmosis, major difficulty in encountered in testing the validity of ORR. According to Eq. (3.79), formerly discussed in Section 3.3, the heat flux equations can be written as

$$J_Q = L_{21}(-v\Delta P/T) + L_{22}(-\Delta T/T^2) \tag{3.135}$$

For the estimation of L_{21}, one needs $(J_Q)_{\Delta T=0}$ the quantity of heat necessary to maintain the system at constant temperature, when the fluid is passed through the membrane by pressure difference. These studies are difficult since $(J_Q)_{\Delta T=0}$ is too small to make the measurements useful [27]. Accordingly, test of ORR in such cases is very difficult.

It has been claimed, however, that ORR is satisfied within 5–8 per cent in the case of liquid He II [30, 31].

3.5. Thermo-osmosis of gases and gaseous mixtures

Number of experimental studies on thermo-osmosis of gases have been reported in literature [32]. The steady-state relation is found to be satisfied in all cases studied within a limited range of non-equilibrium. A typical experimental set-up used for studying thermo-osmosis is shown in Fig. 3.9 [33].

Unglazed porcelain M (porous plate used in organic preparations) was fixed with Araldite. The gases, carbon dioxide and nitrogen used in these studies were passed through concentrated sulphuric acid and P_2O_5 to remove moisture. The apparatus was first evacuated and the gas under investigation was introduced. The cell was again evacuated and filled with the gas. The process of evacuation and filling of the cell was repeated several times to ensure complete removal of the air. The mercury level in the two limbs of the differential manometer, M_l could be adjusted to a definite height with help of the mercury reservoir. Stopcocks N were then closed. After sometime, the pressure of the gas in the cell was noted. The chamber A was heated with a constantan heater, H. Difference of temperature on the two faces of the membrane was measured by copper constantan thermocouple T. Difference of pressure was noted with a cathetometer. Attainment of steady state takes a long time as shown in Fig. 3.10 for a typical run.

For isothermal permeability measurements, differential manometer M_l was replaced by an ordinary manometer, which was in communication with both the compartments A and B. After evacuation, the gas was introduced into one compartment. This created a difference of pressure on the two sides of the membrane, which decreased with time.

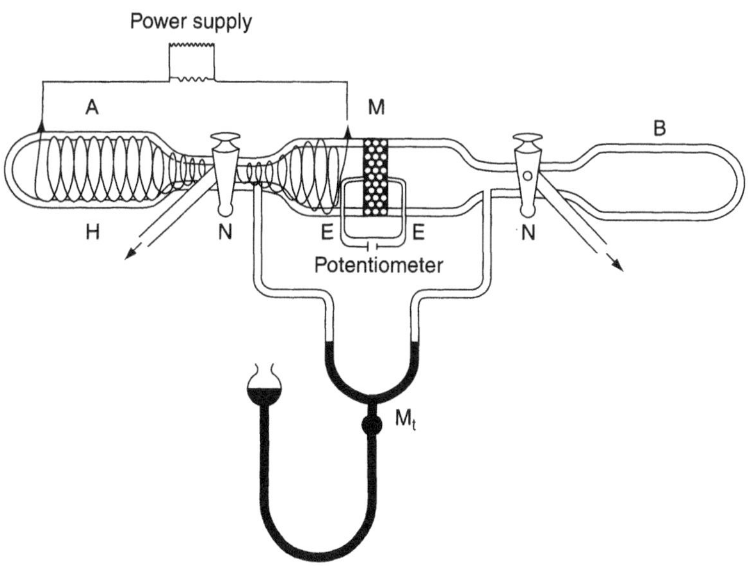

Figure 3.9. Experimental set-up for the measurement of thermo-osmotic concentration difference.

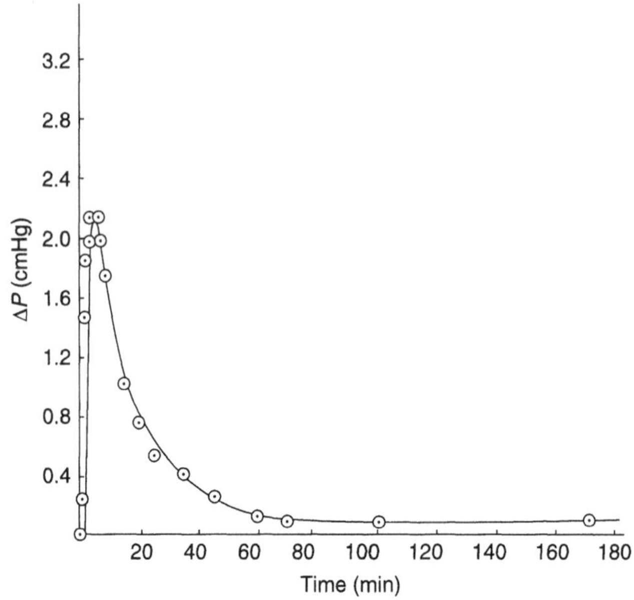

Figure 3.10. Attainment of steady state.

ΔP was noted at different time intervals. The decay of pressure as a function of time is shown in Fig. 3.11. If h is the distance traversed at time t and A is the cross-sectional area of the membrane, the rate of flow J, is given by $J = A(dh/dt)$

The isothermal permeability of a gas may be estimated by plotting J against ΔP. A typical plot is shown in Fig. 3.12 for CO_2 from which isothermal permeability of CO_2 was found to be $7.4 \times 10^{-4}\,cm^2\,(atm\,s)^{-1}$ at 33°C.

Typical thermo-osmotic pressure data for CO_2 using unglazed porcelain membrane at various $\Delta T/T_1T_2$ are presented in Fig. 3.13. The result are in conformity with the steady-state relation. This also implies that heat of transport Q^* is independent of mean temperature within the range of investigation. Heat of transport was also found to be independent of difference of temperature up to $\Delta T = 130$°C. It shows that, so far as thermo-osmosis of gases is concerned, thermodynamic predictions have a wide range of validity.

Rastogi and Singh [34] studied for the first time phenomenon of thermo-osmosis in gaseous mixtures. Measurements of thermo-osmotic pressure difference were carried out using the experimental set-up already shown in Fig. 3.9 for mixtures of CO_2 and O_2 of different composition through unglazed porcelain membrane. Measurement were also carried out using

(i) H_2O and D_2O vapour mixtures [28];
(ii) Ternary mixtures of O_2, C_2H_4 and CO_2 [35]; and
(iii) Quaternary mixtures of CO_2, N_2, C_2H_4 and O_2 [36]

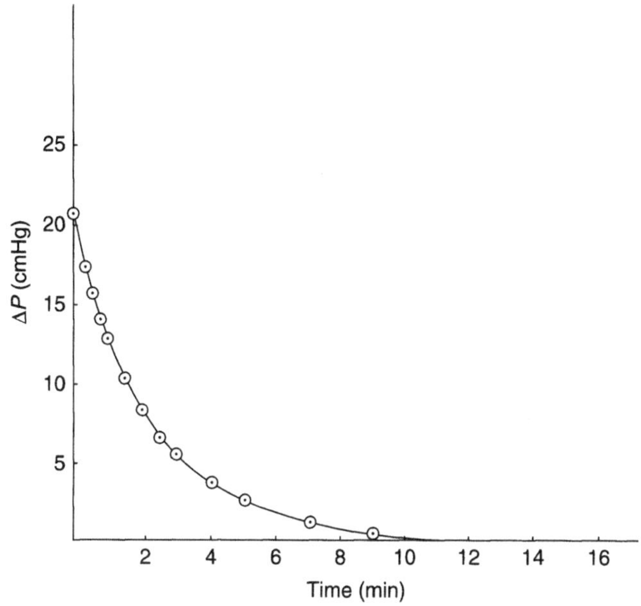

Figure 3.11. Isothermal decay of pressure difference as a function of time.

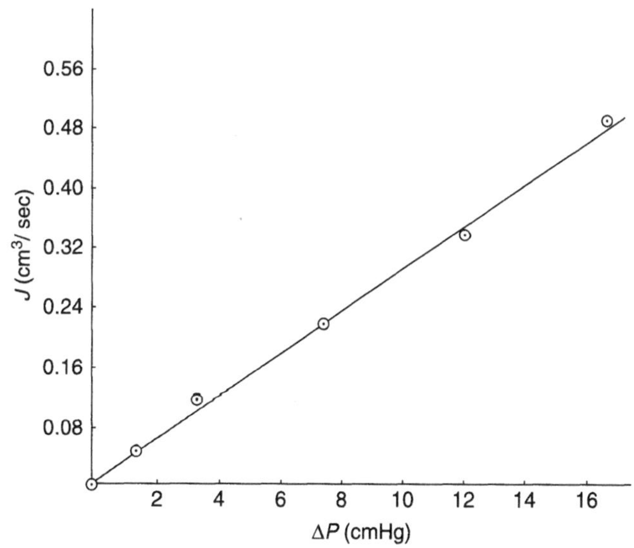

Figure 3.12. Estimation of isothermal permeability.

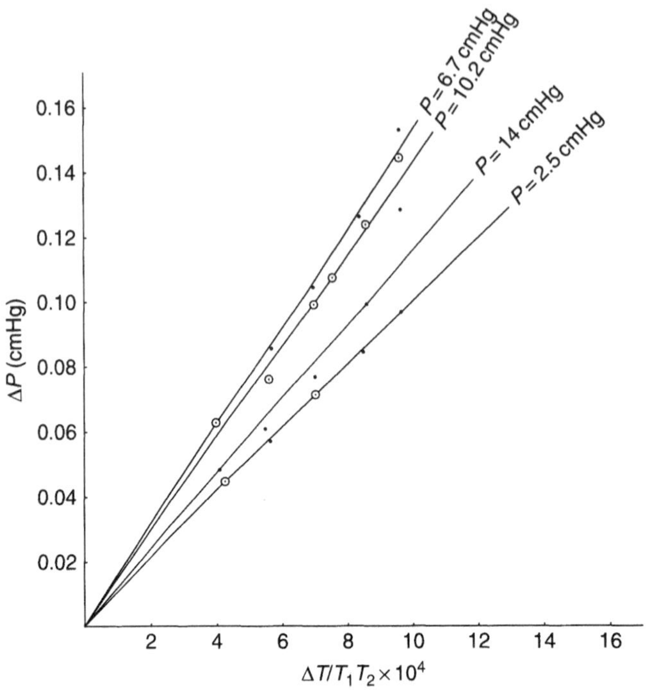

Figure 3.13. Test steady-state relation Eq. (3.88) for CO_2.

The steady-state relations were found to be valid in a wide range. Heats of transport were estimated using the experimental results on the basis of thermodynamic theory of irreversible processes to examine the dependence of heat of transport on the composition of the mixtures. It was found that the heat of transport of the mixture, Q_m^* is related to mass fraction as predicted by Eq. (3.129).

3.5.1. Thermo-osmotic concentration difference

This is a unique non-equilibrium phenomenon similar to thermal diffusion. A steady concentration difference is created when a temperature difference is maintained across the membrane. For liquid mixtures, experimental studies on thermo-osmotic concentration difference have been reported by few workers [17, 37]. In gaseous mixtures it has been demonstrated by Rastogi and Rai [38]. Concentration difference of the order of 1×10^{-3} per degree in the binary gaseous mixtures of CO_2 and C_2H_4 has been reported. They used a modified form of the experimental set-up employed earlier [34] for the measurement of thermo-osmotic pressure difference, and is shown in Fig. 3.9. Both the compartments of the cell were first evacuated. One of the gases was then admitted into the cell and corresponding pressure was noted. Thereafter, the second gas was admitted and the corresponding total pressure was noted to know the partial pressure of the gas. The partial pressures of the two gases were then used to obtain composition of the gaseous mixture. A difference of temperature across the ungazed porcelain membrane, M was maintained using a constantan heater, H. For the measurement of temperatures at the two faces of the membrane, copper constantan thermocouples were used and the resulting difference of pressure was recorded with a cathetometer. After noting the thermo-osmotic pressure difference, the stopcocks were closed. The composition of the gaseous mixture in each of the compartments was determined by refractive index measurements using a Rayleigh Interferometer in the following manner.

The interferometer was first calibrated in term of fringes shifted at atmospheric pressure. A sodium flame ($\lambda = 5893\,\text{Å}$) was used as the source of monochromatic light. The number of fringes shifted was plotted against the scale reading to produce a calibration curve. The refractive index, n_r was then calculated with the help of the following equation.

$$(n_r - 1) = m/P_1 - P_2 \times 76\,\lambda T/273\,L$$

where

m = number of fringes shifted;
T = temperature of the cell;
λ = wave length of sodium light;
P_1 = pressure in the first chamber;
P_2 = pressure in the second chamber; and
L = length of the cell.

A least square analysis showed that the results are best fitted by the equation

$$(n_r - 1) \times 10^5 = 113.2\,C_1 + 68.9$$

C_1 denotes mass fraction of component 1. This equation was subsequently used for the estimation of concentration in each chamber. Experimental results were then utilized to estimate the heat of transport of component gases utilizing the theory discussed in Section 3.2 Steady-state relation (3.97) is found to be valid in a wide range. Further, Eq. (3.129) relating heat of transport of gaseous mixtures and the heat of transport of individual gases is also found to be satisfied. Experimentally estimated heat of transport is found to be in reasonable agreement with calculated values based on mixture of Knudsen gases.

3.6. Thermo-osmosis in biological systems

Thermo-osmosis in biological systems has received scant attention [39], although it is realized that fluxes of thermal energy through biological membranes can occur owing to the different rates of metabolic activity in different components separated by such membranes. However, the temperature differences are likely to be small so that thermo-osmotic transport may not be of serious consequence for biological processes. Gaeta *et al.* [40] investigated thermo-osmotic transport in coenocytic alga valonia utricularis by following incorporation and efflux of tritiated water (^3H_2O).

References

1. L.Z. Waldman, Naturforschung, 29 (1947) 358.
2. R.P. Rastogi and G.L. Madan, Trans. Faraday Soc., 62 (1966) 3325.
3. S.R. DeGroot, The Thermodynamics of Irreversible Processes, North Holland Publishing Co, Amsterdam, 1951.
4. K.G. Denbigh, Thermodynamics of the Steady State, Methuen & Co. Ltd, London, 1951.
5. I. Prigogine, Introduction to the Theories of Irreversible Processes, C.C. Thomas Publishers, Springfield, 1955.
6. S.R. deGroot and P.Mazur, Non-equilibrium Thermodynamics, North Holland Publishing Co, Amsterdam, 1962.
7. D.D. Fitts, Non-equilibrium Thermodynamics, McGraw Hill Book Co, New York, 1962.
8. R. Hasse, Thermodynamics of Irreversible Processes, Addison-Wesley, Massachusetts, 1968.
9. B.N. Srivastava, R.P. Rastogi and A.S. Verma, Physica., 20 (1954) 639.
10. R.P. Rastogi and R.C. Srivastava, Proc. Phys. Soc., 67A (1954) 639.
11. R.P. Rastogi and R.C. Srivastava, Trans. Faraday Soc., 51 (1955) 343.
12. R.P. Rastogi and A.P. Rai, J. Phys. Chem., 70 (1974) 2693.
13. R.P. Rastogi and A.P. Rai, J. Membr. Sci., 7 (1980) 39.
14. E.A. Mason, R.B. Evans and G.M. Walson, J. Phys. Chem., 35 (1960) 6.
15. R.D. Present, Kinetic Theory of Gases, McGraw-Hill Book Co, New York (1958).
16. L.B. Ticknor, J. Phys. Chem., 62 (1958) 1483.
17. R. Haase and C. Steinest, Z. Phys. Chem., Frankf Asusgabe, 21 (1959) 270.
18. K.G. Denbigh and G. Raumann, Proc. Roy. Soc., A20 (1951) 518.
19. R. Bearman, J. Phys. Chem., 61 (1957) 708.
20. M. Crowe, Trans. Faraday Soc., 53 (1957) 692.
21. R.P. Rastogi, R.L. Blokhra and R.K. Agrawal, Trans. Faraday Soc., 60 (1964) 585.
22. R.P. Rastogi and K. Singh, Trans. Faraday Soc., 60 (1964) 1386.
23. K.F. Alexander and K. Wirtz, Z. Physik., 195 (1950) 165.
24. L.B. Ticknor, J. Phys. Chem., 55 (1951) 684.
25. M.S. Dariel and O. Kedem, J. Phys. Chem., 79 (1975) 336.
26. J.I. Menganl and A. Aquilar, J. Memb. Sci., 4 (1978) 209.
27. R.P. Rastogi and K. Singh, Trans. Faraday Soc., 62 (1966) 1754.
28. R.P. Rastogi, P.C. Shukla and B. Yadav, Biochem. Biophys. Acta., 249 (1971) 454.
29. M. Tasaka, T. Mishra and O. Sekignchi, J. Memb. Sci., 54 (1990) 191.
30. D.F. Brewer and D.O. Edwards, Proc. Phys. Soc. (London), 71 (1958) 117.
31. D.G. Miller, in Transport Phenomena in Fluids, H.J.M. Hanley (ed.), Marcel Dekker, New York, 1969, Ch. 11.
32. R.P. Rastogi and B. Mishra, J. Sci. Ind. Res., 41 (1982) 8.

33. R.P. Rastogi, K. Singh and H.P. Singh, J. Phys. Chem., 73 (1969) 2798.
34. R.P. Rastogi, H.P. Singh, J. Phys. Chem., 74 (1970) 1946.
35. R.P. Rastogi and A.P. Rai, J. Phys, Chem., 70 (1974) 269.
36. R.P. Rastogi, A.P. Rai and M.L. Yadav, Indian J. Chem., 12 (1974) 1273.
37. H. Vnik and S.A.A. Chiste, J. Memb. Sci., 1 (1976) 149.
38. R.P. Rastogi and A.P. Rai, J. Memb. Sci., 4 (1979) 221.
39. P.C. Shukla, Membrane, 9 (1984) 202.
40. F.S. Gaeta, P. Canciglia, A.D. Acunto and D.G. Mita, J. Memb. Sci., 16 (1983) 339.

Chapter 4

ELECTRO-OSMOTIC PHENOMENA

4.1. Introduction

The electro-osmotic phenomena are of considerable interest from the standpoint of biological transport, which has been stimulated by the need to understand permeability phenomena and secretion processes [1–5]. Electro-osmotic transport in ion-exchange membranes has received attention because of its technical importance in desalination [6]. A number of applications of electro-osmotic effects which include the use of these effects to pump fluids, generate electricity, measure pressure of flow have been proposed and investigated during recent years. Experimental studies of devices for underwater ultrasonic signal generators, receivers [7] and pressure transducers [8] employing the electro-osmotic principle have also been made.

Electro-osmotic effects are of intrinsic interest also, since they give rise to a number of steady-state phenomena which can be conveniently studied experimentally. These afford examples where the power of non-equilibrium thermodynamics can be easily demonstrated. This is a unique phenomenon where non-linear transport processes have been experimentally observed and studied.

4.2. Non-equilibrium thermodynamics of electro-osmotic phenomena

We shall first outline the theory of electro-osmosis and streaming potential based on thermodynamics of irreversible processes [9–14]. Let us consider two chambers separated by a very thin membrane, so that it merely serves as a dividing surface which we need not consider as a separate phase.

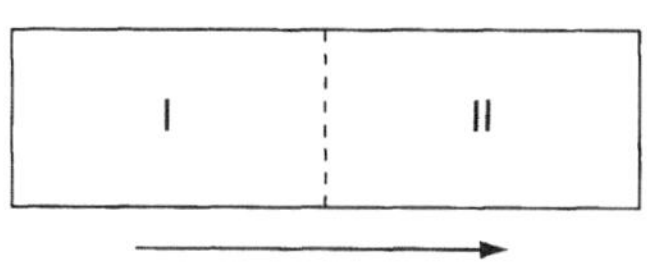

Electric current: Chambers I and II contain a fluid with n components carrying electrical charge e_k ($k = 1, 2, \ldots, n$) per unit of mass. We assume that both the chambers are at the same temperature T and the concentrations are uniform in each chamber. M_k^{I} and M_k^{II} are the masses of component k in chambers I and II, respectively. Similarly ϕ^{I} and ϕ^{II} are the potentials in the two chambers. In the system under discussion, we consider the transport of matter and electricity from one chamber to the other. In the process, both mass and change would be conserved. Thus we must have

$$\mathrm{d}M_k^{\mathrm{I}} + \mathrm{d}M_k^{\mathrm{II}} = 0 \quad (k = 1, 2, \ldots n) \tag{4.1}$$

and

$$\sum_{k=1}^{n} e_k \,\mathrm{d}M_k^{\mathrm{I}} + \sum_{k=1}^{n} e_k \,\mathrm{d}M_k^{\mathrm{II}} = 0 \tag{4.2}$$

Superscripts I and II refer to chambers I and II.

Now our object would be to evaluate entropy production $\mathrm{d_i}S$ due to irreversible processes inside the system, which are simply the transport of matter and electricity. The total entropy production $\mathrm{d}S$ due to internal as well as external factors would be given by

$$T\,\mathrm{d}S = T\,\mathrm{d}S^{\mathrm{I}} + T\,\mathrm{d}S^{\mathrm{II}} \tag{4.3}$$

or

$$T\,\mathrm{d}S = \mathrm{d}U^{\mathrm{I}} + P^{\mathrm{I}}\,\mathrm{d}V^{\mathrm{I}} - \sum_{k=1}^{n} \mu_k^{\mathrm{I}}\,\mathrm{d}M_k^{\mathrm{I}} + \mathrm{d}U^{\mathrm{II}} + P^{\mathrm{II}}\,\mathrm{d}V^{\mathrm{II}} - \sum_{k=1}^{n} \mu_k^{\mathrm{II}}\,\mathrm{d}M_k^{\mathrm{II}} \tag{4.4}$$

where we have assumed the validity of the Gibbs equation for entropy production even outside equilibrium. This assumption is always used in non-equilibrium thermodynamics and it has been shown that this is justified for near equilibrium situations. The term U represents the internal energy; P the pressure; V the volume; and μ_k the chemical potential of species k. From the first law of thermodynamics, we have

$$\mathrm{d}U = \mathrm{d}U^{\mathrm{I}} + \mathrm{d}U^{\mathrm{II}} = \mathrm{d}Q - P^{\mathrm{I}}\,\mathrm{d}V - P^{\mathrm{II}}\,\mathrm{d}V + (\phi^{\mathrm{I}} - \phi^{\mathrm{II}})I\,\mathrm{d}t \tag{4.5}$$

where I is the current given by

$$I = -\sum_{k=1}^{n} e_k\,\mathrm{d}M_k^{\mathrm{I}}/\mathrm{d}t = \sum_{k=1}^{n} e_k\,\mathrm{d}M_k^{\mathrm{II}}/\mathrm{d}t \tag{4.6}$$

and $\mathrm{d}Q$ is the heat absorbed from the surroundings. From Eqs. (4.4) and (4.5) we get

$$T\,\mathrm{d}S = \mathrm{d}Q - \sum_{k=1}^{n} \left(\mu_k^{\mathrm{I}} + e_k\phi^{\mathrm{I}} - \mu_k^{\mathrm{II}} - e_k\phi^{\mathrm{II}}\right)\mathrm{d}M_k^{\mathrm{I}} \tag{4.7}$$

Now $dS = d_eS + d_iS$, where d_eS is the reversible entropy change due to interaction with the surroundings, and hence σ, the rate of entropy production, is given by

$$\sigma = \frac{d_iS}{dt} = -\sum_{k=1}^{n} \Delta(\mu_k + e_k\phi)\frac{dM_k}{dt} \tag{4.8}$$

where Δ represents the difference of a quantity in chambers I and II. When the temperature and concentration remain fixed

$$\Delta\mu_k = v_k \Delta P \tag{4.9}$$

where v_k is the specific volume of component k. Hence

$$\sigma = -\sum_{k=1}^{n} v_k \frac{dM_k^{I}}{dt}\frac{\Delta P}{T} - \sum_{k=1}^{n} -e_k \frac{dM_k^{I}}{dt}\frac{\Delta\phi}{T} \tag{4.10}$$

Equation (4.10) enables us to spot out the fluxes and forces in the system by remembering that entropy production is the sum of the product of fluxes J_i and forces X_i, i.e.

$$\sigma = \Sigma J_i X_i \tag{4.11}$$

Thus, it follows that the volume flow J and the current I are given by

$$J = -\Sigma v_k \frac{dM_k^{I}}{dt} \tag{4.12}$$

$$I = -\sum_{k} e_k \frac{dM_k^{I}}{dt} \tag{4.13}$$

The corresponding forces are $\Delta P/T$ and $\Delta\phi/T$. The explicit transport equations in this linear range would be written as

$$J = L_{11}\left(\frac{\Delta P}{T}\right) + L_{12}\left(\frac{\Delta\phi}{T}\right) \tag{4.14}$$

$$I = L_{21}\left(\frac{\Delta P}{T}\right) + L_{22}\left(\frac{\Delta\phi}{T}\right) \tag{4.15}$$

where Ls are the phenomenological coefficients, governed by the following conditions

$$L_{12} = \mathrm{L}_{21} \text{ (Onsager reciprocity relation)} \tag{4.16a}$$

$$L_{11}\, L_{22} > L_{12}\, L_{21} \tag{4.16b}$$

on account of positive definite character of entropy production.

In the above discussion we have assumed that the barrier or membrane thickness is infinitesimally small. It can be seen that Eq. (4.10) can be deduced even when the membrane has a finite thickness Δx:

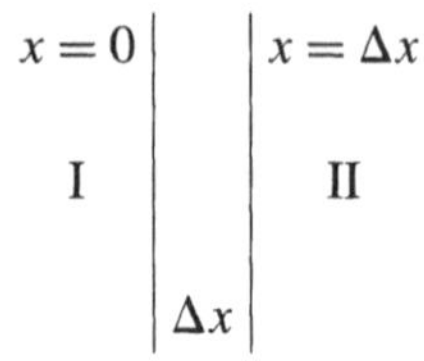

The flows passing through the membrane would be perpendicular to the surface and will have the same value at all points on the surface, provided the membrane is homogeneous. Since the internal entropy production occurs on passage to the membrane, we can also evaluate the net entropy production by integration over the thickness of the membrane by using the following expression:

$$T_\sigma = \int_0^{\Delta x} T\theta \,\mathrm{d}x \tag{4.17}$$

where θ is the entropy production within a volume element of unit area and of thickness $\mathrm{d}t$. It can be shown that

$$\theta = J(-\mathrm{grad}\ P) + I(-\mathrm{grad}\ \phi)$$

so that

$$T_\sigma = J\int_0^{\Delta x} -\frac{\partial P}{\partial x}\,\mathrm{d}x + I\int_0^{\Delta x} -\frac{\partial \phi}{\partial x}\,\mathrm{d}x = J\Delta P + I\Delta\phi \tag{4.18}$$

For simplicity we have considered that all gradients are along the x-coordinate. The above formulation has been extended to the cases of composite membranes [15] and ion-exchange membranes [16, 17].

4.2.1. Non-equilibrium steady states

Equations (4.14) and (4.15) easily yield the following relations for various steady states.

Electro-osmosis:

$$(J)_{\Delta P=0} = L_{12}\frac{\Delta\phi}{T};\quad (J/I)_{\Delta P=0} = \frac{L_{12}}{L_{22}} \tag{4.19}$$

Electro-osmotic pressure:

$$(\Delta P)_{J=0} = -\frac{L_{12}}{L_{11}}\,\Delta\phi \tag{4.20}$$

Streaming potential:

$$(\Delta\phi)_{I=0} = -\frac{L_{21}}{L_{22}}\,\Delta P \tag{4.21}$$

Streaming current:

$$(I)_{\Delta\phi=0} = \frac{L_{21}}{T}\,\Delta P; \quad (I/J)_{\Delta\phi=0} = \frac{L_{21}}{L_{11}} \tag{4.22}$$

These phenomena are interrelated on account of Onsager relation. Thus

$$\left(\frac{\Delta P}{\Delta\phi}\right)_{J=0} = -\left(\frac{I}{J}\right)_{\Delta\phi=0} \tag{4.23}$$

$$\left(\frac{\Delta\phi}{\Delta P}\right)_{I=0} = -\left(\frac{J}{I}\right)_{\Delta P=0} \tag{4.24}$$

It would be interesting to compare the magnitudes of σ for each of the steady states [18]. On substituting the values of J and I from Eqs. (4.14) and (4.15) in (4.10), we have

$$\sigma = L_{11}\left(\frac{\Delta P}{T}\right)^2 + (L_{12}+L_{21})\,\frac{\Delta P\Delta\phi}{T^2} + L_{22}\left(\frac{\Delta\phi}{T}\right)^2 \tag{4.25}$$

We can plot σ as a function of $\Delta P/T$ and $\Delta\phi/T$ on a three-dimensional diagram to yield an elliptical paraboloid with vertex at $\sigma = 0$, which corresponds to the equilibrium state (Fig. 4.1). The major and minor axes would be inclined to the axis representing $\Delta P/T$ and $\Delta\phi/T$ by an angle θ' given by

$$\theta' = \frac{3\pi}{2} + \frac{1}{2}\tan^{-1}\frac{2L_{12}}{L_{11}-L_{12}} \tag{4.26}$$

For a steady state when $J = 0$ and $\Delta\phi =$ constant, the entropy production would be represented by a section of the paraboloid, which would be a parabola with coordinates of vertex given by

$$\sigma = \left(\frac{\Delta\phi}{T}\right)^2\left(\frac{L_{11}L_{22}-L_{12}L_{21}}{L_{11}}\right); \quad \left(\frac{\Delta P}{T}\right) = -\frac{L_{12}}{L_{11}}\frac{\Delta\phi}{T} \tag{4.27}$$

This value of σ would correspond to steady-state value, since it is minimum according to Prigogine's theorem. The values of $\sigma_{\min}$ for different steady states are summarized in Table 4.1.

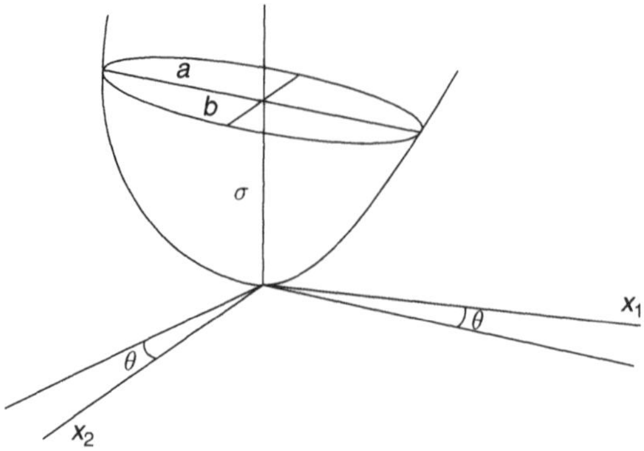

Figure 4.1. Geometrical interpretation of entropy production.

Table 4.1. Values of σ for different steady states.

Steady state	Value of σ
Streaming potential	$\left(\frac{L_{11}L_{22}-L_{12}L_{21}}{L_{22}}\right)\left(\frac{\Delta P}{T}\right)^2$
Electro-osmotic pressure	$\left(\frac{L_{11}L_{22}-L_{12}L_{21}}{L_{11}}\right)\left(\frac{\Delta \phi}{T}\right)^2$
Electro-osmosis	$\left(\frac{L_{21}^2+L_{11}L_{22}-L_{21}(L_{21}+L_{12})}{L_{22}\,(L_{11}L_{22}-L_{21}L_{12})}\right)I^2 \sim L_{22}\left(\frac{\Delta \phi}{T}\right)^2$, when $L_{11}L_{22} >> L_{21}L_{12}$
Streaming current	$\left(\frac{L_{21}^2+L_{11}L_{22}-L_{21}(L_{21}+L_{12})}{L_{11}\,(L_{11}L_{22}-L_{21}L_{12})}\right)J^2 \sim L_{11}\left(\frac{\Delta P}{T}\right)^2$, when $L_{11}L_{22} >> L_{21}L_{12}$

It can be seen that the entropy production values for various steady states are not the same. The value of σ for streaming potential and electro-osmotic pressure would be the same when $(\Delta P)^2/L_{22} = (\Delta\phi)^2/L_{11}$. It also follows that since the entropy production is positive, $L_{11}L_{22}$ must be greater than $L_{12}L_{21}$.

4.3. Theories based on models of membranes

In the development of the thermodynamic theory of the electro-osmotic phenomena, it is not necessary to examine how electro-osmosis or streaming potential occurs. The theory is independent of the nature of the membrane or its character. It yields results in terms of phenomenological coefficients. It should be noted that these coefficients do depend on membrane characteristics. The Onsager reciprocal relation would be valid for the same membrane. In order to get the complete picture, including the reason for the occurrence of these phenomena and the factors on which the phenomenological coefficients depend, we shall discuss the theories based on the specific models of the membrane.

Membranes are essentially of two types so far as their electrical nature is concerned: (i) uncharged type and (ii) charged type, for example ion-exchange membranes. In the former, mobile charges come from the diffuse double layer at the solid–liquid interface, while in the latter, counter-ions are the mobile species. For example, in zeolites or cation-exchange resins, the cations are mobile, whereas the anionic body is not. Such membranes in contact with electrolytic solutions are preferentially permeable to cations. The ion-exchange membranes may be either homogeneous or heterogeneous. The former are coherent ion-exchanger gels, while the latter consist of colloidal ion-exchanger particles embedded in a binder, such as polystyrene or polyethylene.

In terms of mechanical structure, the membrane may be thought of as a bundle of capillaries or channels having a characteristic pore size and length. These may be arranged as (i) a parallel array, i.e. parallel to each other along the x-axis, or (ii) a series array. Membranes having parallel array of capillaries are the simplest and the average number of channels and pore size can be estimated from electro-osmotic [19]. The net flow would be the sum of flows from individual channels, while in the latter, the fluxes must be the same throughout the constituent layers. Thus, for a parallel array, having n capillaries,

$$J = \sum_{i=1}^{n} a_i j_i$$

where j_i is the flow through L_{ik}, a single capillary, and a_i the fractional area given by $A_i/\Sigma A_i$, where A_i is the area of cross-section of the ith capillary. The phenomenological coefficients L_{ik} would be given by

$$L_{ik} = \sum_{i=1}^{n} a_i l_{ik}$$

Similarly, for a series array having membranes α, β, γ, ..., we must have

$$J^{\alpha} = J^{\beta} = J^{\gamma} = \ldots$$

A typical example of a double membrane system is the epithelial cell bounded on either side with liquid of different permeability properties [20].

If directional characteristics are considered, the membrane may be either isotropic or anisotropic. Examples of anisotropic membranes are cellular membranes including axonal membranes composed of different layers [21]. It has been pointed out that sandstones [22] and other rocks, sintered metals, sintered glass and unglazed ceramic bodies may be anisotropic with respect to permeability. In the following, we shall consider the case of uncharged membrane or capillary.

4.3.1. Theories of electro-osmosis in uncharged membranes

It is believed that in uncharged membranes, electro-osmotic flow occurs on account of the existence of electrical double layer at the solid–liquid interface as shown in Fig. 4.2, where the solid is supposed to be negatively charged. The positive charges move towards the negative electrode when an electric field is applied. The potential near the wall would have a fixed value (φ_0). The potential would drop as we move away from the wall in the manner shown in Fig. 4.3, on account of the diffuse nature of the double layer.

The term δ represents the thickness of the so-called Helmholtz double layer, which is fixed, and the potential at the slip-plane is called the ζ-potential. The origin of charges in the double layer is not clearly known, but it is believed to arise on account of preferential adsorption of ions in many cases. The ions in the double layer are relatively

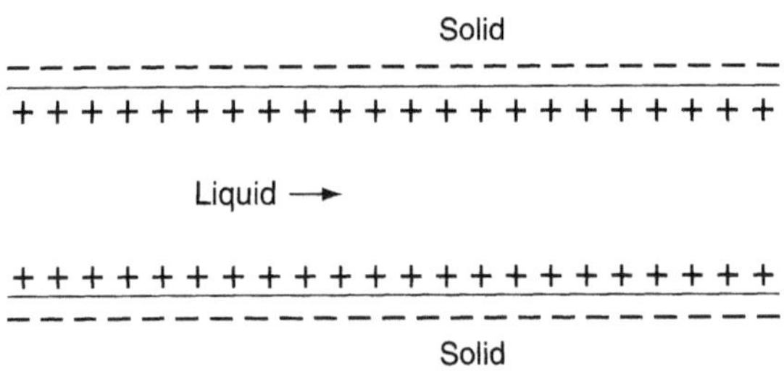

Figure 4.2. Mechanism of electro-osmosis in the capillary.

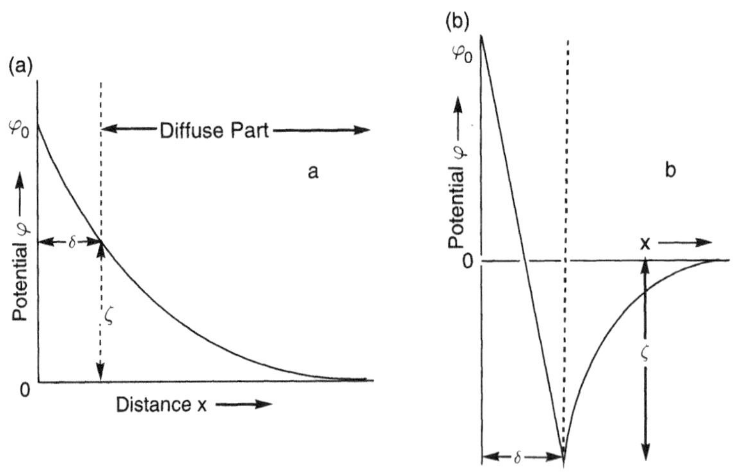

Figure 4.3. Potential distribution (a) when the adsorbed ion is of the same sign as the charge of the outer phase and (b) when a counter-ion is adsorbed.

immobile. The electrolytic transport in the diffuse part of the double layer gives rise to flow of ions in one direction which is responsible for unidirectional flow of solvent.

Theories of electro-kinetic phenomena have been advanced on the basis of Helmholtz double layer model of a parallel plate condenser [23]. This model has been improved by Guoy and Chapman [24], which yields an explanation of (i) the qualitative difference between φ_0 and ζ-potential and (ii) the sensitivity of ζ-potential to the concentration of the non-potential determining ions. However, the refinements were unable to explain this frequently observed inversion of sign of ζ with increasing concentration of certain electrolytes. Further, the theory of Guoy and Chapman leads to capacitance values of this double layer which in concentrated solutions is too high by one order of magnitude. These drawbacks are removed when Stern's correction is applied [25]. But Stern's theory is basically qualitative in character, which is primarily due to the schematic manner in which the charge is divided into adsorption and diffusion components. The use of Langmuir adsorption isotherm introduces considerable limitations in the theory. Thus the problem of true distribution of potential in the double layer is still an open question.

In spite of the above difficulties, a simple theory [26–28] based on Helmholtz model yields a microscopic picture which is useful in understanding the role of pore size and channel length along with the electrical characteristics of this interface in electro-kinetic phenomena. Whereas the macroscopic theory based on irreversible thermodynamics does not depend on any model, the theory discussed below would be valid provided the situation conforms to the model. Both approaches are complementary in understanding the phenomena.

We shall consider a capillary of radius r and length l, filled with a fluid. We assume the existence of double layer similar to a parallel plate condenser (Fig. 4.2). The radius of the capillary is considerably greater than δ, the thickness of the electrical double layer, so that at $x = r$ (at the axis of the capillary), $\varphi = 0$ and $\mathrm{d}\varphi/\mathrm{d}x = 0$.

The application of potential gradient causes the fluid to move. In the steady state, the electrical force would just counterbalance the viscous force. At a distance $\mathrm{d}x$ from the wall,

$$\text{Electrical Force} = \frac{\Delta\phi}{l}\rho\,\mathrm{d}x \tag{4.28}$$

and

$$\text{Viscous Force} = \eta\left(\frac{\mathrm{d}V}{\mathrm{d}x}\right)_{x+\mathrm{d}x} - \eta\left(\frac{\mathrm{d}V}{\mathrm{d}x}\right)_{x} = \eta\left(\frac{\mathrm{d}^2V}{\mathrm{d}x^2}\right)\mathrm{d}x \tag{4.29}$$

where η is the viscosity, $\mathrm{d}V/\mathrm{d}x$ the velocity gradient and ρ the charge density. From Poisson equation

$$\rho = -\frac{\varepsilon}{4\pi}\frac{\mathrm{d}^2\phi}{\mathrm{d}x^2} \tag{4.30}$$

where ε is the dielectric constant. On equating the electrical force and the viscous force and using Eq. (4.30), we have

$$\frac{d^2V}{dx^2} = -\frac{\Delta\phi}{l}\frac{\varepsilon}{4\pi\eta}\frac{d^2\phi}{dx^2} \tag{4.31}$$

which on integration yields

$$\frac{dV}{dx} = -\frac{\Delta\phi}{l}\frac{\varepsilon}{4\pi\eta}\left(\frac{d\phi}{dx}\right) + \text{constant} \tag{4.32}$$

Since the velocity gradient is zero along the axis of the capillary, $dV/dx = 0$, when $x = r$. Further, according to our model $d\phi/dx = 0$, when $x = r$, so that using these boundary conditions we find that the constant of integration is zero, so that

$$\frac{dV}{dx} = -\frac{\Delta\phi}{l}\frac{\varepsilon}{4\pi\eta}\left(\frac{d\phi}{dx}\right) \tag{4.33}$$

On further integration we have

$$V = -\frac{\Delta\phi}{l}\frac{\varepsilon}{4\pi\eta}\phi + \text{constant} \tag{4.34}$$

At the "plane of shear", we have $\phi = \zeta$. Since at this plane no slippage occurs, $V = 0$. Using these boundary condition we get

$$V = -\frac{\Delta\phi}{l}\frac{\varepsilon\zeta}{4\pi\eta} \tag{4.35}$$

so that

$$J = \pi r^2 V = -\frac{r^2\varepsilon\zeta}{4\eta l}\Delta\phi \tag{4.36}$$

Comparing Eqs. (4.19) and (4.36) we obtain

$$\frac{L_{12}}{T} = \frac{r^2\varepsilon\zeta}{4\eta l} \tag{4.37}$$

In the integration of Eq. (4.31) it has been assumed that η and ε are independent of $d\phi/dx$, the local field strength at any point in the double layer. This assumption may be incorrect, since the local electrical field in the double layer would be very high (of the order of 10^5 V cm^{-1}), if there is a fall of 100 mV in a distance of 100 Å. Such a high field would tend to reduce the dielectric constant. More rigorously, we should write

$$\frac{L_{12}}{T} = \frac{r^2}{4l}\int^{\phi_0} \frac{\varepsilon}{\eta}\, d\phi$$

It should be noted that when $r >> \delta$, the charge separation in the capillary occurs only in a thin layer near the wall, so that the bulk of the fluid is electrically neutral. This would not be the case when $r \sim \delta$, since the radial distribution of the field charge would be strongly non-uniform [29]. Further, the above theory has the following additional limitations: (i) the exact location of the slip-plane is not known; (ii) the exact magnitudes of η and ε in the double layer are uncertain; and (iii) the true distribution of the potential in the double layer is not known.

A theory of streaming current produced on account of fluid flow through the capillary can also be developed on the above lines. Let us consider the region between the capillary and the slip-plane. The excess charge e per cm^2 within the distance δ will be transported along the wall with the moving fluid. The speed of the transport of the charge will be just equal to the velocity of the fluid u_δ. The streaming current I would be given by

$$I = 2\pi r\, u_\delta\, e \tag{4.38}$$

The classical equation for laminar flow for a cylindrical capillary yields

$$u_\delta = \frac{1}{4\eta}\frac{\Delta P}{l}\left[r^2 - (r-\delta)^2\right] \sim \frac{1}{2\eta}\frac{\Delta P}{l} r\delta, \quad \text{since } \delta << r \tag{4.39}$$

For Helmholtz double layer, $\delta = \varepsilon\zeta/4\pi e$. On making proper substitutions,

$$I = (r^2\varepsilon\zeta/4\eta l)\,\Delta P \tag{4.40}$$

On comparing Eqs. (4.22), (4.37) and (4.40), we find that

$$\frac{L_{12}}{T} = \frac{L_{21}}{T} = \frac{r^2\varepsilon\,\zeta}{4\eta l} \tag{4.41}$$

so that the above theory is consistent with Onsager's reciprocal relations.

In order to complete the macroscopic description we shall evaluate L_{11}/T and L_{22}/T. If the pore size is much greater than the mean free path, this flow of the fluid through a circular pore of radius r would be governed by Poiseulle's equation [30]

$$J = \frac{\pi r^4}{\delta\eta l}\Delta P \tag{4.42}$$

so that

$$\frac{L_{11}}{T} = \frac{\pi\, r^4}{\delta\eta\, l} \tag{4.43}$$

Further, if x is the specific conductance of the fluid, the conductance of the fluid in the capillary is given by

$$L_{22}/T = \pi r^2 x/l \tag{4.44}$$

When surface conductance x_δ of the solid is comparable to the specific conductance of the liquid

$$\frac{L_{22}}{T} = \pi r^2 \left(k + \frac{2x_\delta}{r}\right) \Big/ l \tag{4.45}$$

If we consider a composite membrane having a parallel array of n capillaries, the right-hand side of Eqs. (4.40), (4.42), (4.43) and (4.44) would have to be multiplied by n. The phenomenology for ultra fine capillaries would be slightly different and has been worked out by a number of workers [31–33].

4.3.2. Theory of charged membranes

The phenomenon of electro-osmosis has also been observed in ion-exchange membranes [6], which are characterized by a high concentration of fixed charges and a correspondingly high concentration of counter-ions. Pore radii are much smaller, approaching molecular dimensions, so that $r << \delta$. A large percentage of counter-ions are mobile. During electro-osmosis, these counter-ions move, carrying solvent along with them. The slip-plane occurs at the surface of each fixed ion or, if it is hydrated, at the surface of primary hydration sheath. In general, ion-exchange membranes are "perm-selective", permitting migration of the ions of the same type, i.e. the co-ions cannot be easily transferred. However, several membranes have been reported where one ion possesses a significantly higher mobility than the other permeable ion species in the membrane phase [34].

A theory of electro-osmotic phenomena suited to charged membranes has been developed by Schmid and co-workers [35–37]. It is assumed that an identical composition is maintained on the two sides of the membrane, so that there is no concentration gradient. The distribution of mobile species within the pores is also assumed to be uniform. The ionic fluxes J_i are given by

$$J_i = -Z_i\overline{C}_i\overline{U}_i\frac{\Delta\phi}{l} + \overline{C}_i\left\{-\frac{\Delta P}{\rho_0 l} + w\overline{U}_0\frac{\Delta\phi}{l}\right\} \tag{4.46}$$

where Z_i is the valency of the ion; $\overline{C}_i$ the concentration of the ith ion; U_i the mobility of the ion in the membrane; ρ_0 the specific flow resistance; and ω the sign of fixed charges. The quantity inside the brackets on the right-hand side represents r_c, the rate of motion of the centre of gravity of the pore liquid. If $\overline{x}$ is the concentration of charges per unit volume, we have the following relation on account of the electro-neutrality condition:

$$\Sigma_i Z_i\overline{C}_i + \omega\overline{x} = 0 \tag{4.47}$$

The current I would be given by

$$I = F\,\Sigma Z_i J_i \tag{4.48}$$

where F is the Faraday constant. It follows that the streaming current and the streaming potential would be given by

$$(I)_{\Delta\phi=0} = \frac{w\overline{x}F}{\rho_0 l}\Delta P \tag{4.49}$$

$$(\Delta\phi)_{I=0} = \frac{w\overline{x}F}{\rho_0 \overline{x}}\Delta P \tag{4.50}$$

where $\overline{x}$ is the specific electrical conductivity of the membrane.

The volume flow J is the product of r_c and the fractional pore area $\overline{\varepsilon}$. Hence

$$(J)_{\Delta P=0} = w\overline{U}_0\frac{\Delta\phi}{l} = \frac{wF\overline{x}}{\rho_0 l}\Delta\phi \tag{4.51}$$

where $\overline{U}_0 = F\overline{x}/\rho_0\varepsilon$. It follows that

$$[(I)_{\Delta\phi=0}/\Delta P] = [(J)_{\Delta P=0}/\Delta\phi]$$

showing that the theory is in agreement with thermodynamic theory. If the pore radius is much greater than the mean free path, $l/\rho_0 = r^4/\delta\eta e$; we can then compare the results for charged and uncharged membranes, showing that $e/(1/2)r^2$ would be equivalent to ωFX. In the theory of charged membranes, no effort is made to correlate ωFX with the dielectric constant or other microscopic parameters.

4.4. Experimental test of thermodynamic theory

The four electro-kinetic effects described by Eqs. (4.19)–(4.22) can be experimentally realized. Much of the earlier data are of doubtful validity [38–40], since it is not known whether the pore geometry remained intact during the experiments of different kinds, as from Eqs. (4.40)–(4.44) it follows that the phenomenological coefficients are sensitive functions of pore geometry. Usually experiments are performed with porous plug which may be considered as a bundle of capillaries. Blocking of capillaries is a serous experimental difficulty. There is a possibility of incomplete wetting of the capillaries due to various factors, including the presence of greasy material. If the blocking and incomplete wetting of the capillaries occur during the experiment, obviously the character of the plug would change during the process and the validity of the results would be doubtful. Complications may also arise on account of (i) the action of the permeant on the capillary interface and (ii) heating of the fluid in the pores. When solutions of electrolytes are used as permeant, it is doubtful whether ζ-potential would remain invariant, in spite of the accumulation of products of electrolysis, unless very small current is passed. In recent experiments [13, 41], electronically regulated power supplies have been used for maintaining high potential differences with minimum possible passage of current. To avoid heating near the pores, magnetic stirring near the electrodes has been resorted to. The use of sinusoidal varying quantities in measuring electro-kinetic properties has been developed by Cooke. The technique with the help of alternation current offers the possibility of elimination of electrode difficulties when using electrolyte solutions. However, the technique is not easy. The earlier measurements [8, 42, 43] were made at low $\Delta\phi$ and low ΔP. In a recent work, ΔP has been extended to a few centimetres of mercury and $\Delta\phi$ has been pushed further to 600 V.

For studying electro-osmosis described by Eq. (4.19), either the bubble technique or the capillary method [44, 45] has been employed. In the former, difficulties due to evaporation of the permeant are avoided. However, the capillary along which the bubble moves has to be perfectly horizontal. The latter method is relatively simple. In principle, a definite potential gradient is applied along the membrane or the capillary, and the movement of the fluid in an external capillary is noted. Instead of a fixed potential gradient, one may apply a fixed current. The electro-osmotic pressure is measured by the pressure difference method. In order to measure the streaming potential, the fluid is allowed to flow through the membrane under a definite pressure gradient, which is maintained constant during the experiment with specially designed pressure heads. The streaming potential is measured with an electrometer, since the résistance of the plug is usually high. The streaming current is measured in a similar fashion.

Considerable amount of work has been done on both charged and uncharged membranes [46–53]. We shall discuss some of the recent data in the context of the conclusion of the thermodynamic theory developed in an earlier section. It may be noted that Eqs. (4.14) and (4.15) are used as axioms in the development of the phenomenological theory. The only justification for these is that these are consistent with the theory developed in the subsequent section. A direct test of the validity of the linear phenomenological equation has been performed recently. For pyrex sinter with water as permeant, $(J)_{\text{total}}$, $(J)_{\Delta\phi=0}$ and $(J)_{\Delta P=0}$ have been measured individually. It is found that up to 60 V, the relationship $(J)_{\text{total}} = (J)_{\Delta\phi=0} + (J)_{\Delta P=0}$ is satisfied within the limits of experimental error. The linear phenomenological equation for mass transport in the electro-osmosis of liquids has also been verified in a similar manner [54]. The data for volume flow for polystyrene sulphonic acid ion-exchange membrane also satisfy the linear phenomenological equation [55].

Previous data on electro-osmotic measurements in support of Onsager relation have been reviewed recently. Only the experimental data of Rutgers and de Smet [42], Lorenz [43] and Miller [56] offer reasonable support. However, Rutger's data refer to capillaries only and no idea can be formed regarding the range of ΔP and $\Delta\phi$ up to which Onsager relation is valid. Moreover, in the case of electrolyte solutions, some complications would arise in making a comparison on account of osmotic effects [57]. Lorenz used a composite membrane made up of quartz powder and pyrex sinter for support. Slight changes in packing of the plug are bound to vitiate the results. Further, the observed streaming current and streaming potential versus ΔP curves do not pass

Table 4.2. Verification of Onsager reciprocal relation [12, 13].

Membrane	L_{11}/T ($\times 10^{-6}\,\text{cm}^5\,\text{s}^{-1}\,\text{dyne}^{-1}$)	L_{12}/T ($\times 10^{-6}\,\text{cm}^3\,\text{A}\,\text{J}^{-1}$)	L_{21}/T ($\times 10^{-6}\,\text{cm}^3\,\text{A}\,\text{J}^{-1}$)
I	3.33×10^{-6}	2.5×10^{-6}	2.61×10^{-6}
II	0.96×10^{-6}	1.13×10^{-6}	1.10×10^{-6}
III	1.27×10^{-6}	0.96×10^{-6}	0.92×10^{-6}
IV	0.92×10^{-6}	0.72×10^{-6}	0.70×10^{-6}
V	1.10×10^{-6}	1.07×10^{-6}	1.03×10^{-6}
VI	1.06×10^{-6}	0.97×10^{-6}	0.90×10^{-6}

through the origin as expected. Lorenz has tested Onsager reciprocal relation for a very limited range, since ΔP has been varied up to approximately 30 cm of acetone and $\Delta\phi$ up to 8.5 V only. On the other hand, Cooke used pyrex–water system for study. In the absence of hydrodynamic permeability data for the system, it is difficult to comment on the results, since it is known that such a system normally has a poor reproducibility [41]. Clear confirmation of Onsager relation is obtained for pyrex–acetone system since the experimental difficulties are minimum for this system [13]. Typical values of the phenomenological coefficients for a few membranes are given in Table 4.2.

On substituting appropriate values for pyrex–acetone system, $L_{11}\,L_{22}$ is found to be of the order of $2 \times 10^{-6}\,\text{cm}^6\,A^2\,J^{-2}$, whereas $L_{12}\,L_{21}$ is equal to $6.5 \times 10^{-8}\,\text{cm}^6\,A^2\,J^{-2}$. Thus, $L_{11}\,L_{22} >> L_{12}\,L_{21}$ and the inequality is proved. The current required for a flow of $1\,\text{cm}^3$ s of acetone would be of the order of 25.6 mA, since $I = J(L_{22}/L_{11})$. The entropy production for various steady states can be estimated by using the relations given in Table 4.1. For streaming potential and streaming current $\sigma = 10^{-11}\,\text{J}\,\text{deg}^{-1}\,\text{s}^{-1}$, when ΔP is of the order of 10^4 dynes. For electro-osmosis and electro-osmotic pressure, σ is of the order of $3 \times 10^{-4}\,\text{J}\,\text{deg}^{-1}\,\text{s}^{-1}$, when $\Delta\phi = 100\,\text{V}$.

The efficiency of energy conversion may also be estimated in view of the interest in the engineering applications of electro-kinetic phenomena during recent years [58–61]. One may define the efficiencies of energy conversion E_e and E_s, which are related to coupling in electro-osmosis and streaming potential, as follows:

$$E_e = -J\Delta P/I\Delta\phi; \quad E_s = -I\Delta\phi/J\Delta P \tag{4.52}$$

Calculations show that $E_s \sim 10^{-7}$, while $E_e << 1$ for pyrex–acetone sinter. In view of the extremely low efficiency, these cannot be utilized for energy conversion.

In the linear region where the fluxes are linearly related to forces, σ has a minimum value in the stationary state. However, according to Prigogine and Glansdorff, $-(\text{d}\sigma)_J = -\Sigma J_i\,\dot{X}_i \geq 0$, the equality holding for the stationary state [62]. The $\dot{X}_i$ term is the time derivative of force X_i. In the case of electro-osmosis

$$(\text{d}\sigma)_J = \left(L_{11}\frac{\Delta P}{T} + L_{12}\frac{\Delta\phi}{T}\right)\left(\frac{\Delta\dot{P}}{T}\right) \tag{4.53}$$

For pyrex–water system, it turns out that $(\mathrm{d}\sigma/\mathrm{d}t)_J$ is the order of $-4 \times 10^{-5}\,\mathrm{J\,deg^{-1}\,s^{-2}}$, when $\Delta\phi = 60\,\mathrm{V}$, which confirms the above generalization.

Water transference number t_w is an important parameter in the study of ion-exchange membranes. The value of t_w is given by

$$t_w = \left(\frac{J}{I}\right)_{\Delta P=0} \frac{F}{18} = \frac{L_{12}}{L_{22}} \frac{F}{18} \tag{4.54}$$

It follows that within the domain of validity of linear thermodynamics of irreversible processes, t_w of should be independent of current density. Actually such a behaviour was observed by earlier workers [29, 63]. However, abnormalities at very low current densities have been noted, the reason for which is obscure. Marked dependence of electro-osmotic water transport on current has also been observed at higher current densities [48], which may be due to non-linear effects discussed below.

Both the thermodynamic theory and the macroscopic theory lead to the conclusions that (i) $(J)_{\Delta P=0}$ is proportional to $\Delta\phi$ and (ii) when $J = 0$, ΔP is proportional to $\Delta\phi$. Whereas the thermodynamics clearly shows that this may be true when departures from equilibrium are not significant, the theories based on specific model do not set any limit to their validity. The non-linear region has been studied extensively only recently by Rastogi and co-workers. Some typical results [64] presented in Figs. 4.4–4.6 show that the departure from linearity exceeds the experimental uncertainty beyond $\Delta\phi > 150\,\mathrm{V}$ and $\Delta P > 20\,\mathrm{cm}$. However, the streaming potential varies linearly with pressure difference within the range studied (Fig. 4.7).

The electro-osmotic data in the experimentally studied non-linear region are found to satisfy the following non-linear transport equation [12, 65]

$$J = L_{11}X_1 + L_{12}X_2 + L_{112}X_1X_2 + \tfrac{1}{2}L_{122}X_2^2 + \tfrac{1}{2}L_{1122}X_1^2X_2 + \tfrac{1}{2}L_{1122}X_1X_2^2 \tag{4.55}$$

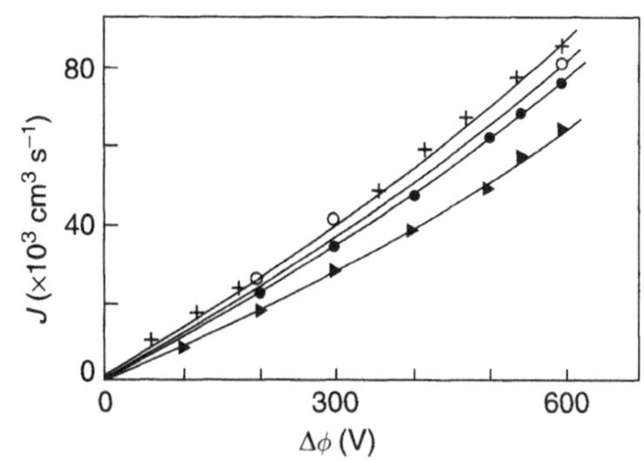

Figure 4.4. Dependence of volume flow (J) on potential difference ($\Delta\phi$) (o, membrane I; •, membrane II; ▲, membrane III; and x, membrane IV) [64].

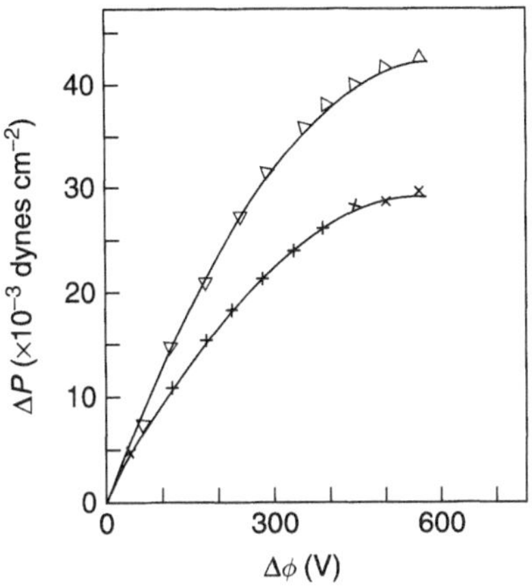

Figure 4.5. Dependence of electro-osmotic pressure (ΔP) on potential difference ($\Delta\phi$) (x, membrane IV; Δ, membrane V) [64].

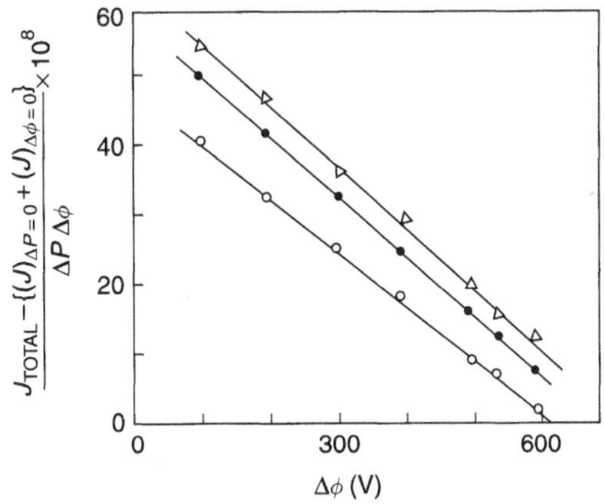

Figure 4.6. Estimation of L_{1122} and L_{112} using Eq. (4.11) (o, membrane I; •, membrane II; Δ, membrane III) [64].

It would be worthwhile to discuss the reason for departure from non-linearity at this stage [66] as change in structural factors r, n and l and the non-constancy of L_{ik} may cause deviation from linear behaviour. Since L_{ik} refers to mean temperature, pressure and potential prevailing in the membrane, these can vary if the gradients across the membrane are non-linear. In electro-osmotic phenomena, there is interaction between flow due to a pressure gradient and the flow due to potential gradient. The former gives rise to high-velocity gradient near the axis of the capillary and very low velocity gradient near the wall. In the case of electro-osmotic flow, it primarily takes place near the wall. The two flows would be independent, so long as there is no interaction near the wall and the net flow would be the sum of the two. The moment

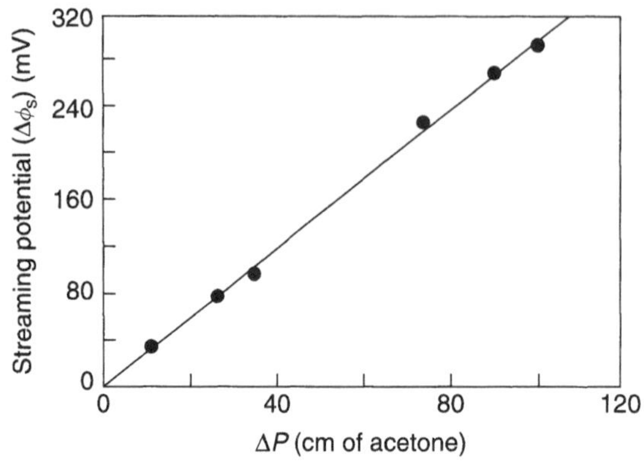

Figure 4.7. Estimation of L_{21} from streaming potential measurements [64].

interaction starts, the flow may be expected to be non-linear. In physical terms, this may mean that the charge distribution near the interface would be disturbed when the flow exceeds a certain limit, thus probably affecting the position of the slip-plane. Extensive experimental data have accumulated [67, 68] involving different types of membranes, e.g. (i) charged, (ii) uncharged, (iii) series (kaolinite + crysotile), (iv) parallel, (v) weak cation- and anion-exchange membranes, (vi) strong cation- and anion-exchange membrane, (vii) liquid membrane, cellulose-supported membrane (phospho-lipid-cholesterol and lecithin-cholesterol). Fluids involved include (i) polar organic liquids, (ii) mixtures of two such liquids, (iii) aqueous solution of electrolyte and (iv) mixtures of solution of two electrolytes. These provide reliable information about the domain of validity of various postulates.

Study of domain of validity of linear laws is of particular interest in the context of biological systems. For example, even Poiseulle's law, which has a much larger domain of validity otherwise, is reported to have significant deviations in physiological systems. The effect of increase of arterial pressure on flow through the various tissues of the body is greater than one would expect on the basis of Poiseulle's law [69].

4.5. Concluding remarks

Experimental studies for a variety of membranes have been reviewed in detail in several reviews [67, 68]. These show the following:

Phenomenological flux equations for both coupled and uncoupled flows are satisfied in a wide range.

Onsager reciprocity relations are also satisfied in the range where phenomenological relations for coupled flows are satisfied.

Steady-state relation for (i) electro-osmotic pressure and (ii) streaming potential is also satisfied in the same range.

Build-up and decay of steady electro-osmotic pressure or streaming potential is exponential with respect to time, obeying the following relations [70, 71]:

$$\Delta P = (\Delta P)_\infty [1 - \exp(-t/r_1)]$$

$$\Delta P = (\Delta P)_0 \exp(-t/r_2)$$

where t denotes time while r_1 and r_2 denote relaxation time for build-up and decay. Further $(\Delta P)_0$ is the pressure difference when $t = 0$ and $(\Delta P)_\infty$ is the pressure difference at the steady state during build-up.

References

1. H.A. Abramson, Electro-kinetic Phenomena and Their Application to Biology and Medicine, American Chemical Society Monograph, Chemical Co. Inc., New York, 1934.
2. H.N. Christensein, Biological Transport, W.A. Benjamin Inc., New York, 1962.
3. G.H. Bourne, Division of Labour in Cells, Academic Press, New York, 1962.
4. R.M. Freidenberg, The Electrostatics of Biological Cell Membranes, North Holland Publishing, Amsterdam, 1967.
5. E.J. Embros (ed.), Cell Electrophoresis, Little Brown, Boston, 1965.
6. H. Helfrich, Ion-Exchange, McGraw-Hill Book, Toronto, 1962.
7. E. Yeager, H. Ditrick and F. Hovorka, J. Acoust. Soc. Am., 25 (1953) 456.
8. C.E. Cooke, J. Chem. Phys., 23 (1955) 2299.
9. S.R. de Grout and P. Mazur, Non-equilibrium Thermodynamics, North Holland, Amsterdam, 1962.
10. I. Prigogine, Thermodynamics of Irreversible Processes, Wiley, New York, 1967, Ch. 7.
11. A. Katchalsky and P.F. Curran, Non-equilibrium Thermodynamics in Biophysics, Harvard Univ. Press, Cambridge, MA, 1965.
12. R.P. Rastogi, K. Singh and M.L. Srivastava, J. Phys. Chem. Ithaca, 73 (1969) 46.
13. R.P. Rastogi and K.M. Jha, Trans. Faraday Soc., 62 (1966) 585.
14. P. Mazur and J. Th. G. Overbeek, Rec. Trav. Chem., 70 (1951) 83.
15. O. Kedem and A. Katchalsky, Trans. Faraday Soc., 59 (1963) 1931.
16. A.J. Staverman, Trans. Faraday Soc., 48 (1952) 176.
17. K.S. Spiegler, Trans. Faraday Soc., 54 (1958) 1409.
18. R.P. Rastogi and R.C. Srivastava, Physica, 27 (1961) 265.
19. R.P. Rastogi, K. Singh and S.N. Singh, Indian J. Chem., 6A (1968) 466.
20. S.I. Rapoport, J. Theor. Biol., 19 (1968) 247.
21. A. Katchalsky, in The Neuro-sciences, G.C. Quarton, T. Melmechuk and F.O. Schmitt (eds), Rockefeller Press, New York, 1957.
22. P.C. Carman, Flow of Gases Through Porous Media, Academic Press, New York, 1956.
23. H. Helmholtz, Wid. Ann. N.F., 7 (1879) 339.

24. G. Guoy and D.L. Chapman, Philos. Mag., 25(6) (1913) 475.
25. O. Stern, Z. Elektrochem., 30 (1924) 508.
26. J. Th. G. Overbeek, in Colloid Science, Vol. 1, H.R. Kruyt (ed.), Elsevier, Amsterdam, 1952, Ch. 5.
27. A. Sheludko, Colloid Chemistry, Elsevier, Amsterdam, 1966.
28. J.T. Davis and E.K. Rideal, Interfacial Phenomena, Academic Press, New York, 1961.
29. D. Mackay and P. Mears, Trans. Faraday Soc., 55 (1959) 1221.
30. A.S. Ramsay, A Treatise on Hydromechanics, Part II, Hydrodynamics, G. Bell and Sons, London, 1957.
31. L. Dresner, J. Phys. Chem., 67 (1963) 1635.
32. R.J. Gross and J.F. Osterle, J. Phys. Chem., 49 (1968) 228.
33. B. Billard, Applied Mathematics, Preprint No. 4, Univ. of Queensland, Brisbane, 1968.
34. H.P. Gregor and D.M. Wetstone, Discuss. Faraday Soc., 219 (1956) 162.
35. G. Schmid and H. Schwarz, Z. Elektrochem., 56 (1952) 35.
36. G. Schmid, Z. Elektrochem., 56 (1952) 181.
37. R. Schlögl, Discuss. Faraday Soc., 21 (1956) 46.
38. D.R. Briggs, J. Phys. Chem. Ithaca, 32 (1928) 641.
39. R. Dubois and H.R. Alexander, J. Phys. Chem, Ithaca, 40 (1936) 543.
40. H.B. Bull, J. Phys. Chem. Ithaca, 39 (1933) 577.
41. R.P. Rastogi and K.M. Jha, J. Phys. Chem. Ithaca, 70 (1966) 1017.
42. A.J. Rutgers and M. de Smet, Trans. Faraday Soc., 43 (1967) 102.
43. P.B. Lorenz, J. Phys. Chem. Ithaca, 57 (1953) 430.
44. N. Lakshminarayaniah, Chem. Rev., 65 (1965) 526.
45. N. Lakshminarayaniah and V. Subrahmanyan, J. Phys. Chem. Ithaca, 72 (1968) 1253.
46. A.G. Winger, R. Ferguson and R. Kunin, J. Phys. Chem. Ithaca, 60 (1956) 556.
47. R.J. Stewart and W.J. Graydon, J. Phys. Chem. Ithaca, 61 (1957) 164.
48. A. Depic and J.G. Hills, Discuss. Faraday Soc., 21 (1956) 150.
49. A.S. Tombalkian, H.J. Barton and W.F. Graydon, J. Phys. Chem., 66 (1962) 1006.
50. A.S. Tombalkian, M. Worsley and W.F. Graydon, J. Am. Chem. Soc., 88 (1966) 661.
51. Y. Kobatake, M. Yuasa and H. Fujita, J. Phys. Chem., 72 (1968) 1752.
52. Y. Toyoshima, M. Yuasa and H. Fujita, Trans. Faraday Soc., 63 (1967) 2814.
53. Y. Toyoshima, Y. Kobatake and H. Fujita, Trans. Faraday Soc., 63 (1967) 2828.
54. R.P. Rastogi and K. Singh, Trans. Faraday Soc., 62 (1966) 1754.
55. A.S. Tombalkian, J. Phys. Chem., 72 (1968) 1566.
56. D.G. Miller, Chem. Rev., 60 (1960) 15.
57. O. Kedem and A. Katchalsky, Trans. Faraday Soc., 59 (1963) 1918.
58. J.R. Dixon and F.W. Schafer, J. Chem. Educ., 43 (1966) 180.
59. F.A. Morrison and J.F. Osterle, J. Chem. Phys., 43 (1965) 2111.
60. R.J. Gross and J.F. Osterle, J. Chem. Phys., 49 (1968) 228.
61. O. Kedem and S.R. Caplan, Trans. Faraday Soc., 61 (1965) 1897.
62. P. Glansdorff and I. Prigogine, Physica, 20 (1954) 773.
63. C.W. Carr, R. McClintock and K. Solner, J. Electrochem. Soc., 109 (1962) 251.
64. R.P. Rastogi, J. Sci. Ind. Res., 28 (1969) 284.
65. R.P. Rastogi, K. Singh and S.N. Singh, J. Phys. Chem., 73 (1969) 1593.
66. R.P. Rastogi and K. Singh, Trans. Faraday Soc., 63 (1967) 2917.

67. R.C. Srivastava and R.P. Rastogi, Transport Mediated by Electrified Interfaces, Elsevier, Amsterdam, 2003.
68. R.P. Rastogi, R.C. Srivastava and S.N. Singh, Chem. Rev., 93 (1993) 1945.
69. A.C. Guyton, Text-Book of Medical Physiology, Saunders, Philadelphia, 1981, Ch. 14, p. 152
70. R.C. Srivastava and R.P. Rastogi. Transport Mediated by Electrified Interfaces, Elsevier, Amsterdam, 2003, pp. 70–75.
71. R.P. Rastogi, G.P. Mishra, P.C. Pandey, K. Bala and K. Kumar, J. Colloid Interface Sci., 217 (1999) 275.

Chapter 5

NON-EQUILIBRIUM PHENOMENA IN CONTINUOUS SYSTEMS

5.1. Introduction

In the earlier chapters, transport phenomena involving a barrier have been discussed from the angle of (i) basic understanding of the physico-chemical phenomena and (ii) test of the linear thermodynamics of irreversible processes. Similar phenomena in continuous systems such as thermal diffusion (Soret effect)/Dufour effect are of equal physico-chemical interest.

In a system without a membrane or a barrier, temperature gradient can give rise to concentration gradient and the phenomenon is called thermal diffusion. Conversely, a concentration gradient can give rise to the temperature gradient. The phenomena is known as Dufour effect. The latter effect was discovered in gases by Waldman [1] and in liquids by Rastogi and Madan [2]. These non-equilibrium phenomena also occur in both liquids and gases where coupling between heat flux and mass flux is involved. Non-equilibrium thermodynamics gives good deal of insight into the phenomena from macroscopic point of view. Accordingly, it is intended to discuss theory and experimental studies of the phenomena in this chapter. An attempt would be made to assess to what extent the thermodynamic theory is consistent with kinetic theories and experimental data.

Similar phenomena such as diffusion potential and thermal diffusion potential in systems where ion transport is involved are also of considerable interest. Coupling of flow of ions relative to solvent is involved in the development of diffusion potential, while in the case of thermal diffusion potential, coupling of flow of ions and energy flow is involved. In such situations, the effective transference number as compared to Hittorf transference number is affected. Interesting experimental results have been reported in the context of galvanic cells (thermo-cells), in which the two electrodes are not at the same temperature where results have been interpreted in terms of thermodynamics of irreversible processes [3].

5.2. Theory: thermodynamic considerations

In discontinuous systems, intensive state variables have the same value throughout in each sub-system. But the same are discontinuous at the boundary of homogeneous regions. On the other hand in continuous systems, the incentive state variables are functions of both time and space coordinates.

Systems involving thermal diffusion and Dufour effect are continuous systems without a barrier. For investigating continuous systems, the local variation of properties has to be considered. We shall first consider a general case where mass flux, heat flux and chemical reactions are occurring [4, 5].

If ρ_k is the density (mass per unit volume of species/component k), then we have for its local variation

$$-\frac{\partial \rho_k}{\partial t} = -\mathrm{div}\rho_k v_k + \sum_{j=1}^{r} \upsilon_{kj} J_c \tag{5.1}$$

$j =$ number of reactions $(1, \ldots, r)$
$\delta/\delta t =$ local time derivative

The local change on the left-hand side in equation (5.1) is equal to negative divergence of a flow plus a source term gives the production and destruction of the substance k. The divergence of flow has the simple meaning of giving per unit volume, the excess of the flow which leaves a small volume to the flow which penetrates into it.

We note that

$$\rho = \Sigma \rho_k = \text{total density} \quad \text{and} \quad v = \text{centre of mass velocity} \tag{5.2}$$

Furthermore, we have

$$\sum_{k=1}^{n} \nu_{kj} = 0, \quad \mathrm{j} = 1, 2, ..\mathrm{r} \tag{5.3}$$

since mass is conserved in each chemical reaction.

Now summing over all substances/components k and making use relations (5.2) and (5.3), we get the following relation for the conservation of mass:

$$\delta/\delta t = -\mathrm{div}\rho v = -(\partial \rho\, v_k/\partial x + \partial \rho\, v_k/\partial y + \partial \rho\, v_k/\partial z) \tag{5.4}$$

The force equation can be written as

$$\rho\, \mathrm{d}v/\mathrm{d}t = -\mathrm{grad}\, P + \sum_{k=1}^{n} F_k\, \rho_k \tag{5.5}$$

where P is the pressure and F_k the external force per unit of mass on substance k.

The energy equation for the energy per unit mass with the exclusion of the barycentric kinetic energy is

$$\rho \mathrm{d}(1/2v^2 + u)/\mathrm{d}t = -\mathrm{div}(\rho_v + J_q) + \sum F_k v_k \rho_k \tag{5.6}$$

Here, J_q is the flow of heat (can be seen as part of energy flow).

According to second law of thermodynamics,

$$T\,dS/dt = dU/dt + P\,dV/dt - \sum_k \mu_k\,dM_k/dt \tag{5.7}$$

where the differentials are substantial derivatives with respect to centre of gravity movement. S is total entropy, U the total energy and μ_k the chemical potential of species k.

It can also be written entirely involving intensive quantities, viz.

$$T\,ds/dt = du/dt + P\,dv/dt - \sum \mu_k\,dc_k/dt \tag{5.8}$$

where $s = S/M$, $u = U/M$, $v = V/M$ and concentration $c_k = M_k/M = \rho_k$.

Now, multiplying Eq. (5.5) by V and subtracting Eq. (5.5) from Eq. (5.6), we get

$$\rho\,dU/dt = -P\,\mathrm{div}\,V - \mathrm{div}\,J_q + \sum F_k\,J_k \tag{5.9}$$

where mass flux of component k

$$J_k = \rho_k(V_k - V) \tag{5.10}$$

and

$$\sum_{k=1}^{n} J_k = 0 \tag{5.11}$$

On introducing Eq. (5.11) in Eq. (5.9), one obtains

$$d_u/dt = -\rho d_v/dt - \rho^{-1}\mathrm{div}J_q + \rho^{-1}\sum_k F_k\,J_k \tag{5.12}$$

The entropy balance equation can be rewritten as

$$\rho T\frac{ds}{dt} = -\mathrm{div}\,J_q + \sum_k F_k\,J_k + \sum \mu_k\,\mathrm{div}\,J_k - J_c\sum \mu_k v_k \tag{5.13}$$

An alternative form of the above equation is

$$\rho\,ds/dt = -\mathrm{div}(J_q - \sum \mu_k\,J_k)/T + J_q X_u + \sum J_q X_k + A\,J_c = -\mathrm{div}\,J_s + \sigma \tag{5.14}$$

noting that negative divergence of entropy flow $J_s = (J_q - \Sigma\,\mu_k\,J_k)/T$.

Hence, entropy production with source strength is given by

$$\sigma = (J_q\,X_u + \sum_k J_k\,X_k + A\,J_c)/T \tag{5.15}$$

where $X_u = -\text{grad}\, T/T$; $X_k = F_k - T$ grad (μ_k/T) and

$$A = -\sum \mu_k V_k$$

We now consider the phenomena of thermal diffusion, Soret effect and Dufour effect in a specific system. We consider an isotropic system consisting of two components 1 and 2. The concentration and temperature are non-uniform in the system. The pressure is supposed to be different at all points in the system so that mechanical equilibrium is rapidly established. There are no viscous forces and we shall neglect the viscous phenomena. Furthermore we assume that no chemical reactions are occurring. Since the system is non-uniform, local quantities have to be considered.

On combining the balance equation for energy assuming the absence of velocity gradients, the first law of thermodynamics and assuming Gibbs equation for entropy production for the case of local equilibrium, σ the entropy production per unit volume per unit time due to the occurrence of irreversible processes in the system is given by

$$\sigma = \frac{1}{T^2} J_q \text{ grad } T - \frac{1}{T} \sum_{k=1}^{n} j_k \{T \text{ grad } (\mu_k/T)\} \tag{5.16}$$

In Eq. (5.16) J_q and J_k are the heat flux and mass flux defined as follows:

$$\rho\, \mathrm{d}c_k/\mathrm{d}t = -\text{div}\, J_k; \quad \rho \mathrm{d}q/\mathrm{d}t = -\text{div}\, J_q \tag{5.17}$$

c_k is the mass fraction of component k and $\mathrm{d}q$ is the heat added per unit mass. It may be noted that it is difficult to estimate J_q.

If we define a new flux J'_q by

$$J'_q = J_q - \sum_{k=1}^{n} h_k J_k \tag{5.18}$$

where h_k is the specific enthalpy of component k, Eq. (5.16) can be rewritten as

$$\sigma = -1/T^2 J'_q \text{ grad } T - 1/T \sum_{k=1} J_k (\text{grad}\, \mu_k)_T \tag{5.19}$$

so that the new force corresponding to flux J_k is (grad $\mu_k)_T$ which can be explicitly known.

It may be noted that in the system, both the flows are not independent since

$$J_1 + J_2 = 0$$

On using the Gibbs–Duhem equation, it follows that

$$\text{grad}(\mu_1 - \mu_2) = (\mathrm{d}\mu_1)_{T,P}/c_2$$

Now we can write the phenomenological relations as follows:

$$J'_q = -L_{qq}(\text{grad } T/T^2) - L_{q1}(\text{grad}\,\mu_1)_{T,P}/c_2T$$
$$J_1 = L_{1q}(\text{grad } T/T^2) - L_{11}(\text{grad}\,\mu_1)_{T,P}/c_2T$$

With the Onsager relation $L_{q1} = L_{1q}$, Ls are called phenomenological coefficients, Here,

$$L_{qq} = \lambda T^2$$
$$L_{q1} = D''\rho\, c_1 c_2\, T^2$$
$$L_{iq} = D'\rho\, c_1 c_2\, T^2$$
$$L_{11} = D\rho\, c_2 T/(\partial\mu_1/\partial_{C_1})_{T,P}$$
$$S_T = D'(\partial\mu_1/\partial_{C_1})_{T,P}$$

where

D'' = Dufour coefficient
D' = thermal diffusion coefficient
λ = thermal conductivity
D = diffusion coefficient
S_T = Soret coefficient

Thermal diffusion factor $\alpha = D'/D$. It may be noted that if Onsager relation is satisfied, $D'' = D'$.

5.3. Experimental studies in gaseous systems

Theoretical and experimental studies on thermal diffusion is liquids and gases have been earlier reviewed by Grew [6]. However, the theoretical interpretation is largely based on Chapman Enskog Kinetic theory [7, 8]. In the context of non-isothermal diffusion, Cary [9] has reported that Onsager reciprocity relations are satisfied within experimental error of 8%. Experimentally, one finds in gases, in the context of Dufour effect, temperature difference ΔT of the order of 1°C. These lead to $D'' = D'$, confirming Onsager reciprocity relation [10].

Studies on Dufour effect for a number of gases and gaseous mixtures have also been reported by Rastogi and Madan [11]. Experimental set-up used for the measurement of Dufour effect is shown in Fig. 5.1.

ΔT generated is <1°C. Values of thermal diffusion factor estimated and discussed are in agreement with kinetic theory as regards order. The value of calculated Soret coefficient from the data is also found to be in agreement as regards order with the experimental values obtained by other workers.

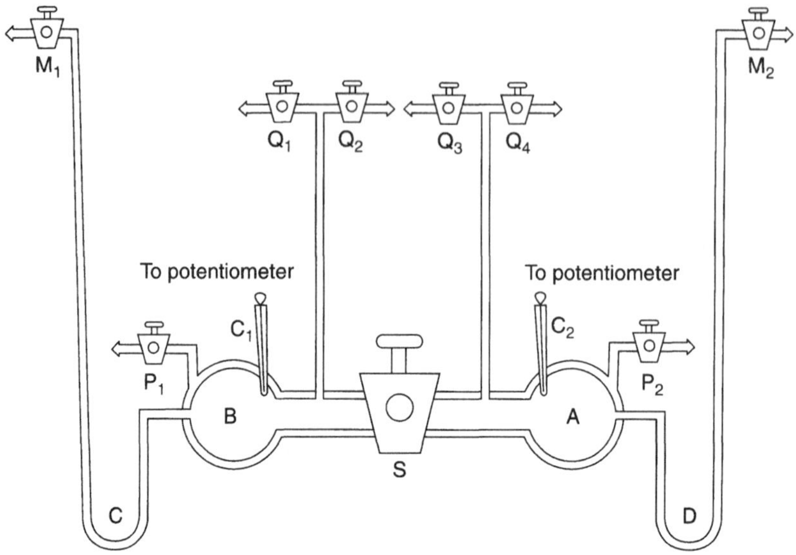

Figure 5.1. Experimental cell for the measurement of Dufour effect in gases. A, B, double-walled glass bulbs of capacity 250 ml; S, stopcock having a bore of 2 cm diameter.

5.4. Dufour effect in liquid mixtures

Rastogi and Madan [2] investigated Dufour effect in liquids for the first time. They employed a simple experimental set-up as shown in Fig. 5.2 for the purpose.

Later on, improved technique was employed by Rastogi and Yadava [12, 13] for investigation of Dufour effect in several liquid mixtures. Diffusion coefficient was also measured. D'' estimated from the data has been compared with those for thermal diffusion coefficients estimated from the known data on Soret coefficient using measured values of diffusion coefficients. Although ΔT is found to be $\approx$0.3°C, the values of D'' and D' are on reasonable agreement.

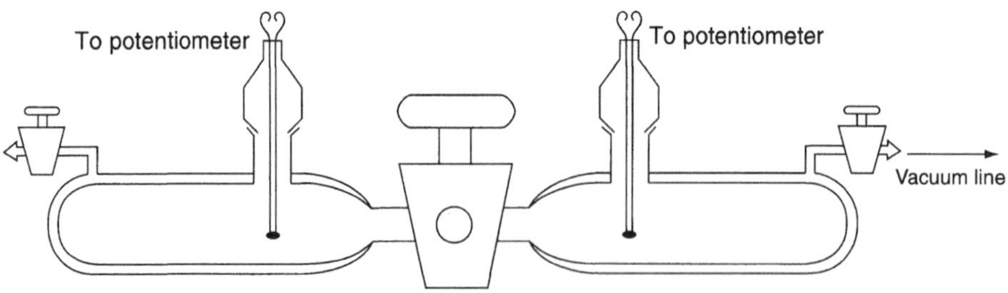

Figure 5.2. Apparatus for the measurement of Dufour effect in liquids.

5.5. Thermal diffusion potential

The passage of electric current through an aqueous solution in the presence of temperature, potential and concentration gradient involves interesting phenomena where three forces are coupled. This leads to development of (i) diffusion potential and (ii) thermal diffusion potential. This also affects transport numbers.

Although heat of transport of gases has been given due attention, it is equally interesting to study the characteristic properties of transfer and its dependence on temperature, concentration and presence of other ions [14, 15].

Within the range of validity of Debye–Huckel theory, the heat of transfer of all ions should approach the same value in the medium of same ionic strength. For binary electrolytes, we have in the stationary state [16]

$$Z_1 F \operatorname{grad} \varphi + (\operatorname{grad} \mu_1)_{T,P} + (Q_1^0/T) \operatorname{grad} T = 0 \tag{5.20}$$

$$Z_2 F \operatorname{grad} \varphi + (\operatorname{grad} \mu_2)_{T,P} + (Q_2^0/T) \operatorname{grad} T = 0 \tag{5.21}$$

where Z and μ are the valency and chemical potential of the two ions. Q_1^0 and Q_2^0 are the heat of transport of the two ions. Heat of transport is the heat absorbed in compartment I and liberated in compartment II when one mole of fluid is transferred from I and II. For ideal solutions, these reduce to

$$Z_1 F \operatorname{grad} \varphi + (RT/m)\mathrm{d}m + (Q_1^*/T) \operatorname{grad} T = 0 \tag{5.22}$$

$$Z_1 F \operatorname{grad} \varphi + (RT/m)\mathrm{d}m + Q_2^*/T \operatorname{grad} T = 0 \tag{5.23}$$

because $Z_1 = Z_2$. In the above equations m denotes molality. Combining Eqs. (5.22) and (5.23), we get

$$Q_1^* = Q_2^* \tag{5.24}$$

The asterisk denotes that the quantities are at the ideal state.

For investigating the effect of the nature of anions on the heat of transport of silver ions, the cells $Ag/AgNO_3/Ag{:}Ag/AgCH_3OO/Ag$ were studied.

From the theory [16], it follows that

$$\{(\mathrm{d}\varphi/\mathrm{d}T)^{\mathrm{II}}\}_{t\to\infty} - \{(\mathrm{d}\varphi/\mathrm{d}T)^{\mathrm{I}}\}_{t\to\infty} = \{(Q_2^*)_{\mathrm{I}} - (Q_2^*)_{\mathrm{II}}\}/Z_2 FT \tag{5.25}$$

where subscript I represents the first cell and II the second cell. Z_2 is the valency of the ion with which the electrode is reversible. F is the Faraday and T is the absolute temperature. $(\mathrm{d}\varphi/\mathrm{d}T)_{t\to\infty}$ is the thermo-electric power in the steady state. The thermo-electric power in the two cells using the medium of same ionic strength would be equal, and heat of transport of the silver ions would also be equal in both the cases. To test this point, the thermo-cell as given in Fig. 5.3 was used [17].

Thermo-electric power for each cell was measured for different molar concentrations of $AgNO_3$ for cell I and silver acetate for cell II. The results are plotted Fig. 5.4.

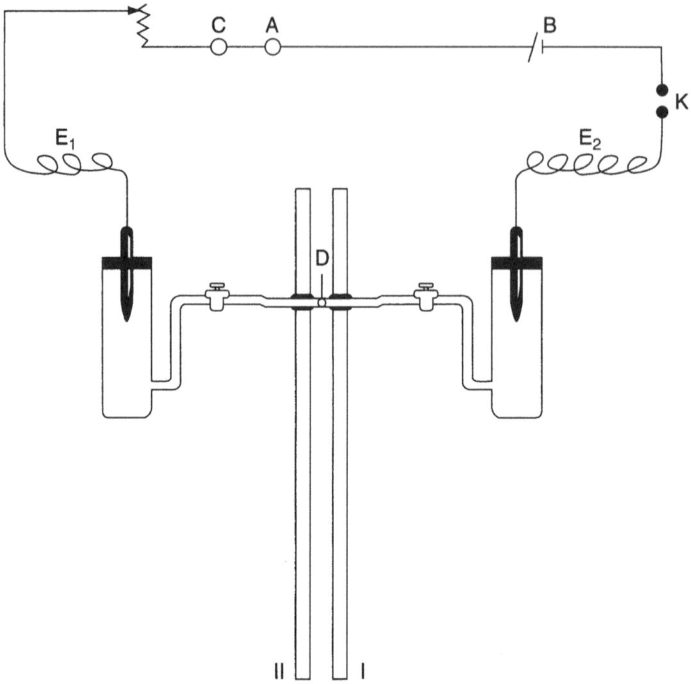

Figure 5.3. Apparatus (thermo-cell) for measuring transport of electricity in the presence of temperature gradient. C, coulometer; A, ammeter; B, battery; K, key; E_I, E_2, electrodes; D, sintered glass disc; I, thermostat maintained at temperature T_I; II, thermostat maintained at temperature T_{II}.

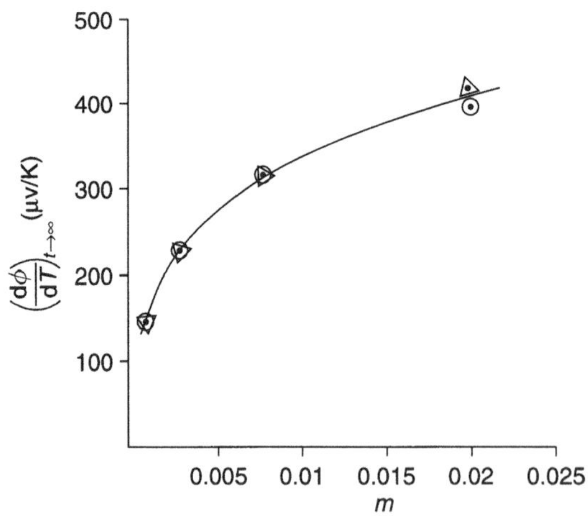

Figure 5.4. Effect of concentration (molar) and the nature of the anion on heat of transfer of silver ion; o-o $AgNO_3$; Δ-Δ CH_3COOAg.

Both the curves overlap showing that the heat of transfer of the silver ion is independent of the nature of anion.

Electrical transport in the presence of temperature gradient can be better understood by the application of concepts involved in irreversible thermodynamics. We consider the passage of an electric current through aqueous solution of electrolyte. If the charge carried by a cation is e_1 and that by an anion is e_2 and the respective mass flows are J_1 and J_2, then the t_1 current carried by the cations is given by

$$t_1 = e_1 J_1/(e_2 J_2 + e_1 J_1) \tag{5.26}$$

When the transport of electricity occurs at zero gradient of temperature and chemical potential, t_1 is equal to the Hittorf's transport number t_{1h} given by [18]

$$t_{1h} = (e_1 J_1/e_2 J_2 + e_1 J_1)\, \mathrm{d}T/\mathrm{d}x = 0; \quad \mathrm{d}\mu/\mathrm{d}x = 0 \tag{5.27}$$

Rastogi and co-workers [17, 18] have investigated that influence on magnitude of transport number in presence of temperature and chemical potential gradient. The results have been interpreted in terms of linear relations between fluxes and forces. Results suggest that the sign of difference between measured transport number and Hittorf's transport number depends on the sign of Soret coefficient [3, 4].

In the above cases, although experiments have not provided a comprehensive test of non-equilibrium thermodynamics, it can be seen that results yield interesting information, when these are interpreted in terms of linear relations between fluxes and forces.

Although non-equilibrium thermodynamics has serious limitations in view of limited domain of validity of Gibbs equation, but as a first approximation, it leads to interesting results for systems close to equilibrium. It provides a base for further extension of studies for systems far from equilibrium and even beyond.

5.6. Electric potentials generated at crystal interface

Development of electric potentials has been reported when (i) water containing an electrolyte as an impurity crystallizes [19], (ii) an electrolyte crystallizes from its supersaturated solution [20, 21], (iii) when an electrolyte dissolves in a solvent (water and non-aqueous solvents) [22] and (iv) molten electrolyte crystallizes [23]. With respect to (i) Workman and Reynolds suggested that preferential incorporation of ions in the ice lattice is an important factor responsible for the generation of freezing potentials. This view has been supported by other workers [24]. Alternatively, it has been suggested that these potentials might depend on dipole orientation, proton conductance and ionic entrapment in the ice lattice and ionic diffusion. It may be noted that freezing potentials are not developed when pure water is allowed to freeze.

Systematic experimental and theoretical studies on generation of precipitation and dissolution potentials have been undertaken by Rastogi and co-workers [22, 25–27].

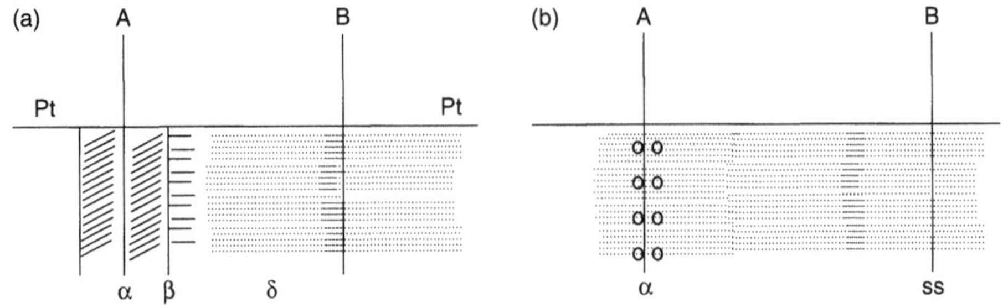

Figure 5.5. (a) Schematic diagram of cell for recording the dissolution potential of electrolytes: A, B, platinum electrodes; α, solid phase (electrolyte embedded on the electrode undergoing dissolution); β, layer of solution in the immediate vicinity of the solid phase; δ, solvent phase. (b) Schematic diagram of cell for the measurement of precipitation potentials: A, B, platinum electrodes; α, precipitating solid electrolyte; ss, super-saturated solution of electrolyte.

The experimental set-up used for the purpose of recording precipitation and dissolution potentials are schematically recorded (Fig. 5.5a and b).

It has been shown that precipitation or dissolution potential = observed potential − Nernst potential − diffusion potential − phase potential − thermo-emf.

Each type of potential has been measured using elaborate experimental set-up. For uni-univalent electrolytes like KCI, KBr and KI, the estimated values of precipitation and dissolution potentials are found to be in the range of 20–100 mV with a negative sign. Experimental values of some typical electrolytes are recorded in Table 5.1.

Thermodynamic theory based on non-equilibrium thermodynamics yields the following expression for the two types of potentials for 1:1 electrolyte:

1. Dissolution potential

$$(\Delta\phi)_{\text{diss}} = -RT/F[(U^{+} - U^{-})/(U^{+} + U^{-1})] \ln a^{\beta}/a^{S}\, U \qquad (5.28)$$

where

U^{+} = ionic mobility of cation
U^{-} = ionic mobility of cation
a^{β} = activity in β-phase
a^{s} = activity of saturated solution.

Table 5.1. Dissolution and precipitation potentials of some electrolytes.

Electrolyte	Medium	True dissolution potential	True precipitation potential
NaCl	Water	−80 mV	−54 mV
	D_2O	−42 mV	−49 mV
KCl	Water	−97 mV	−56 mV

2. Precipitation potential

$$(\Delta\phi)_{\mathrm{ppt}} = -RT/F[(U^+ - U^-)/(U^+ + U^-)]\ln a^{\mathrm{ss}}/a^* \qquad (5.29)$$

where

a^{ss} = activity of supersaturated solution
a^* = activity of saturated solution
R is the gas constant, T is the temperature and F is the Faraday.

During the crystallization process, following processes are expected to occur.

(a) Diffusion of ions from the bulk to the growing interface.
(b) Partial or complete desolvation of the ions moving towards the growing crystal lattice.
(c) Subsequent adsorption or desorption of these ions on the lattice. In fact the lattice energy and hydration energy act as the source of precipitation and dissolution potentials.

Thermo-chemical measurements confirm this inference.

Equations (5.28) and (5.29) lead to the conclusion that

(a) The sign of dissolution/precipitation potential would depend on the sign of $(U^+ + U^-)$.
(b) The potentials would depend on temperature.
(c) Magnitude of potentials would depend on the concentration of the medium.
(d) Decay of dissolution potentials would occur in the same manner.

The experimental results are in agreement with the above prediction [28].

A crucial test was made in the context of dissolution and precipitation potential of KI in water and DMF. In one case $U^+ < U^-$ while in the other case $U^+ > U^-$. Accordingly, one would expect difference in the sign of the potential in the two cases. Experimental results agree with this prediction [26]. Similarly, in solution of mixed electrolytes, the order of mobilities of cations and anions can be reversed as compared to the case of solution of a single electrolyte [26]. Effort has also been made to have a quantitative test of the theory. There is agreement between experimental and theoretically computed values of potentials as regards sign and order [29].

References

1. L.Z. Waldman, Z. Naturf., 29 (1947) 358.
2. R.P. Rastogi and G.L. Madan, J. Chem. Phys., 43 (1965) 4179.

3. K.G. Denbigh, The Thermodynamics of Steady State, Methuen and Co. Ltd, New York, 1951.
4. S.R. De Groot, Thermodynamics of Irreversible Processes, North-Holland Publishing Company, Amsterdam, 1952, Chapter VII.
5. S.R. De Groot, and P. Mazur, Non-equilibrium Thermodynamics, North-Holland Publishing Company, Amsterdam, 1962, p. 263, 273.
6. K.E. Grew, in Transport Phenomena in Fluids, H.J.M. Hanley (ed.), Marcel Dekker, New York and London (1969) Chapter 10.
7. S. Chapman and T.G. Cowling, The Mathematical Theory of Non-uniform Gases, Cambridge University Press, 2nd ed., 1952.
8. J.O. Hirschfelder, C.F. Curtiss and R.B. Bird, Molecular Theory of Gases and Liquids, Wiley, New York, 1974.
9. J.W. Cary, J. Phys. Chem., 67 (1963) 126.
10. S.R. De Groot and P. Mazur, Non-equilibrium Thermodynamics, North-Holland Publishing Company, Amsterdam, 1962, Chapter XI, p. 279.
11. R.P. Rastogi and G.L. Madan, Trans. Faraday Soc., 62 (1966) 3322.
12. R.P. Rastogi and B.L.S. Yadava, J. Chem. Phys., 51 (1969) 2826.
13. R.P. Rastogi and B.L.S. Yadava, J. Chem. Phys., 52 (1970) 2791.
14. H.J.V. Tyrrell and G.L. Hollis, Trans. Faraday Soc., 48 (1952) 893.
15. J.N. Agar and J.C.R. Turner, Proc. R. Soc., 255A (1960) 307.
16. R. Hasse, Trans. Faraday Soc., 49 (1953) 724.
17. R.P. Rastogi, R.C. Srivastava and R.L. Blokhra, Physica, 26 (1960) 1167.
18. R.P. Rastogi, R.L. Blokhra and K. Singh, Physica, 29 (1963) 841.
19. E. Workman and S. Reynolds, Phys. Rev., 74 (1948) 709; 78 (1950) 254.
20. R.P. Rastogi, R.K. Das and B.P. Batra, Nature, 191 (1961) 764.
21. H.L. Girdhar, K. Gobil and R.P. Matta, J. Electroanal. Chem., 123 (1981) 389.
22. R.P. Rastogi, R.D. Shukla and S. Bhagat, J. Electrochem. Soc., 121 (1974) 1563.
23. R.P. Rastogi and R.D. Shukla, Indian J. Chem., 6 (1968) 611.
24. P. Heinmets, Trans. Faraday Soc., 58 (1962) 788.
25. R.P. Rastogi and R.D. Shukla, J. Appl. Phys., 41 (1970) 2787, See also J. Sci. Ind. Res. (INDIA) 29 (1970) 317.
26. R.P. Rastogi and S.A. Khan, J. Electrochem. Soc., 129 (1980) 1989.
27. R.P. Rastogi and S.A. Khan, J. Electrochem. Soc., 132 (1983) 1327.
28. R.P. Rastogi and P.C. Pandey, Indian J. Chem., 24A (1985) 449.
29. R.P. Rastogi and P.C. Pandey and A.K. Tripathi, J. Indian Chem. Soc., 53 (1986) 76.

Chapter 6

ELECTROPHORESIS AND SEDIMENTATION POTENTIAL

6.1. Introduction

Electrophoresis and sedimentation potential also offer a test of predictions of thermodynamics of irreversible processes, provided these are supplemented by classical analysis of the data. Few measurements of sedimentation potential have been reported [1] and the theories due to Kruyt [2], Debye and Huckel [3] and Henry [4] are not in complete agreement. The thermodynamics of irreversible processes [5] may be helpful since the theory does not depend on any model. In the present chapter it is intended (i) to test linear phenomenological relations, (ii) to test the Onsager's reciprocal relation and (iii) to examine the validity of conflicting theories of electrophoresis.

6.2. Thermodynamic theory

We consider a system having a neutral fluid between two electrodes, 1 cm apart. Insoluble solid particles of a particular size are allowed to fall under gravity in the medium in a tube of radius r. The potential energy is converted into electrical energy and a sedimentation current flows. Alternatively, if an electric field is applied, particles move and give rise to electrophoresis.

Our system contains two components, viz. fluid and glass particles. C_k, the change in concentration of component k, is given by

$$\rho \, \mathrm{d}C_k/\mathrm{d}t = -\mathrm{div} J_k (k = 1, 2) \tag{6.1}$$

where ρ is the density and J_k the flow of matter per unit area. It is convenient to define it with respect to velocity of movement of centre of mass $\overline{V}$, i.e.

$$J_k = \rho_k (V_k - \overline{V}) \tag{6.2}$$

where ρ_k is the density of component k. The absolute flow J_k^0 may be defined as

$$J_k^0 = \rho_k V_k \tag{6.3}$$

As shown by de-Groot, Mazur and Overbeek [6] the entropy production σ is

$$T\sigma = \sum_{k=1}^{n} J_k^0 X_k \tag{6.4}$$

where the forces X_k are given by

$$X_k = e_k E + g - \text{grad}\, \mu_k \tag{6.5}$$

e_k is the charge per unit mass of species k, E the electrical field, g the gravitational field per unit mass and μ_k the chemical potential of species k. At constant temperature, Eq. (6.5) reduces to

$$X_k = e_k E + g - v_k \text{ grad } P \tag{6.6}$$

where v_k is the partial specific volume of component k. Since the total volume flow through a section vanishes,

$$\sum_{k=1}^{n} v_k J_k^0 = 0 \tag{6.7}$$

Hence for a two-component system

$$T\sigma = J_1^0\{e_k E + (1 - v_1/v_2)g\} \tag{6.8}$$

where the subscript 1 refers to pyrex particles and subscript 2 to the medium. We choose electric current density I and mass flow J as the two fluxes defined as

$$I = e_1 J_1^0 \tag{6.9}$$

$$J = J_1^0(1 - v_1/v_2) \tag{6.10}$$

The corresponding forces are E and g, respectively. The phenomenological relations can be written as

$$I = L_{11}E + L_{12}g \tag{6.11}$$

$$J = L_{21}E + L_{22}g \tag{6.12}$$

because of Onsager relation $L_{12} = L_{21}$. From Eqs. (6.11) and (6.12),

$$L_{12} = -(E/g)_{I=0} L_{11} \tag{6.13}$$

$$L_{21} = (J/E)_{g=0} \tag{6.14}$$

Here $E = \Delta\phi/l$, where $\Delta\phi$ is the potential difference and l the distance between the two electrodes. L_{11} is related to conductivity and hence L_{12} can be estimated from the measurements of sedimentation potential. For estimating L_{21}, one can measure J which is related to electrophoretic velocity V_e in cm s^{-1}, in the following manner:

$$J = m(1 - \rho_2/\rho_1)V_e \tag{6.15}$$

where m is the mass of particles suspended between two electrodes. If r is the radius of the particles, n the number of particles suspended between the two electrodes and V the volume, m would be equal to $4n\pi r^3\rho_1/3V$ provided the particles are assumed to be spherical. Since m and V_e are known experimentally, L_{21} can be calculated.

6.3. Comparison with Helmholtz double layer theory

The electrophoretic velocity is given from double layer theory [2] as

$$V_e = D\zeta E/4\pi\eta \tag{6.16}$$

where η is the viscosity of the medium, D its dielectric constant and ζ the zeta potential of interface. Similarly for the sedimentation potential [2],

$$E = \frac{D\zeta}{4\pi\eta\lambda}\frac{n}{V}\frac{4}{3}\pi r^3(\rho_1 - \rho_2)g \tag{6.17}$$

where λ is the specific conductivity. From Eqs. (6.16) and (6.17), it follows that

$$\frac{V_e}{E} = E\lambda \bigg/ \left[\frac{n}{V}\frac{4}{3}\pi r^3(\rho_1 - \rho_2)g\right] \tag{6.18}$$

and from Eqs. (6.13), (6.14), (6.16) and (6.17),

$$L_{12} = L_{21} = \frac{D\zeta}{4\pi\eta}\frac{n}{V}\frac{4}{3}\pi r^3\rho_1\left(1 - \frac{\rho_2}{\rho_1}\right) \tag{6.19}$$

This shows that the kinetic theory is consistent with the thermodynamic theory. Experimental proof of Onsager relation would be obtained if the experiments on sedimentation potential and electrophoretic velocity are performed for the particles of same size and shape.

There is some controversy regarding the numerical factor 4 in Eq. (6.16). According to rigorous theory of Debye and Huckel [3], it is 6, when the electrophoretic drag due to double layer is also considered. However, Henry [4] has shown that it lies between 4 and 6. Thus Eq. (6.16) should be written as

$$(V_e/E)_{g=0} = D\zeta/f\pi\eta \tag{6.20}$$

where the value of f lies between 4 and 6.

6.4. Test of theory: experimental studies

Exhaustive experimental studies on electrophoresis and sedimentation potential have been reported by Rastogi and Mishra [7] with a view to test the thermodynamic theory of the phenomena. The experimental setup involving (i) sedimentation column and (ii) optical assembly are shown in Figs. 6.1 and 6.2. Both sedimentation potential and electrophoretic velocity were measured with these setups.

As indicated in Section 6.2, linear relation between V_e and $\Delta\phi$ can be tested experimentally by plotting V_e against $\Delta\phi$ using Eq. (6.15). The slope yields L_{12} while intercept gives L_{22} after some work. The observed rate of movement of boundary V_e in cm min^{-1} is equal to sum of $(V_e)_{g=0}$ and V_g, the sedimentation velocity due to gravity. Similarly L_{12} and L_{22} can be estimated using streaming current measurements. Rastogi and Mishra [7] experimentally studied the following three systems: (i) pyrex–water, (ii) pyrex–acetone and (iii) quartz–water. Typical results on electrophoretic velocity measurements for specific systems are plotted in Fig. 6.3.

Experimental results show that for pyrex–water system L_{21} is constant, at least up to 150 V. On the other hand for pyrex–acetone system, linear phenomenological relation is found to be valid up to 450 V. Onsager reciprocity relation was tested in each case and the results are given in Table 6.1 [7]. The results are in agreement with those of earlier workers [8].

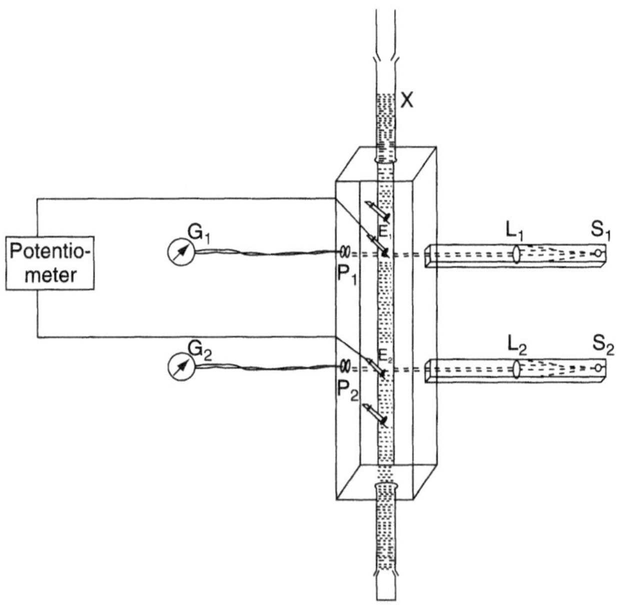

Figure 6.1. Assembly for measurement of sedimentation potential and sedimentation rate [7]. P_1/P_2 = photo-cells, G_1/G_2 = scalamp galvanometers, S_1/S_2 = light sources which are at the focii of two lenses L_1 and L_2.

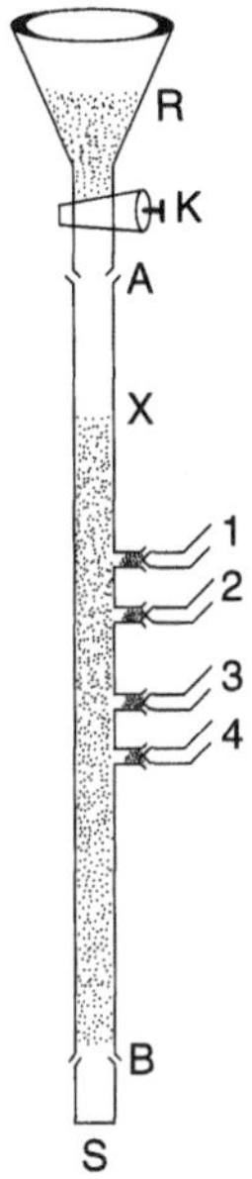

Figure 6.2. Sedimentation columns [7] 1, 2, 3, 4, quick fit joints.

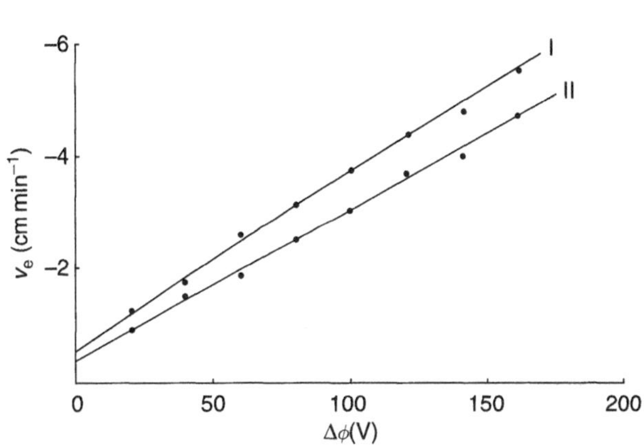

Figure 6.3. Plot of V_e versus $\Delta\phi$.

Table 6.1. Verification of Onsager reciprocity relation (temp. 35°C).

System	L_{12} ($\times 10^{13}$ Å g cm^{-1} erg^{-1})	L_{21} ($\times 10^{13}$ Å g cm^{-1} erg^{-1})
Pyrex–water	2.9	2.68
Pyrex–acetone	1.32	1.15
Quartz–water	2.26	1.96

In conclusion we can see that all the postulates of irreversible thermodynamics are found to be in agreement with experimental data over a wide range in the case of electro-kinetic phenomena (electro-osmosis, streaming potential, electrophoresis and sedimentation potential). Similarly, thermo-osmosis provides support to the thermodynamic theory. However, test of Onsager reciprocity relation is difficult in this case. Thus, steady-state thermodynamics in the liner range for membrane phenomena (systems consisting two subsystems separated by a barrier) has a wide range of applicability. The only limitation is the range of validity of Gibbs equation. The limitation is observed in biological membranes, which are thinner and have much smaller pores. Nevertheless, even for such membranes, for getting an insight into the nature of physical processes in such systems, steady-state thermodynamics has proved to be a valuable guide.

In continuous systems, the situation is not so clear. On account of experimental difficulties, exhaustive tests as in the above case have not been performed. Nevertheless, experimental results on thermal diffusion and Dofour effect/Soret do support the theoretical predictions although in a much limited range, again on account of limited range of validity of Gibbs equation.

The limitation of Gibbs equation and later efforts to extend the applicability through extended thermodynamics would be discussed in the appendices.

Steady-state thermodynamics in the linear range provides a good glimpse of the non-equilibrium region close to equilibrium. The utility of steady-state thermodynamics is illustrated in the case of electro-kinetic phenomena in Parts Two and Three in the regions more and more distant from equilibrium (non-linear steady state, bistability, oscillations, pattern formation) including complexity and complex phenomena.

References

1. A.W. Adamson and A.P. Gast, Physical Chemistry of Surfaces, Sixth edition, Wiley, New York, 1997.
2. H.R. Kruyt, Colloid Science, Vol. I, Elsevier, Amsterdam, 1952, pp. 204, 207, 221.
3. P. Debye and E. Hickel, Phys. Z., 25 (1924) 204.
4. D.C. Henry, Proc. Roy. Soc. A., 133 (1931) 106.
5. S.R. de-Groot and P. Mazur, Non-equilibrium Thermodynamics, North Holland, Amsterdam, 1962, p. 364.
6. S.R. de-Groot, J.Th.G. Overbeek and P. Mazur, J. Chem. Phys., 41 (1952) 764.
7. R.P. Rastogi and B.M. Mishra, Trans. Faraday Soc., 63 (1967) 584, 2926.
8. A.J. Rutgers and M. de Smet, Trans. Faraday Soc., 41 (1945) 764.

Part Two

NON-LINEAR STEADY STATES – DISSIPATIVE STRUCTURE (TIME ORDER AND SPACE ORDER)

Chapter 7

NON-LINEAR STEADY STATES

7.1. Introduction

In Part One, we have discussed theory and related experiments concerning steady states in the range of validity of linear phenomenological equations. We shall call such states as linear steady states. We come across stable steady states even beyond the domain of validity of linear non-equilibrium thermodynamics (LNT) where the flux equations are non-linear.

We will first discuss the form of non-linear flux equations for electro-osmosis and streaming current, mass and heat flux in thermal diffusion and chemical reactions from experimental and theoretical angle.

7.2. Non-linear flux equations in electro-kinetic phenomena

Volume flux of methanol and methanol–water mixtures through Dowex-1 (Cl^-) form has been investigated when pressure difference and potential difference is simultaneously imposed. The following third-order non-linear phenomenological equation has been found to be adequate for the description of volume flux data:

$$J_v = L_{11}(\Delta P/T) + L_{12}(\Delta\phi/T) + \tfrac{1}{2}L_{122}(\Delta\phi/T)^2 + L_{112}(\Delta P/T)(\Delta\phi/T)$$
$$+ \tfrac{1}{2}L_{1122}(\Delta P/T)(\Delta\phi/T)^2 + \tfrac{1}{2}L_{1112}(\Delta P/T)^2(\Delta\phi/T) \qquad (7.1)$$

Applicability of the above equation has been tested by using the flux data when [1]

(a) ΔP is kept constant and $\Delta\phi$ is varied; and
(b) $\Delta\phi$ is kept constant and ΔP is varied.

The first order and higher order coefficients were estimated from the data in order to check the consistency. These are recorded in Table 7.1

There is good agreement as regards the first three coefficients. However, in the case of the last there is only agreement as regards order.

Table 7.1. Phenomenological coefficient for Dowex-1(01)/methanol systems.

Phenomenological Coefficients	$\Delta P = \text{constant}$	$\Delta\phi = \text{constant}$
$L_{11}T^{-1}\,\text{cm}^5\ 300^{-1}\,\text{dyne}^{-1}$	2.80×10^{-4}	2.80×10^{-4}
$L_{12}T^{-1}\,\text{cm}^3\ 800^{-1}\,\text{v}^{-1}$	1.95×10^{-6}	2.00×10^{-6}
$L_{112}T^{-2}\,\text{cm}^5\ 800^{-1}\,\text{v}^{-1}\,\text{dyne}^{-1}$	-0.25×10^{-6}	-0.25×10^{-6}
$L_{122}T^{-2}\,\text{cm}^3\ 800^{-1}\,\text{v}^{-2}$	0.10×10^{-8}	0.03×10^{-8}
$L_{1112}T^{-3}\,\text{cm}^7\ 800^{-1}\,\text{v}^{-1}\,\text{dyne}^{-2}$	0.25×10^{-8}	0.06×10^{-8}
$L_{1122}T^{-3}\,\text{cm}^5\ 800^{-1}\,\text{v}^{-2}\,\text{dyne}^{-1}$	0.10×10^{-8}	0.06×10^{-8}

Experimental data on electro-osmosis over a wide range of $\Delta\phi$ and ΔP for a variety of systems such as (i) pyrex-sinter/acetone; (ii) quartz plug/acetone; (iii) pyrex sinter/methanol; (iv) pyrex sinter/methyl-ethyl ketone; (v) quartz/methanol; and (vi) quartz/methyl-ethyl ketone have been accumulated which further confirms the above non-linear equation for volume flux [1, 2].

Non-linear flux equations for electro-osmotic flow, as well as streaming current are also experimentally obtained. On the basis of number of studies with different membranes, it turns out that

$$(J_v)_{\Delta P=0} = L_{12}(\Delta\phi/T) + 1/2\,(L_{122}\,\Delta\phi/T)^2 \tag{7.2}$$

which has been tested for number of systems such as pyrex sinter/polar solvent, quartz/polar solvent, Amberlite (IRC-50 Ba^{2+}) form/methanol–water, Amberlite IRC-50(Al^{3+}) form/methanol–water and Zeokarb-225(Na^+) form/methanol–water [1, 3, 4, 5].

Results for particular systems are plotted in Fig. 7.1

In some cases such as electro-osmosis of 0.1 NaCl or KCl solution exponential relation between $(J_v)_{\Delta P=0}$ and $\Delta\phi$ i.e.,

$$(J_V/\Delta\phi)_{\Delta\phi=0} = L_{12}\,e^{b\Delta\phi} \tag{7.3}$$

has been reported [6]. In Eq. (7.3), b is a constant. However, for more dilute solutions such as 0.01N or 0.001N NaCl or KCl solution, linear relation is found to be valid. This indicates that for higher concentration of electrolytes, non-linearity would be more pronounced

Electrophoretic velocity also depends on $\Delta\phi$ in a non-linear manner (Fig. 7.2).

For higher values of ΔP and streaming current $(I)_{\Delta\phi=0}$, experimental studies obey the following non-linear equation involving higher powers of fluxes:

$$(I)_{\Delta\phi=0} = L_{21}(\Delta P/T) + 1/2\,L_{211}(\Delta P/T)^2 \tag{7.4}$$

Typical results for ion-exchange membranes such as Zeokarb 225 (Na^+) form/ methanol–water are shown in Fig. 7.3.

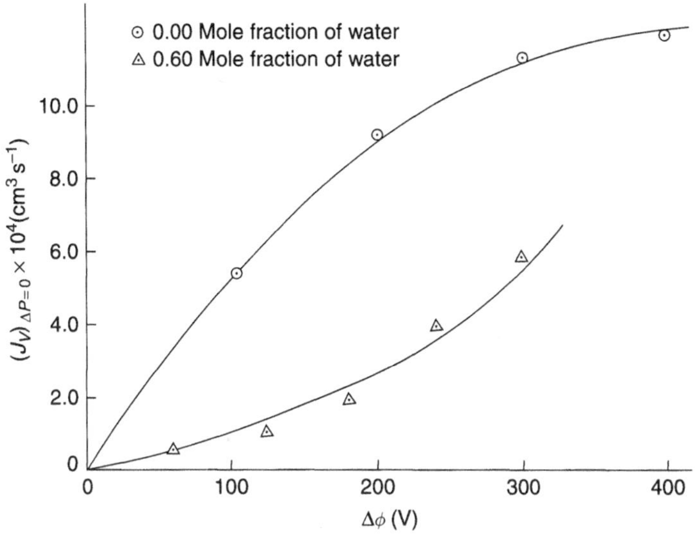

Figure 7.1. Dependence of electro-osmotic flux $(J_V)_{\Delta P=0}$ on potential difference $\Delta\phi$ for Zeokarb 225 (Na^+ form)/methanol–water system.

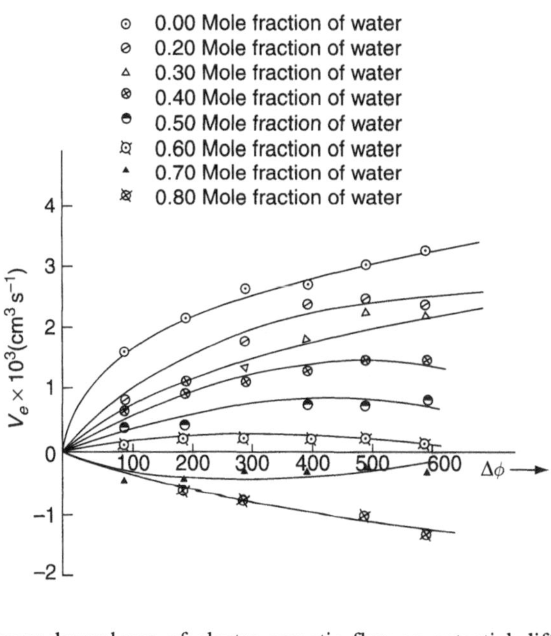

Figure 7.2. Non-Linear dependence of electro-osmotic flux on potential difference for IRC-50 (H^+ form)/methanol–water system.

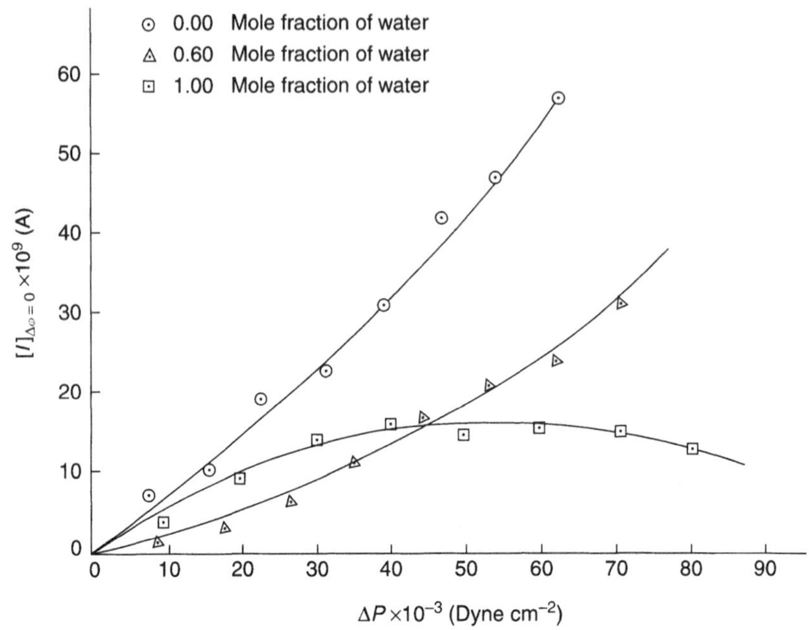

Figure 7.3. Dependence of streaming current on $(I)_{\Delta\Phi=0}$ on pressure difference for Zeokarb 225(Na^+ form)/methanol–water system.

7.3. Non-linear steady states

7.3.1. Electro-osmotic pressure

Electro-osmotic pressure per unit potential difference in the non-linear range varies with pressure difference in the following manner:

$$(\Delta P/\Delta\phi)_{J_v=0} = L_{12}/L_{11} - 1/2L_{122}/L_{11}(\Delta\phi/T) \tag{7.5}$$

Typical results for pyrex/acetone system are recorded in Fig. 4.4 [7]. In the linear steady state, electro-osmotic pressure varies linearly with potential difference.

7.3.2. Streaming potential

Streaming potential per unit pressure difference in the non-linear region is given by a similar relation

$$(\Delta\phi/\Delta P)_{I=0} = -L_{21}/L_{22} - 1/2\,L_{211}/L_{22}(\Delta P/T) \tag{7.6}$$

Streaming potential versus pressure difference curve is non-linear as indicated by Fig. 7.4.

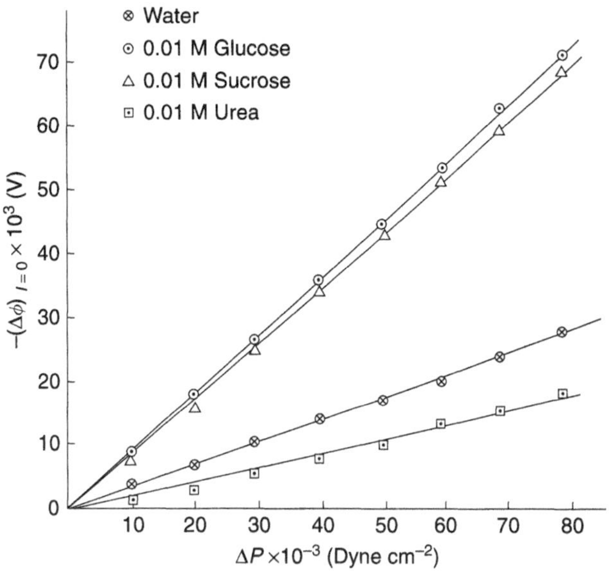

Figure 7.4. Dependence of streaming current $(I)_{\Delta\Phi=0}$ on pressure difference for Estrone/aqueous solutions of non-electrolytes.

In case of ion-exchange membranes, complicated non-linear behaviour is observed in Fig. 7.5. A typical case is Zeokarb 225 (Ba^{2+}) form/methanol–water system [8].

In Zeokarb 225 (Al^{3+}) form/methanol–water system, indication of bistability without hysteresis is obtained as indicated in Fig. 7.6.

Some other important features of ion-exchange membranes have been observed; these are given below [9, 10, 11]:

(a) In case of ion-exchange membranes, both electro-osmotic flux as well as electrophoretic velocity have been found to be non-linear in $(\Delta\phi)$, where even the change of direction of the flux occurs. Experimental results are mutually consistent since the sign of inversion occurs at a particular value of potential difference.
(b) Sign inversion is obtained in case of weak-cation exchange membranes.

7.4. Interpretation of second-order coefficients in the light of double layer theory

7.4.1. Interpretation of L_{122}

The origin of second-order coefficients, particularly L_{122} and L_{211}, can be easily understood on the basis of Helmholtz double layer theory. It may be noted that local

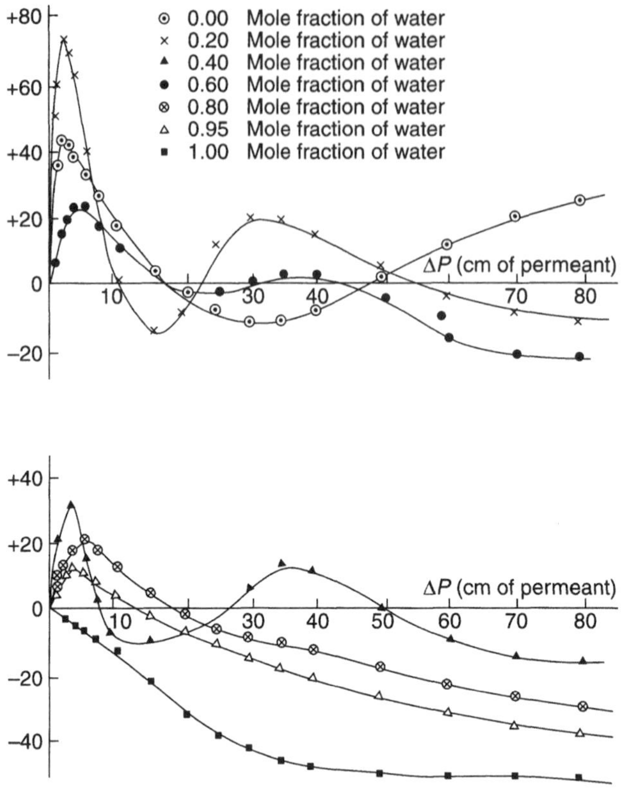

Figure 7.5. Dependence of streaming potential on pressure difference for Zeokarb-225 (Ba^{++} form/methanol–water system.

equilibrium is disturbed by increasing the magnitude of ΔP and $\Delta\phi$. The former affects the thickness of the electrical double layer while the latter affects the charge density ultimately influencing the effective zeta potential.

We would first try to understand the circumstances in which L_{122} becomes significant [12]. For this purpose let us consider the structure of the electrical double layer (see Fig. 4.2).

The effective electro-kinetic potential (zeta potential) would be given by

$$\zeta_{\text{eff}} = \Delta\psi + \Delta\chi \tag{7.7}$$

where $\Delta\psi$ = charge contribution potential

$$\Delta\chi = \text{dipolar potential}$$

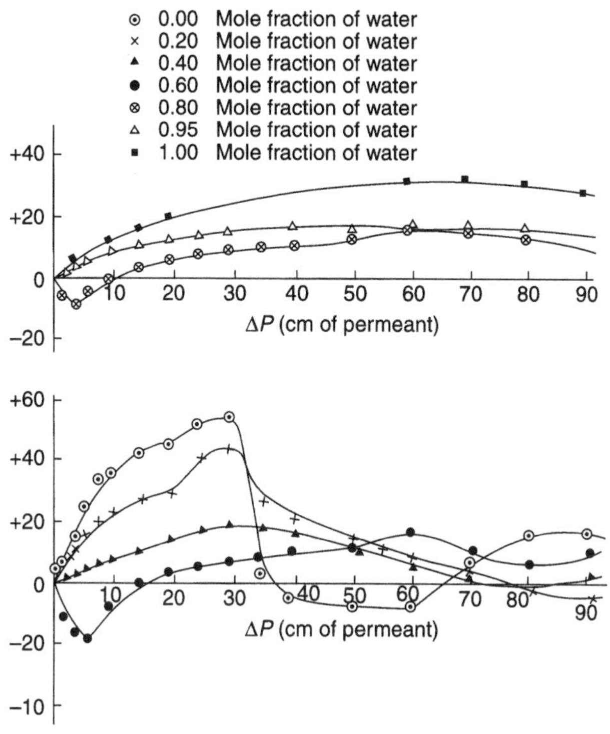

Figure 7.6. Dependence of streaming potential on pressure difference for Zeokarb-225 (Al^{+++} form)/methanol–water system.

It can be shown that $\Delta\chi$ would be given by

$$\Delta\chi = \frac{\mu N_T}{4\pi\varepsilon_0}\frac{\bar{\mu}\vec{X}(1-C)}{kT} \tag{7.8}$$

where, N_T = total number of dipole on the interface including flip-up and flip-down states at the charge concerned;

$\vec{X}$ = electric field strength;
C = number of dipoles which interact with a particular dipole;
μ = dipole moment vector;
ε_0 = dielectric constant of the medium;
k = Boltzmann constant;
T = temperature.

When the idea of effective zeta-potential is incorporated in the double layer theory of electro-osmosis [12], it can be shown that

$$L_{122} = a^2\ N_T\ \mu(1-C)\ \overline{\mu}/8\pi\eta r^2 kl \tag{7.9}$$

where a = radius of the capillary;
η = viscosity;
l = length of the capillary.

The theory explains the sign reversal observed during electro-osmosis of methanol–water mixture through Zeokarb 222 (a weak cation-exchange membrane). The effective zeta potential can be zero or undergo sign reversal depending on the magnitude of $\Delta\chi$ H_20 and $\Delta\chi$ CH_3OH, which depend on the field strength. Effect can be zero or undergo sign reversal depending on the circumstances. This explains how the zeta potential changes with the composition of alcohol–water mixtures and the electric field as observed by [13].

7.4.2. Interpretation of L_{211}

Non-linearity arising on account of second order coefficient L_{211} in electro-osmosis primarily occurs on account of stress on the double layer formed at the interface when pressure difference is applied. As pressure difference increases, strain would also increase [14]. If d_0 is the thickness of the double layer before application of the stress, outer Helmholtz plane (OHP) would be pushed by an angle θ when stress is applied. On account of this the effective thickness of the double layer would be $d_0 \cos\theta$. The lateral and longitudinal stress would be given by

$$\text{Lateral Strain} = \frac{\overrightarrow{d_0 - d_0\cos\theta}}{d_0} \tag{7.10}$$

Since stress is equal to ΔP, modulus α of rigidity of double layer would be given by

$$\alpha = \overrightarrow{\Delta P}\frac{d_0\sigma'}{\overrightarrow{d_0 - d_0\cos\theta}} = \overrightarrow{\Delta P}\frac{\sigma'}{\mathrm{i}(1-\cos\theta)} \tag{7.11}$$

where σ' is called Poisson's ratio and i is the unit vector, so that

$$\cos\theta = 1 - \overrightarrow{\Delta P}\frac{\sigma'}{ik} \tag{7.12}$$

The charge contribution potential $\Delta\psi_{\text{eff}}$ would be given by

$$\Delta\psi_{\text{eff}} = \frac{4ed_0\cos\theta}{\varepsilon} \tag{7.13}$$

where e is the charge density. On substituting the value of $\cos\theta$ using the earlier equation, we get

$$\Delta\psi_{\text{eff}} = \Delta\psi_0\left(1 - \overrightarrow{\Delta P}\frac{\sigma'}{ik}\right) \tag{7.14}$$

where $\Delta\psi_o$ is equal to $4\pi e\, d_o\, \zeta$. Accordingly

$$\xi_{\text{eff}} = \Delta\psi_0 \left(1 - \overrightarrow{\Delta P}\frac{\sigma'}{ik}\right) \tag{7.15}$$

According to Helmholtz double layer theory, streaming current $(I)_{\Delta\phi=0}$ is given by [15]

$$(I)_{\Delta\phi=0} = \frac{a^2 \varepsilon \xi_{\text{eff}}}{4\eta e}\overrightarrow{\Delta P} \tag{7.16}$$

On substituting the value of ξ_{eff}, in the above equation, we get

$$(I)_{\Delta\phi=0} = \frac{a^2 \varepsilon \overrightarrow{\Delta P}\Delta\psi_0}{4\eta e} - \frac{a^2 \varepsilon \Delta\psi_0}{4\eta e}\frac{\sigma'}{\text{i}x}(\overrightarrow{\Delta P})^2 \tag{7.17}$$

Thus,

$$\frac{L_{21}}{T} = \frac{a^2 \varepsilon \Delta\psi_0}{4\eta e} \quad \text{and} \quad \frac{L_{211}}{T_2} = -\frac{a^2 \varepsilon \Delta\psi_0}{4\eta e}\frac{\sigma'}{\text{i}x} \tag{7.18}$$

Vector i for water (permeant) would be positive corresponding to direction of flip-up dipole [↑] while it is negative for methanol (permeant) for which the direction of flip-down dipole is [↓]. Thus, for water, L_{211} would be negative, while for methanol it would be positive. The available experimental data support the conclusion [16].

7.5. Non-linear transport equations in gaseous medium

Theoretical studies had been undertaken [17, 18] for analysing transport equations for mass flux as well as energy flux in continuous systems although experimental studies are not available for the purpose.

The state of a gas is characterized by the distribution function f which depends on the peculiar velocity, the radius vector and the time t. For a stationary gas, the distribution function is Maxwellian and is denoted by f_o, which is given by

$$f_o = n\,(m/2\pi kT)^{3/2}\ \exp[-mc^2/2kT] \tag{7.19}$$

where n = number density i.e. total number of molecules per unit volume, m = mass of one molecule, c = mass velocity, k = Boltzman constant, and T represents temperature.

In a non-stationary situation, the distribution function can be written as

$$f = f_o(1 + \phi(1) + \phi(2) + \cdots) \tag{7.20}$$

where $f_o\ \phi(1)$ and $f_o\ \phi(2)$ are the second and third approximation to the distribution function, respectively. The kinetic processes can be understood provided the time dependence of f can be explicitly characterized [19]. In the first approximation, $f = f_o$. For kinetic derivation of transport equations, the first approximation cannot be used. Linear transport equations for mass flux and energy flux are obtained when f is put equal to $f_o\ (1+\phi(1))$ i.e. when second approximation to the distribution is used. These have considerable importance within the framework of thermodynamics of irreversible processes in the linear range when third approximation to the distribution function is used i.e. $f = f_o\ [1+\phi(1)+\phi(2)]$ [20]. The third approximation for a simple gas has been worked out by Burnett [21], while that for mass flux in a mixed gas has been investigated by Chapman and Cowling [22]. Rastogi and Misra [15] have also used the theory of Chapman and Cowling for non-uniform gases involving third approximation to distribution function in order to examine the magnitude and tensorial character of non-linear terms. It is observed that the terms containing products of gradients and higher order space derivatives of barycentric velocity appear in transport equations, but higher powers of a single force do not occur in contrast with cases of electro-osmosis and thermo-osmosis discussed in Part One. Calculations show that second order terms are negligible in the transport equations for mass as well as energy flux as compared to other terms.

An assessment of the relative magnitudes of various terms in transport equation for mass flux has been made by Chapman and Cowling [22]. The estimation of relative significance of various terms [18] in the mass flux and energy flux equations, for a mixture of hydrogen and helium, at a mean temperature 300° K, gradient of mass fraction ranging from 0.99 to 0.09 cm^{-1} and grad T ranging from 10 to 40° cm^{-1} confirmed the conclusion of Chapman and Enskog. In case of heat flow, the second order terms are much smaller so that they do not materially contribute to the heat flux.

This is also true for mass flux. Only those terms are important which are associated with the force X_v corresponding to symmetrical non-divergent part of the viscous pressure tensor and space derivative of X_v. These can become important in capillaries. It seems that linear transport equations have larger domain of validity than expected.

An examination of the derived flux equations [18] leads to the following conclusions: (i) Onsager reciprocity relation (ORR) is obeyed. (ii) All coefficients are scalar. (iii) Higher powers of single force do not occur. (iv) Space derivatives of forces occur in the transport equations. (v) In none of the $X_i\ X_j$ terms, the tensorial order of X_i and X_j term is the same, (vi) All the X_iX_j terms have the same tensorial order as the fluxes. (vii) Non-linearity arises on account of gradient of barycentric velocity.

The above conclusions have serious limitations, which are as follows:

(i) The theory is applicable to dilute gases since only binary collisions are considered.

(ii) The applicability of the theory is limited to mono-atomic gases since molecular collisions are considered to be elastic, which implies that (a) Only translational energy of the molecules is taken into account; (b) Only spherically symmetric molecules are considered; and (c) Molecular collisions are described by classical mechanics.

7.6. Non–linear flux equations and non-linear steady states in chemical reactions

7.6.1. Non-linear flux equations

In case of chemical reactions, linear relation is obtained between reaction rate J and affinity A of the reaction when $A/RT << 1$, where R is the gas constant and T is the temperature. In case of coupled reactions in specific cases, linear phenomenological relations between reaction rates and affinities are obtained. In the linear range, ORR are found to be valid. The non-linear rate equations involve higher powers of affinities. However, linear steady state are not easily experimentally obtained. Steady state in chemical reactions requires steps involving negative and positive feedback, which ultimately generate non-linear rate equations. However, in non-linear region, chemical reactions under specific conditions can exhibit the phenomena of steady states in continuous stirred tank reactor (CSTR) as discussed in Chapter 9 on chemical oscillations. In the following subsections, we shall discuss (i) Linear (rate) and non-linear flux equation for a single reversible reaction; and (ii) Linear phenomenological relation and non-linear flux equation for coupled reactions.

7.6.2. Linear flux equation for a single reversible reaction

Let us consider the reaction $A \rightleftharpoons B$ for illustration. Affinity A of the reaction in a general case is defined by

$$A = -\Sigma\, \nu_i \mu_i \tag{7.21}$$

where ν_i = stoichiometric number of species ν

μ_i = chemical potential species ν_i ; μ_i is counted positive for the reactants on the left hand side and negative for the reactants on the right hand side. Thus, A in the present case is equal to $\mu_A - \mu_B$. The rate of reaction would be given by

$$-\mathrm{d}C_A/\mathrm{d}t = k_1\, C_A - k'_1\, C_B \tag{7.22}$$

where k_1 and k'_1 are the rate constants of forward and back reactions, respectively. C_A and C_B are the concentrations of A and B, respectively. Chemical potentials of A and B would be given by

$$\mu_A = \mu_A{}^{\mathrm{o}} + RT\ \ln\ C_A; \quad \overline{\mu}_A = \mu_A{}^{\mathrm{o}} + R\ \ln\ \overline{C}_A \tag{7.23}$$

$$\mu_B = \mu_{\mathrm{B}}{}^{\mathrm{o}} + RT\ \ln\ C_B; \quad \overline{\mu}_B = \mu_B{}^{\mathrm{o}} + RT\ \ln\ \overline{C}_B \tag{7.24}$$

where $\mu_A{}^{\mathrm{O}}$ and $\mu_B{}^{\mathrm{O}}$ are constants; $\overline{\mu}_A$ and $\overline{\mu}_B$ are the chemical potentials of A and B at equilibrium; $\overline{C}_A$ and $\overline{C}_B$ denote the concentration of A and B at equilibrium. So that

$$\mu_a - \overline{\mu_A} = RT \ln \frac{C_A}{\overline{C_a}} \quad \text{and} \quad \mu_B - \overline{\mu_B} = RT \ln \frac{C_B}{\overline{C_B}} \tag{7.25}$$

and further,

$$C_A = \overline{C_A} \exp\left(\frac{\mu_A - \overline{\mu_A}}{RT}\right) \quad \text{and} \quad C_B = \overline{C_B} \exp\left(\frac{\mu_B - \overline{\mu_B}}{RT}\right) \tag{7.26}$$

Substituting the values of C_A and C_B from in Eq. (7.22) using Eq. (7.26), one gets

$$\frac{\mathrm{d}C_A}{\mathrm{d}t} = k_1 \overline{C_A} \exp\left(\frac{\mu_A - \overline{\mu_A}}{RT}\right) - k_1' \overline{C_B} \exp\left(\frac{\mu_B - \overline{\mu_B}}{RT}\right) \tag{7.27}$$

Noting that at equilibrium $k_1\,\overline{C}_A = k'_1\,\overline{C}_B$, one obtains

$$-\frac{\mathrm{d}C_A}{\mathrm{d}t} = k_1 \overline{C_A} \left[\exp\left(\frac{\mu_A - \overline{\mu_A}}{RT}\right) - \exp\left(\frac{\mu_B - \overline{\mu_B}}{RT}\right)\right]$$

$$= k_1 \overline{C_A} \left[\left(\frac{\mu_A - \overline{\mu_A}}{RT}\right) + \left(\frac{\mu_B - \overline{\mu_B}}{RT}\right) + \cdots\right]$$

$$j = \frac{\mathrm{d}C_A}{\mathrm{d}t} = k_1 \overline{C_A} \left[\left(\frac{A}{RT}\right) + \frac{1}{2}\left(\frac{A}{RT}\right)^2 + \cdots\right] \tag{7.28}$$

since, $\overline{\mu}_B = \overline{\mu}_A$. When $(A/RT) << 1$, $\mathrm{d}C_A/\mathrm{d}t = k_1\,\overline{C}_A\,(A/RT)$. Thus, we obtain the linear equation. However, when this is not so, we get non-linear equation (7.28)

In order to have an idea of the range upto which linear relation is valid, for a typical case of isomerisation reaction between NH_4CNS (ammonium thio-cyanate) and thio-urea viz.

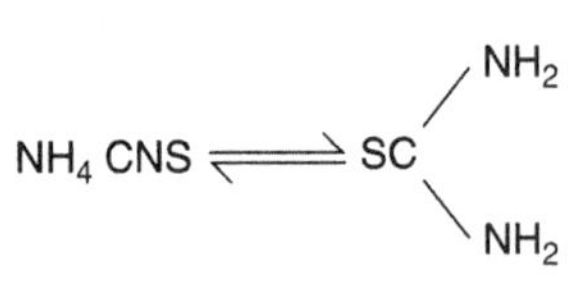

for which experimental data have been examined [23]. Experimental results clearly indicate the non-linear relation between reaction rate and affinity as predicted by Eq. (7.28).

7.6.3. Phenomenological relations in coupled reactions

Let us consider the series of reactions

$$A \rightleftharpoons B$$

$$B \rightleftharpoons C$$

$$C \rightleftharpoons A$$

where coupling takes place. Let J_1, J_2 and J_3 be the reaction rates of each step. Following the usual procedure described in part I, the entropy production σ can be expressed as

$$T\sigma = J_1 \; A_1 + J_2 \; A_2 + J_3 \; A_3 \tag{7.29}$$

where A_1, A_2 and A_3 are the affinities of corresponding reaction steps. It can be shown easily that

$$A_1 + A_2 + A_3 = (\mu_B - \mu_A) + (\mu_C - \mu_B) + (\mu_A - \mu_B) = 0 \tag{7.30}$$

Further it can be shown that

$$\left(-\frac{dx_A}{dt}\right)_{\text{eff}} = -\left(-\frac{dx_A}{dt}\right)_1 + \left(-\frac{dx_C}{dt}\right)_2 = J_1 - J_3 \tag{7.31}$$

$$\left(-\frac{dx_C}{dt}\right)_{\text{eff}} = -\left(-\frac{dx_C}{dt}\right)_3 + \left(-\frac{dx_B}{dt}\right)_2 = J_2 - J_3 \tag{7.32}$$

where the subscript eff denotes the effective rate and subscripts 1, 2, and 3 denote the quantities for the respective steps. It can be easily shown that since $X_A + X_B + X_C =$ constant,

$$J_1 = J_2 = J_3 \tag{7.33}$$

illustrating the condition of microscopic reversibility [24]. Using Eq. (10), Eq. (9) can be rearranged as follows

$$T\sigma = (J_1 - J_3) \; A_1 + (J_2 - J_3) \; A_3 \tag{7.34}$$

where $(J_1 - J_3)$ and $(J_2 - J_3)$ are the new fluxes. Significance of $(J_1 - J_3)$ and $(J_2 - J_3)$ can be understood as follows.

$$J_1 = k_1 \; X_A^{\text{o}} \left[\left(\frac{A_1}{RT}\right) + \begin{array}{l}\text{higher order terms}\\ \text{involving higher powers}\\ \text{of forces}\end{array} \right]$$

$$J_2 = k_2\ X_B^{\mathrm{o}}\left[\left(\frac{A_2}{RT}\right) + \text{higher order terms}\right]$$

$$J_3 = k_3\ X_C^{\mathrm{o}}\left[\left(\frac{A_3}{RT}\right) + \text{higher order terms}\right]$$

$$= k_3 X_C^{\mathrm{o}}\left[\left(\frac{A_1 + A_2}{RT}\right) + \text{higher order terms}\right]$$

and therefore,

$$J_1 - J_3 = k_1\ X_A^{\mathrm{o}}\left(\frac{A_1}{RT}\right) + k_3 X_C^{\mathrm{o}}\left(\frac{A_1 + A_2}{RT}\right) + \text{higher order terms}$$

$$J_2 - J_3 = k_2\ X_B^{\mathrm{o}}\left(\frac{A_2}{RT}\right) + k_3 X_C^{\mathrm{o}}\left(\frac{A_1 + A_2}{RT}\right) + \text{higher order terms}$$

or,

$$J_1 - J_3 = (k_1\ X_A^{\mathrm{o}} + k_3\ X_C^{\mathrm{o}})\left(\frac{A_1}{RT}\right) + k_3\ X_C^{\mathrm{o}}\left(\frac{A_2}{RT}\right) + \text{higher order terms}$$

$$J_2 - J_3 = \left(\frac{k_2\ X_B^{\mathrm{o}} + k_3\ X_C^{\mathrm{o}}}{RT}\right)\left(\frac{A_2}{RT}\right) + k_3\ X_C^{\mathrm{o}}\left(\frac{A_1}{RT}\right) + \text{higher order terms}$$

where,

$$L_{11} = (k_1\ X_A^{\mathrm{o}} + k_3\ X_C^{\mathrm{o}});\quad L_{12} = k_3\ X_C^{\mathrm{o}}$$

$$L_{21} = k_3\ X_C^{\mathrm{o}};\quad L_{22} = (k_2\ X_B^{\mathrm{o}} + k_3\ X_C^{\mathrm{o}})$$

It is obvious that ORR is obeyed. Thus, in case of chemical reactions, linear and non-linear flux equations similar to the case of electro-kinetic phenomena are predicted and ORR between first-order cross-coefficients is also obeyed. However, experimental studies so far have not been made, although triangular reaction of the above type for isomerisation of $\Delta\alpha$ – pentenoic acid has been investigated from a different viewpoint [25].

7.6.4. Non-linear steady states in chemical reactions

Steady states, as we have seen in Part One, are obtained when fluxes in opposite directions are involved. In electro-osmosis, hydrodynamic flow is opposed by electro-osmotic flux. In thermo-osmosis, hydrodynamic flow is opposed by thermo-osmotic flux. In case of chemical reactions, such situations can arise when positive feedback is opposed by negative feedback. For example, when autocatalysis is opposed by inhibitory reaction, steady state can be attained. However, the reaction rates are non-linear and have only non-linear steady states in practice. We illustrate this point by the following example.

Let us consider following sequence of reactions in a CSTR.

There is a continuous supply of the reactants A and B into the reactor with the flow rate k_f as indicated by steps (i) and (ii). In the reservoirs, the concentration Ao and Bo are maintained constant

$$
\begin{array}{llll}
A_o & \xrightarrow{k_f} A & \text{(Source of supply of A)} & \text{(i)} \\
B_o & \xrightarrow{k_f} B & \text{(Source of supply of B)} & \text{(ii)} \\
A+B & \longrightarrow 2B & \text{(autocatalysis)} & \text{(iii)} \\
B & \longrightarrow C & \text{(Decay of B)} & \text{(iv)} \\
 & & \text{(half-order kinetics)} &
\end{array}
$$

In the reactor, A and B react and autocatalysis occurs by step (iii) with a rate constant k_1, while decay of B to C occurs via step (iv) with the rate constant k_2. Further, k_f is the flow rate of A and B from the reservoirs. If half-order kinetics for step (iv) is assumed, the kinetic equations (non-linear) are as follows:

$$\mathrm{d}A/\mathrm{d}t = -k_1\ AB + k_f(A_o - A) \tag{7.35}$$

$$\mathrm{d}B/\mathrm{d}t = k_1\ AB - k_2\ B^{1/2} + k_f(B_o - B) \tag{7.36}$$

Steady state would be reached when $\mathrm{d}A/\mathrm{d}t = 0$ or $\mathrm{d}B/\mathrm{d}t = 0$ i.e. when (a) rate of inflow of A by (i) = rate of consumption of A by step (iii); and (b) rate of inflow of B by (ii) = rate of net production of B by combined steps (iii) and (iv).

In Eqs. (7.35) and (7.36), *A* and *B* denote the concentration of respective species. Computer analysis of such a reacting system has recently bee made by Rastogi and Khare [26].

On account of severe limitations, experimental studies on linear steady state could not be undertaken so far. However, the non-linear range experimental studies on bistability in reacting systems (Chapter 8) and chemical oscillations in CSTR do provide convincing examples of stable non-linear steady states under specific circumstances.

7.7. General remarks

Specific features of non-linear steady states (beyond linear steady state) can be summarized as follows:

(a) Linear non-equilibrium thermodynamics is useful for identifying fluxes and forces, which continue to be valid for a wide non-linear range. It is a good tool for extrapolation of concepts to non-linear region.

(b) Non-linear flux equations involve higher powers of gradients but ORR between first order coefficients is satisfied.

(c) The theorem of minimum entropy production for the linear steady state does not hold in the non-linear range.
(d) Non-linear steady states are usually stable. In some cases, bistability (multistability) may appear with and without hysteresis.
(e) Non-linear steady states can conveniently be experimentally investigated in the case of Electro-kinetic phenomena and chemical reactions.
(f) In cases such as above, non-linearity also depends on structure factors (e.g. Helmholtz double layer) and non-linear kinetics.
(g) In case of thermal diffusion in gases, non-linear flux equations of mass and heat transport involve space derivatives of gradient, but higher powers of gradients of single force do not occur, although ORR is satisfied between first order coefficients.

References

1. R.P. Rastogi, S.N. Singh and M.L. Srivastava, J. Phys. Chem., 74 (1970) 2960.
2. M.L. Srivastava, Indian J. Chem., 9 (1971) 48.
3. R.C. Srivastava and R.P. Rastogi, Transport Mediated by Electrified Interfaces, Elsevier, Amsterdam, 2003.
4. R.P. Rastogi, K. Singh and S.N. Singh, J. Phys. Chem., 73 (1969) 1593.
5. R. Sabad, Ph.D. thesis, Gorakhpur University, 1977.
6. R.P. Rastogi, G. Srinivas, R.C. Srivastava, A.P. Mishra, J. Colloid Interface Sci., 175 (1995) 262–275.
7. R.P. Rastogi and K.M. Jha, Trans.Faraday Soc., 62 (1966) 585.
8. R.P. Rastogi, K. Singh, R. Kumar, S.A. Khan, J. Phys. Chem., 81 (1977) 2114.
9. R. Kumar, Ph.D. thesis, Gorakhpur University, 1977.
10. R.P. Rastogi, K. Singh and J. Singh, J. Phys. Chem., 79 (1975) 2574.
11. R.P Rastogi, K. Singh, R. Kumar and R. Sabad, J. Memb. Sci., 2 (1977) 317.
12. R.P. Rastogi and R. Shabad, J. Phys. Chem., 81 (1977) 1953.
13. R.P. Rastogi, K. Singh and J. Singh, J. Phys. Chem, 79 (1975) 2574.
14. R.P. Rastogi and R. Sabad, Indian J. Chem., 21A (1982) 859.
15. R.P. Rastogi, J. Sci. Ind. Res., 28 (1969) 28.
16. R.P. Rastogi, R. Shabad, B.M. Upadhyaya and S.A. Khan, J. Non-equilib. thermodyn., 6 (1981) 271.
17. S. Chapman and T.G. Cowling, The Mathematical Theory of Non-Uniform Gases, Cambridge University Press, Cambridge, 1960.
18. R.P. Rastogi and B.P. Misra, J. Phys. Chem., 74 (1970) 119.
19. I. Prigogine, Non-Equilibrium Statistical Mechanics, Interscience, New York, 1962.
20. S.R. de Groot and P. Mazur, Non-equilibrium Thermodynamics, North-Holland Publishing, Co, Amsterdam, 1962.
21. D. Burnett, Proc. Lond. Math. Soc., 40 (1935) 382.
22. S. Chapman and T.G. Cowling, Proc. Roy. Soc., A179 (1941) 150.

23. R.P. Rastogi, R.C. Srivastava and K. Singh, Trans. Faraday Soc., 61 (1965) 854.
24. A. Katchalsky and P.F. Curran, Non-equilibrium Thermodynamics in Biophysics, Harvard University Press, Cambridge Massachusetts, 1965, p. 94.
25. D.J.G. Ives and R.G. Kerlongue, J. Chem. Soc., 1362 (1940) 1362.
26. R.P. Rastogi and R. Khare, Indian J. Chem., 33A (1994) 1.

Chapter 8

BIFURCATION PHENOMENON AND MULTI-STABILITY

8.1. Introduction

Currently, non-linear science has become a hot area of multi-disciplinary research. Oscillatory reaction and multiple steady state are a typical class of non-equilibrium phenomenon which occurs far from equilibrium [1, 2].

A system under suitable condition may switch from one stable A to either of two stable states B and C, on changing some parameter which is called bifurcation parameter. Successive bifurcation can also take place.

The coexistence between two stable steady states is referred to as bistability. However, it is generally associated with the phenomenon of hysterisis in which a system jumps back and forth between the two brownies of stable states for different critical values from which bifurcation occurs. These phenomena are observed in a number of physical, physiological and biological and social systems.

A variety of bifurcation phenomena in chemical systems have been discovered during the last few decades [3]. Results have been interpreted in terms of methodology considerably improved and refined by developments in non-linear dynamics. Bifurcation is also quite common in transition from one state to another in biological systems indicated by small change in timing or electrical conductivity.

8.2. Dynamical non-linear systems

The systems discussed above belong to the class of non-linear dynamical systems. In such cases, the non-linear equation represents evolution of a solution with time or some variable like time. Such non-linear equations may be (i) algebraic, (ii) functional, (iii) ordinary partial differential equations and (iv) integral equations or a combination of these. Non-equilibrium systems can be defined by the type of equations as defined above involving the bifurcation parameter. The solution may change depending on the particular values of parameter. The solutions change at bifurcation points. Such situations do occur in the form of bistability and oscillations.

When the system moves further away from equilibrium as defined by certain parameters, the solutions of a non-equilibrium system can change their character as the above type of parameter called bifurcation parameter change. At this stage, a new type of non-equilibrium state can emerge. This stage is called bifurcation point. Such states do occur in the form of bistability and oscillations in the type of systems discussed in this communication.

Bifurcation means a division in two, a splitting apart, a change. In dynamical systems, the object of bifurcation theory is to study the changes that maps undergo as parameters change. These changes often involve the periodic point structure, but may also involve other changes as well

We will now consider a simple example to illustrate how variation of a parameter can lead to qualitative change in the solution of an equation. We consider the equation

$$X^2 - a = 0 \tag{8.1}$$

We will now examine the real solutions when the values of a change the roots of equation (8.1):

$$X = \pm\sqrt{a} \tag{8.2}$$

Now, three cases arise.

(a) $a > 0$, the two roots are real
(b) $a = 0$, only one root
(c) $a < 0$, the two roots are imaginary

The real cases (b) and (c) are relevant for our discussion. We can graphically represent the solutions according to Fig. 8.1. The solution is a parabola, when X is plotted against a (Fig. 8.2).

The dotted region represents the imaginary solution (when $a < 0$). The character of the solution changes when a approaches the value zero, which acts as the bifurcation point. When a increases further, the character of the solution changes and two values of x are obtained.

A variety of bifurcation phenomena in chemical systems have been discovered during the last few decades. Results have been interpreted on the basis of the framework originally suggested by Poincare, who used the term bifurcation to describe the "splitting" of

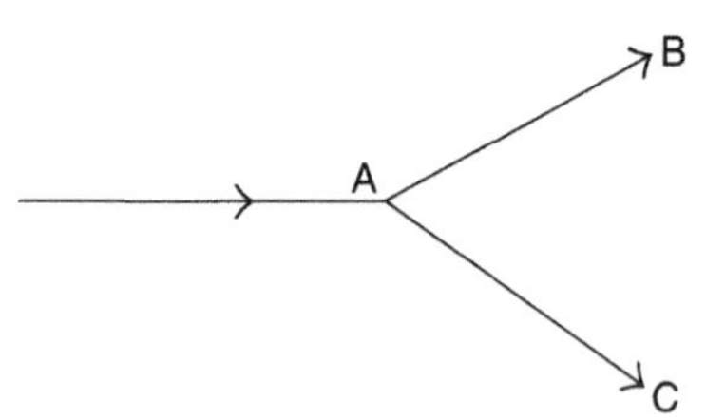

Figure 8.1. Bifurcation phenomenon.

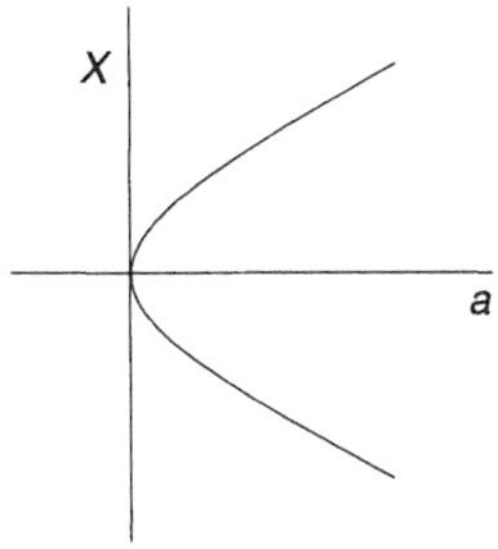

Figure 8.2. Graphical solution of Eq. (8.2).

equilibrium solutions in a family of differential equations. Last two methodologies are considerably being improved and refined by researches in mathematical basis of non-linear dynamics. We will by to present an elementary account of these ideas in this article.

8.3. Typical types of bifurcation

We would start with the simple examples of (a) tangent bifurcation (b) pitchfork bifurcation and (c) Hopf bifurcation [Figs. 8.3 and 8.4].

(a) ***Tangent bifurcation [4]***

Let us consider the graph of function $f(x) = \lambda\, e^x$ where λ is some constant. Depending on the value of λ, the curve would be a tangent at $f(x) = 1$; $x = 1$ passing through the origin as shown below.

Three situations can arise when (i) $\lambda > 1/e$, (ii) $\lambda = 1/e$ and (iii) $\lambda < 1/e$. In case (i) the curve will not touch OC, while in case (iii), the curve will cut OC at two points. On the other hand for case (ii), OC will be tangent to the curve. This type of bifurcation is called tangent bifurcation and λ is called bifurcation parameter.

(b) ***Pitchfork bifurcation***

Consider the differential equation:

$$\dot{x} = dx/dt = \mu x - x^3 \tag{8.3}$$

It is obvious that $dx/dt = 0$ when $x = 0$ or $x = \pm\sqrt{\mu}$. The unique fixed point $x = 0$ existing for $\mu \leq 0$ is stable while for $\mu > 0$ is unstable. Furthermore, for $\mu > 0$, the fixed points $x = \pm\sqrt{\mu}$ are stable. This result can be expressed diagrammatically as follows.

The bifurcation diagram is represented by the curve A′B′C′D′, which looks like a tuning fork. The stable branch A′B′ bifurcates to branches B′C′ and B′D′ at the point B′ which is called the bifurcation point.

There is a formal resemblance between bistability in a reacting system and the pitchfork bifurcation.

Let us illustrate the idea by another simple example. Let there be two variables x and y dependent on time in the following manner:

$$x = A\exp(\lambda t); \quad y = B\exp(\lambda t) \tag{8.4}$$

which happen to be hypothetical solutions of

$$\dot{x} = a_1 x + a_2 y \quad \text{and} \quad \dot{y} = b_1 x + b_2 y \tag{8.5}$$

By linear stability analysis it is found that

$$\lambda^2 - (a_1 + b_2)\lambda + a_1 b_2 - a_2 b_1 = 0 \tag{8.6}$$

(c) ***Hopf bifurcation* [5]**

Any chemical reaction mechanism may be considered as a system of n differential equations of the general form $\partial y/\partial t = f(y, \lambda)$ where y and f are vectors with n components. If $f(y, \lambda)$ represents a non-linear dynamical system, a large multitude of interesting phenomena may occur.

The particular point that connects steady states to limit cycles is commonly called a Hopf bifurcation.

There are two types of Hopf bifurcation: (i) supercritical; (ii) sub-critical.

In the supercritical case, the amplitudes of the oscillations smoothly increase from zero as bifurcation parameter λ moves away from λ_0 into the oscillatory region. The amplitude increase is proportional to $(\lambda - \lambda_0)^{1/2}$.

A sub-critical Hopf bifurcation contains an interval of λ in which the stable steady state is connected by an unstable periodic orbit. Therefore, in addition to the bifurcation point, a turning point exists that connects a stable with an unstable periodic solution.

Chemical systems are expected to be particularly sensitive towards fluctuations at their sub-critical Hopf bifurcation. On tinkering the hard cove of oscillatory mechanism by (i) induction of some additional steps or by (ii) acceleration and deceleration of some steps, one can change the oscillatory features and the reaction mechanism.

Hopf bifurcation and Jug handle-type bifurcation are most commonly observed in chemical systems. In the former case, at the bifurcation point, time period remains finite but amplitude tends to zero. However, in the latter case, both the time period and amplitude have finite values.

Bifurcation features and deformation of limit cycles for NaH_2PO_4 + cyclo hexanone/acetone + Mn^{2+} + BrO_3^- + H_2SO_4 oscillatory reaction system has been

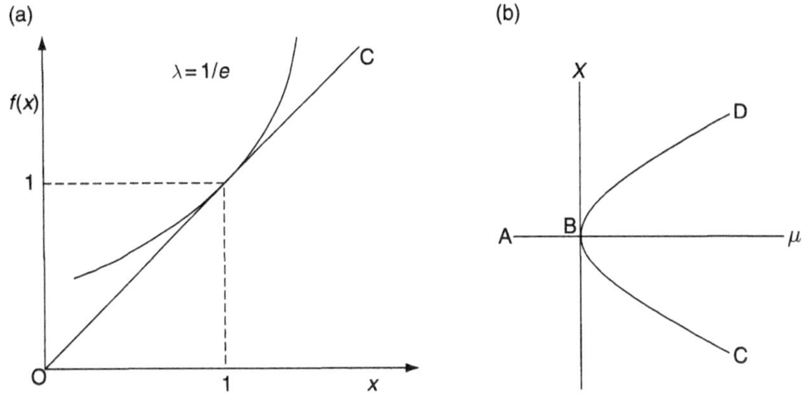

Figure 8.3. (a) Tangent bifurcation, (b) pitchforke bifurcation.

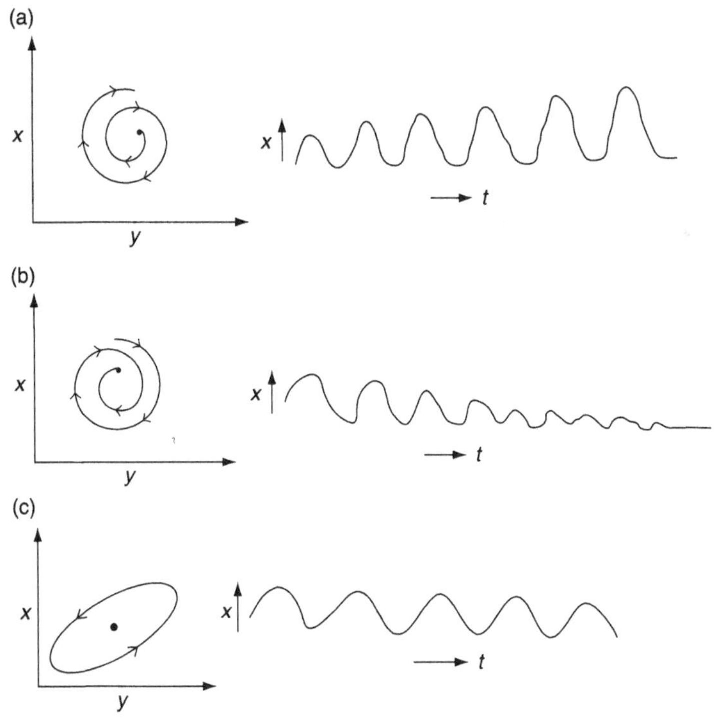

Figure 8.4. (a) Supercritical bifurcation, (b) subcritical bifurcation, (c) undamped oscillations.

reported [6]. Acetone system yields unperturbed limit cycles, whereas in the case of cyclohexanone systems, the limit cycles are deformed and knot is obtained. At the upper limit, Hopf bifurcation is obtained in the case of acetone while Jug Handle bifurcation is obtained in the case of cyclohexanone.

Let us focus our attention on Eq. (8.6) ignoring for a moment how it is derived. If $a_1b_2 - a_2b_1 = +\text{ve}$, then the following situation can occur. The values of λ would be given by

$$\lambda = \frac{1}{2}\left[(a_1 + b_2) \pm \sqrt{(a_1 + b_2)^2 - 4(a_1b_2 - a_2b_1)}\right] \tag{8.7}$$

If $D = +(a_1 + b_2)^2 - 4(a_1b_2 - a_2b_1) > 0$, then the two roots of λ are real. On the other hand when $D < 0$, the roots of λ would be complex conjugate. Now, if

$(a_1 + b_2) > 0$, the singular point is an unstable focus which corresponds to oscillations that increase in amplitude.

$(a_1 + b_2) < 0$, damped oscillations occur; the singular point is a stable focus about which the spiral trajectories in the $x - y$ plane wind up.

$(a_1 + b_2) = 0$, the roots $\lambda_1 = -\lambda_2$, are imaginary. Undamped oscillations take place. The phase trajectories in the $x - y$ plane are concentric ellipses and the singular point is the centre.

Note: The modern approach of studying steady-state multiplicity is to find the set of parameters at which a change (bifurcation) in the number of solutions occurs. The bifurcation loci coalesce in general at certain highly degenerate (singular) points. The behaviour in the vicinity of these point is universal, in the sense that it depends only on the character of singular points, but not on the specific form of the mathematical model. This enables a very general and systematic classification of the local multiplicity features by focusing attention on the singular points. Characteristics of different types of bifurcation have been summarized in Table 8.1.

Table 8.1. Characteristic of bifurcation point.

Bifurcation	Characteristic
Supercritical Hopf bifurcation and inverted Hopf bifurcation	At the birth or death of periodic orbits, the amplitude of oscillations is zero but the time period is finite
Jug handle bifurcation	Both the amplitude and time period of bifurcation oscillations are finite at the critical value of the parameter
Saddle loop bifurcation	The oscillations appear and disappear with finite amplitude but with finite period
Saddle node infinite period bifurcation	The oscillation appears or disappears with a finite amplitude but with an infinite period. If a limit cycle disappears via this bifurcation, then an excitable steady state can appear simultaneously

8.4. Bifurcation from steady state to bistability

A system is bistable if, for a given value of perturbation x, its response y may take two stable values under steady-state conditions. An indicator of bistability is the appearance of hysteresis as a loop in the curve $y = f[x]$. In the case of chemical reactions in continuously stirred tank reactor (CSTR), the perturbation is provided by flow rate and response is observed by noting the concentrations of oscillating species. In case of electro-kinetic phenomena, current, I, provides the perturbation while the corresponding potential difference, $\Delta\phi$, yields the response. Similarly, in the case of magnetic resonance, external magnetic field generates the perturbation while the response is registered by the nuclear field.

It may be noted that bistability can occur even in the absence of hysterisis. Theoretical studies of biochemical and combustion systems have shown that one of the limit points bounding the domain of bistability may not be accessible to the system.

8.5. Bifurcation from steady state to oscillatory state

Departure from stable state to an oscillatory state also involves a particular type of bifurcation. Consider a system governed by following differential equations in two variables x and y:

$$\mathrm{d}x/\mathrm{d}t = ax + by \tag{8.8}$$

$$\mathrm{d}y/\mathrm{d}t = cx + dy \tag{8.9}$$

where a, b, c and d are constants. Putting $x = x_0\,\mathrm{e}^{\lambda t}$ and $y = y_0\,\mathrm{e}^{\lambda t}$, where x_0, y_0 and λ are constant, and following the normal mode analysis, we get relation similar to (8.7)

$$\lambda^2 - (a+b)\lambda + (ad - bc) = 0 \tag{8.10}$$

On solving the above equation we get

$$\lambda\pm = \alpha(\rho) \pm \iota\omega(\rho) \tag{8.11}$$

where $\alpha(\beta) = (a+b)/2$ and $\omega^2 = (a+b)^2 - 4(ad-bc)$ when $4(ad-bc) > (a+b)^2$ and ω is the frequency and ρ is the bifurcation parameter at $\rho = 0$; the solution bifurcates from a steady state solution to periodic solution. The amplitude is proportional to $p^{1/2}$ and time period is $2\pi/\omega$. The bifurcation diagram is as shown in Fig. 8.3.

8.6. Multi-stability

The entropy export to the environment becomes possible if the system is provided continuously with an overcritical amount of the energy. So the system can leave equilibrium or does never reach it. Multi-stability can occur in different types of systems. Some of the typical examples are given in Table 8.2.

Table 8.2. Bistability in different types of systems.

Areas	Phenomena
Animal psychology	Bistable behaviour of an animal under stress (flight or attack)
Economics	Two stable levels of good production (a) Low production with high prices (b) High production with low prices
Nerve physiology	(a) Switching mechanism in nerve cells from rest to action potential (b) Epileptic brain, seizure in epilepsy
Biochemistry	Bistable mechanism in enzyme reaction, protein/biochemical systems
Electronic devices	(a) Multivibrator (flip-flop) (b) Bistable operation of a laser coupled to an external resonator
Spatial bistability	Spatial bistability of two-dimensional turing pattern in a reaction-diffusion system
Magnetic resonance	Bistability of magnetic resonance of conduction electrons
Optics	Electro-opticals absorption in quantum well.

In Chemistry, a large number of chemical reactions exhibiting bistability have been reported [8].

8.7. A simple mathematical model of bistability

In case of bistability, it is logical to expect that if the steady state value of any quantity X is related to certain parameter Y, then a cubic relation of the type

$$(Y-\alpha)(Y-\beta)(Y-\gamma)=X-\alpha\beta\gamma \tag{8.12}$$

can yield three values of Y for a particular value of X in a certain region. To illustrate the point, the computer solution of the equation

$$Y^3-6Y^2+11Y-X=0 \tag{8.13}$$

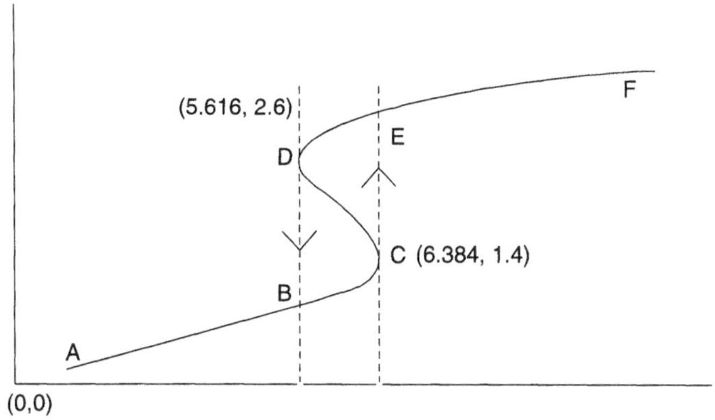

Figure 8.5. Simple model of bistability.

where $\alpha = 1$, $\beta = 2$ and $\gamma = 3$, has been plotted. The curve ABCDEF provides the solution (Fig. 8.5).

It may be noted that along AB and EF, one gets only one value of Y for a particular value of X. There are three values of Y corresponding to states BC, CD and DF when X lies in the region denoted by <−>. These correspond to multiple steady states. In many physico-chemical systems, one of the solutions along CD is not permissible on account of theoretical constraints as in electro-kinetic phenomenon. In such cases, bistability and hysterisis are obtained. The points C and D where the nature of the solution changes is called the bifurcation point.

8.8. A simple model for reacting systems

For discussing the bifurcation phenomena in the case of chemical reactions [1], we consider the following simple set of reactions:

$$\mathrm{A} + 2\mathrm{X} \underset{K_2}{\overset{K_1}{\rightleftharpoons}} 3\mathrm{X}$$

$$\mathrm{X} \underset{K_4}{\overset{K_3}{\rightleftharpoons}} B$$

The system is connected to reservoirs of A and B, so that their concentrations are kept constant. The rate equation is given by

$$\mathrm{d}x/\mathrm{d}t = -k_2X^3 + k_1AX^2 - k_3X + k_4B \tag{8.14}$$

where X, A and B denote the concentrations of the reactants, and k_1, k_2, k_3 and k_4 are the rate constants. In the steady state, $dX/dt = 0$, so that we have for the steady state

$$X_0^3 - aX_0^2 + kX_0 - b = 0 \tag{8.15}$$

where X_0 is the steady-state value of X and

$$a = k_1 A/k_2; \quad b = k_4 B/k_2; \quad k = k_3/k_2$$

If $q = 1/9a^2 + 1/3k$ and $r = 1/27a^3 + 1/6ak + 1/2b$, the following cases arise:

(a) when $q^3 + r^2$ is greater than zero, the equation has one real root and two imaginary roots,
(b) when $q^3 + r^2 = 0$, the equation admits at least two identical roots, and
(c) when $q^3 + r^2$ is less than zero, the equation admits three real roots.

The bistability diagram obtained by plotting X_0 against b is of the type shown in Fig. 8.5.

8.9. Bistability in reacting systems

We discuss below a few representative systems where (i) only bistability occurs and (ii) both bistability and oscillations occur. Bistability results for some typical chemical systems are discussed below. Quite a few different cases [6, 7] have earlier been discussed by Field and Burger [8].

8.9.1. $Mn^{2+} + BrO^{3-}$ system

Experimental studies [9] on $Mn^{2+} + BrO^{3-}$ system in CSTR have clearly established the existence of bistability (hysteresis loops) by (i) varying bromide ion inflow concentration, (ii) varying bromate ion inflow concentration, (iii) varying catalyst inflow concentration and (iv) varying flow rate and measuring the Br potential in the steady state.

8.9.2. Cerous-bromate system

Geiseler and Follner [10] discovered bistability in cerous-bromate system and found that under certain set of constraints, the system in sulphuric acid medium can exist in either one or two stable states. Bar-Eli and Noyes [11] have explained their results on the basis of the following mechanism.

$$BrO_3^- + Br^- + 2H^+ \longrightarrow HBrO_2 + HOBr$$
$$HBrO_2 + Br^- + 2H^+ \longrightarrow HBrO_2 + 2HOBr$$

$$HOBr + Br^- + H^+ \longrightarrow Br_2 + H_2O$$
$$BrO_3^- + HBrO_2 + H^+ \longrightarrow 2BrO_2^- + H_2O$$
$$Ce^{3+} + BrO_2 + H^+ \longrightarrow Ce^{4+} + HBrO_2$$
$$Ce^{4+} + BrO_2 + H_2O \longrightarrow Ce^{3+} + BrO_3^- + H_2O$$
$$2HBrO_2 \longrightarrow BrO_3^- + 2HOBr + H^+$$

By adding a flow term, the above set of equations yields seven rate equations. Using the appropriate computational procedure, these workers reported hysteresis limits and estimated bromide ion concentrations in steady states for various values of initial bromide ion concentration.

For generating oscillations, a sub-system has to be added to the system under consideration. Such a sub-system is provided when malonic acid is added to the system.

8.9.3. Chlorite–iodide reaction system

The reaction between chlorite and iodide has been studied over a wide range of conditions such as varying flow rate, pH, chlorite and iodide concentrations in a CSTR [12]. The system exhibits both bistability and hysteresis. The region of bistablility is consistent with cross-shaped diagram as postulated by Biossonade and De Kepper [13]. Ali and Menzinger [14] have reported inhomogeneity-induced transformation of the well-known bistability hysteresis in the chlorite/iodide reaction into an isola through the destabilization of the high-electric potential branch at low flow rate.

8.9.4. Briggs-Rausscher reaction system

Pacault *et al.* [15] studied this oscillatory reaction in a CSTR and reported complex oscillations, multiple steady states accompanied by a variety of bifurcation and hysteresis phenomena. They proposed a mechanism which was subjected to computer simulation. The $[IO_3]_0 - [I_2]_0$ clearly demonstrates two separate regions of bistability and oscillations. Noyes *et al.* [16] and Furrow *et al.* [17] suggested same mechanism independently by using different methods, thereby supporting both the validity of the mechanism and their approach

8.9.5. Iodate–arsenite reaction

There are two steady states in iodate–arsenite system in a CSTR as observed by De Kepper *et al.* [18, 19]. It is found that there exist two steady states for certain initial values of $[IO_3]_0$ and $[H_3AsO_3]_0$ and the flow rate. The transition from one state to the other may be induced by a suitable transient perturbation, e.g. injection of a small amount of some species into the CSTR or a temporary change in the flow rate.

Overshoots and undershoots are also observed when stepwise changes are made in $[H_3AsO_3]_0$ at a fixed flow rate. On the other hand, no overshoots and undershoots are observed when $[IO_3]_0$ is varied in an analogous manner.

It is believed that when the ratio of initial concentrations $[H_3AsO_3]_0/[IO_3]_0$ is greater than 3, the following reaction takes place

$$5H_3AsO_3 + 2IO_3^- + 2H^+ \rightleftharpoons 5H_3AsO_4 + I_2 + H_2O$$

The subscript 0 denotes the initial concentration. When the ratio of concentrations is given by

$$[H_3AsO_3]_0/[IO_3]_0 < 3.$$

The reaction proceeds as

$$3H_3AsO_3 + IO_3^- + 5I^- \rightleftharpoons 3H_3AsO_4 + 6I^-$$

which produces I^- autocatalytically [20] and which is followed by the reaction

$$5I^- + IO_3^- + 6H^+ \rightleftharpoons 3I_2 + 3H_2O$$

Numerical analysis of bistability in the above system has been attempted by Kepper *et al.* [18] and Papsin *et al.* [19]. In the latter case, rate equations have been generated by considering a set of seven reactions, and the equations have been solved for seven species, assuming $[IO_3^-]_0$ to be constant. In the former case, four rate equations are generated for four unknowns, viz., $[IO_3^-]_0[I_2]$ and $[H_3AsO_3]$. Solutions have been obtained with the use of Gear's algorithm for coupled sets of stiff differential equations.

Interesting results have been obtained by Rastogi *et al.* [20] when the law of conservation of electric charges and electro-neutrality is taken into account in the ionic reaction-diffusion system. The reaction scheme can be written as

$$5A + 2X_1^- + 2X_3^+ \underset{K_2}{\overset{K_1}{\rightleftharpoons}} 5B + C + E$$

$$3A + X_1^- + 5X_2^+ \underset{K_4}{\overset{K_3}{\rightleftharpoons}} 3B + 6X_2^-$$

$$X_1^- + 5X_2^+ + 6X_2^+ \underset{K_5}{\overset{K_4}{\rightleftharpoons}} 3C + 3E$$

which can be simplified to the model reaction system

$$X_1^- + X_3^+ \underset{K_2}{\overset{K_1}{\rightleftharpoons}} A$$

$$X_1^- + X_2^- \underset{K_4}{\overset{K_3}{\rightleftharpoons}} 2X_2^-$$

$$X_1^- + X_2^- + 2X_3^+ \underset{K_6}{\overset{K_5}{\rightleftharpoons}} B$$

where $A = H_3AsO_3$; $B = H_3AsO_4$; $C = I_2$; $X_1 = IO_3^-$ $X_2 = I^-$; $X_3^+ = H^+$

A and B are the products of particular reaction using appropriate dimensionless concentrations X_1, X_2 and X_3 and quantities in terms of dimensionless $[IO_3^-]'[I^-]$, $[H^+]$.

Noting that on account of electro-neutrality $X_1^- + X_2^- = X_3^+$ and for the steady state $X_1^- = 0$, $X_2^- = 0$, $X_3^+ = 0$, the following two solutions are obtained:

(a) $X_1^- = X_3^+ = 1$; $X_2^+ = 0$ and

(b) $X_1^+ = (\alpha/\beta)^{1/2}(2\alpha - 1)/4\alpha \pm 1/4\ \{[2\alpha + 1]^2 - 8\beta\}^{1/2}/\{\alpha\beta\}^{1/2}$

where $\alpha = k_3/k_1$; $\beta = k_2k_3A/k_1^2$

Thus, when $\beta > \alpha$ and $\alpha < 1/2$, one gets only one solution corresponding to a monostable state since negative values of X_2^- are inadmissible. However, when $\beta > \alpha$ and $\alpha < 1/2$ both the solutions (a) and (b) are possible, and bistability occurs.

The influence of KCNS and $NaHCO_3$ the bistability diagram and hysteresis for the iodate–arsenite system has been investigated. On addition of KCNS, the critical point is shifted towards lower residence times, while on addition of $NaHCO_3$ it is shifted towards higher residence times [21]. KCNS produces more of I^-, whereas $NaHCO_3$ changes the concentration of H^+ in the system.

Non-periodic oscillations have been observed in the above system in CSTR. The visual oscillations are accompanied by oscillations in the redox potential and I^-. Oscillations are inhibited by IO_3^-, Cl^- and $NaHCO_3$. These depend on $[IO_3^-]$ [arsenite] and the reciprocal of the residence time. There is an upper limit and a lower limit of $[IO_3^-]$ as well as $[H_3AsO_3]$ in between which oscillations are observed. It is found that these occur in a narrow range of flow rate which is close to the region when the system can switch from the monostable state to a state of bistability [21(b)]. However, for periodic oscillations, different strategies were adopted by Epstein and co-workers [12].

8.9.6. Gallic acid BrO_3^- system [22, 23]

A bistability between an oscillatory branch and a flow branch has been reported for the gallic acid system by using the stirring rate as control parameter, indicating complex role of temperature and stirring rate as the control parameters [23]. Bistability in an uncatalysed oscillator in a CSTR has been observed in initial $[BrO_3^-]_0$ $[Br^-]_0$ concentration space. It occurs between an oscillatory and a flow branch. Hysterisis is also found to occur.

8.10. Spatial bistability

Spatial bistability [24, 25] of two-dimensional Turing patterns in a reaction-diffusion system has recently been observed. Starting from a uniform state and decreasing the [malonic acid] in the chlorite–iodide + malonic acid (CIMA) reaction, first a Turing bifurcation to a quasi two-dimensional striped pattern was observed. With further decrease in [malonic acid], there was transition t < hexagons. The transition from stripes to hexagons was hysteric – there was a range in [malonic acid] in which either stripes or hexagons could be stable.

8.11. Bistability in magnetic resonance

Two types of EPR (electron paramagnetic resonance) detectable bistability have been detected during recent years. The first is related to the EPR spectrometer itself and not to the sample. The physical cause of the phenomenon is the non-linear behaviour of the sample to cavity coupling [26]. The width of the hysteresis loop was found to vary with the filling factor of the cavity. This type was detected by Giordano *et al.* [26] during studies on polypyrrole radical.

The second type of bistability is intrinsic to the conduction electron spins. Under saturation conditions, Overhauser effect is the primary condition for the observation of bistability. Under such a condition, the effective magnetic field, and hence the nuclear field B_n, takes two stable steady-state values for a given value of the external magnetic field B_0.

Bistability of the magnetic resonance of conduction electrons in gallium oxide [27] and small metallic lithium particles [28] has recently been reported. It was observed that EPR spectrum of conduction electrons in β-Ga_2O_3 single crystals exhibits bistability with a magnetic field memory at temperatures up to room temperature. Bistability effects recently generated much interest because of their potential applications in information storage and signal processing.

Conduction electrons in gallium oxide β-Ga_2O_3 exhibit bistability detectable by EPR at X band ($V = 8.4$ GHz; B0, 340 mT) and temperatures higher than 100 K, but the effect is also observable at room temperature.

The bistability effect manifests itself by the occurrence of different line shapes and positions of EPR lines recorded on forward and backward sweeps of the external magnetic field. Giordano *et al.* [26] tried to develop the theory of bistability of conduction electrons. According to them, the Overhausser effect is the primary condition for the observation of the bistability. Furthermore, EPR bistability of the conduction electrons under Overhausser effect will exist if the effective magnetic field β_{eff} and hence the nuclear field B_n takes two stable steady-state values for a given value of the external magnetic field B_0.

8.11.1. Far from equilibrium region

In the equilibrium situation, the thermodynamic forces: (if one consider electro-kinetic phenomena) ΔP, $\Delta\phi$, ΔC and the consequent fluxes J_v. I, etc. are zero. For finite values but not large values of the differences, the linear thermodynamics of irreversible process using linear phenomenological equation adequately describes the electro-kinetic phenomena. When the magnitude of the thermodynamic forces further increases, the non-linear phenomenological equation becomes significant, and for a single fluid, the theoretical framework holds good, provided Gibbs equation is satisfied. On the other hand, under specific conditions, when a difference in concentration of an electrolyte is maintained across the membrane, bistability and temporal oscillations in volume flow, pressure difference, potential difference and resistance have been reported by Teorell [29, 30], Mears and Page [31, 32] and Kobatake and Fujita [33, 34]. Obviously, for such a case, another source of non-linearity is introduced by the involvement of concentration difference.

Since electro-kinetic systems can be maintained in the far-from-equilibrium region to any desired extent by controlling the variables, they constitute a very good system for the study of dynamic instability from experimental angle. A brief discussion of bistability and oscillations in electro-kinetic phenomena from an experimental point of view is given below.

8.12. Bistability in electro-kinetic phenomena

In electro-kinetic phenomena, bistability was studied by several groups in a system of following type [35–38] the system was as follows. The plots of I (current) versus $\Delta\phi$ were found to be typical N-shaped curves (Fig. 8.6). These N-shaped curves have been referred to as flip-flop type.

A typical example of bistability in electro-kinetic phenomena [35–38] is demonstrated in Fig. 8.6, which is essentially the current–voltage curve obtained when solutions of different concentrations of the electrolyte are kept on the two sides of the membrane. Current is passed and the transmembrane potential $\Delta\phi$ is recorded.

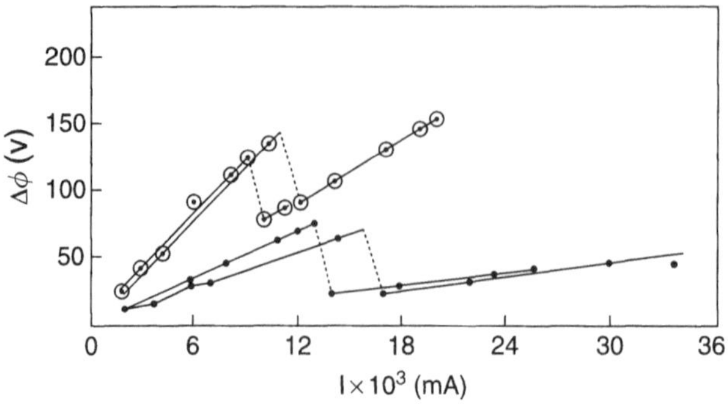

Figure 8.6. Dependance of transmembrane current on applied potential for (a) 0.1 N NaCl/ membrane/0.01 N NaCl at $\Delta P = 0$ and $\Delta P = 6$ cm of H_2O column, (b) 0.1 N KCl/membrane/0.01 N KCl at $\Delta P = 0$ and $\Delta P = 6$ cm of H_2O column.

A similar type of curve is obtained when $\Delta\phi$ is varied and current is recorded. The current increases with the transmembrane potential. On attaining a certain value, it suddenly decreases (discontinuously) and then again increases linearly with slope much smaller than that found in the initial state. An N-shaped current–voltage curve is obtained which is called the flip-flop type. It is obvious that in the region of bistability, there are two values of $\Delta\phi$ corresponding to a particular value of I.

Recent experiments have shown that bistability is not always observed in electro-kinetic phenomena. Whereas bistability is observed in the case of pyrex sinter, this is not so in the case of millipore filter [37]. In the case of pyrex sinter, the equation relating I and $\Delta\phi$ is non-linear and has cubic terms in $\Delta\phi$. Hence, bistability is expected in this case, whereas this is not so in the case of millipore filter, since I versus $\Delta\phi$ equation is only quadratic in $\Delta\phi$.

Another point has to be noted in this connection. Electro-kinetic oscillations are observed in the case of both types of membranes. Thus, it follows that bistability is not a necessary condition for oscillation to occur. The former phenomenon corresponds to a situation where $\Delta\phi$ varies with change in I when ΔP is fixed or zero and ΔC is fixed.

On the other hand, in case of oscillations, one essentially considers the situation when I and ΔC are fixed but $\Delta\phi$ and ΔP are allowed to vary. It means that if the restriction on constancy of ΔP is removed, the system can display oscillations. Such a possibility has been demonstrated [34] by using vertical tubes generating pressure difference ΔP instead of horizontal tubes in the experimental set-up.

The two steady states exist within the narrow range of I, i.e. for the same value of I, it can have two values of ϕ. In other words, there are two steady states, one with lower resistance and the other with higher resistance. The steady state with lower resistance is obtained when the membrane pores got filled with concentrated solution while the latter corresponds when the membrane pores got filled with dilute solutions.

In the case of bistability of electro-kinetic phenomena, dependence of the transmembrane current on the applied potential difference is investigated when the pressure difference across the membrane is held constant [31, 33, 34] and a concentration difference is maintained on the two sides of the membrane. The current (I) increases with the transmembrane potential $\Delta\phi$. On attaining a certain value, it suddenly decreases (discontinuously) and then again increases linearly with slope much smaller than that found in the initial state. An N-shaped current–voltage curve [31, 39] is obtained, which is called the flip-flop type.

So far as the mechanism of bistability is concerned in non-reacting system including membrane process, there is general agreement that it is related to the interplay of the hydrodynamics flux and electrostatic flux, which act in opposite directions [32].

Quantitative theories of the phenomena were developed by Kobatake and Fujita [33, 34] and Meares and Page [32]. The following assumptions are implicit in the theories [35].

(a) The contributions of streaming current and diffusion current (due to concentration gradient alone) are ignored.
(b) Electro-osmotic permeability is assumed to depend only on concentration. However, it is known that it is dependent on potential in the non-linear region.

Kobatake and Fujita [33, 34] developed a theory of the phenomena by deducing continuity equations for electricity and mass flow by combining the Navier–Stokes equation for the mass movement and the concept of thermodynamics of irreversible processes. The steady-state solution of the resulting equations yielded the characteristic N-shaped relation between the electric current (I) and the transmembrane potential ($\Delta\phi$) which involved hysterisis between the transition points. The features of the curves predicted by Kobatake and Fujita [33, 34] compared well with the Franck's experimental results [40] on a sintered glass membrane separating 0.01 and 0.01 N NaCl solutions, with the only difference being that no hysteresis was observed. However, in later experiments, Meares and Page [31] observed both bistability and hysterisis. Kobatake [34] also attempted to study the phenomenon in terms of generalized entropy production as proposed by Glandsdorff and Prigogine [41–43] and has tried to show that the transition between two steady states takes place when the generalized entropy productions of the two states corresponding to the high and low electric resistance of the membrane have the same value. The possibility of multiplicity of steady states can be understood if one examines the characteristic of non-linear flux equation for the H^+ form of the resins [38].

When $I = 0$, one gets a quadratic equation in ΔP and consequently two values of ΔP for the same value of $\Delta\phi$. It should be noted that ΔP is the independent valuable and $\Delta\phi$ is the dependent variable in the experiment, and furthermore steady states are obtained continuously on varying the pressure and hence no instability is involved.

8.13. Bistability in biological systems

Bistability is also observed in physiological systems, and the phenomenon has a formal similarity with electro-kinetic phenomenon. In nerve physiology, the so-called switching mechanism in nerve cells from a rest to action potential corresponds to the above phenomena. Current–voltage curves [44, 45] similar to the above are obtained in biological systems where we have different concentrations of electrolytes across the cell membranes.

The ratio of the values of conductance of the membrane in the two steady states is given in Table 8.3 for different cases. The same for the pyrex membrane sodium chloride system has also been recorded for the sake of comparison.

Excitation of the squid axon membrane in iso-osmotic potassium chloride solution has been investigated [46–48] and is found to be particularly interesting. In this context, current–voltage curves were obtained by keeping (i) the current constant and (ii) the voltage constant. A typical curve [48] is shown in Fig. 8.7 which displays hysteresis.

Table 8.3. Ratio of conductance of the membrane in two steady states [44].

System	Approximate ratio of conductance of solutions on the two sides of the membranes
Valonia cell	1:2
Squid axon in iso-osomotic KCl	1:16
Cell electroplax	1:2
Frog node	1:6

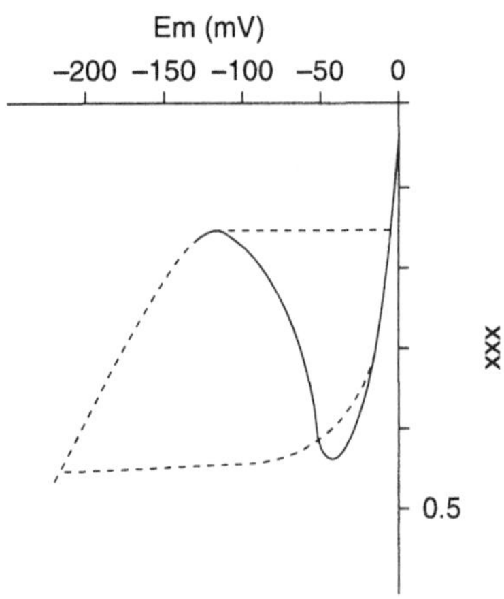

Figure 8.7. Voltage–current curve for squid axon membrane in 0.5 M KCl: ——— potential control, --------- current control [3b].

About 0.5 M KCI iso-osmotic solution was used. Attempt was also made to correlate the phenomena with the time course of an "action potential" in 0.5 M potassium chloride.

The current–voltage and voltage–current curves are not similar in this case unlike those for electro-kinetic phenomena. Presumably, under conditions of voltage control, movement of ions through the membrane might influence the magnitude of current. Similarly, when transition from steady state occurs by perturbation with the help of a hyperpolarizing current, a pulse of depolarizing current is generated and an action potential is developed. Thus, non-steady action potential is developed in entirely different situations.

It is evident that non-equilibrium phenomena will receive more and more attention in the coming years, since these are important from the viewpoint of practical applications also. For example, it has recently been shown that a dramatic enhancement [49] of the production rate may be achieved by external periodic forcing of non-linear chemical reactions that contain thresholds such as bistability.

Two stable states in frog node of Ranvier in 20–40 in 0.5 M potassium chloride solutions are observed by Stampfli [46]. Similar phenomena are also observed in the case of squid action membrane [46, 47]. It should be noted that the biological phenomena could be correlated with the generation of action potential. The value of conductance of membrane when filled with the solutions in the two steady states is given in Table 8.3.

It is also worth to mention here that the bistability phenomena can be easily understood in animal behaviour and social behaviour as mentioned in Section 8.6.

References

1. G. Nicolis and I. Prigogine, Self Organization in Non-equilibrium Systems, John Wiley and sons, New York, 1977.
2. P. Gray and S.K. Scott, Chemical Oscillations and Instabilities in Non-linear Chemical Kinetics, Oxford Science Publication, 1990.
3. (a) P. De Kepper and J. Boissonade Oscillations and Travelling Waves in Chemical Systems, John Wiley and Sons, 1985, p. 230. (b) R.P. Rastogi and A.P. Mishra, Ind. J. Chem., 38A (1999) 859.
4. P. De Kepper, I.R. Epstien and K. Kustin, J. Chem. Soc., 103 (1982) 6121.
5. (a) P. Reseh *et al.*, J. Phys. Chem., 1991, 95, 6270. (b) P.P. Rastogi and G.P. Misra, Indian J. Chem., 29A (1991) 941.
6. P. Gray and S.K. Scott, J. Phys. Chem., 87 (1983) 1835.
7. N. Ganpathisubramaniam and K. Showalter, J. Am. Chem. Soc., 106 (1984) 816; J. Phys. Chem., 80 (1984) 4177.
8. R.J. Field and M. Burger, Oscillations and Traveling Waves in Chemical Systems, Wiley, New York, 1977.
9. W. Geiseler, J. Phys. Chem., 86 (1982) 4394.
10. W. Geiseler and H.H. Follner, Biophys. Chem., 6 (1977) 107.
11. K. Bar-Eli and R.M. Noyes, J. Phys. Chem., 82 (1978) 1352.

12. C.E. Dateo, M. Orban, P. De Kepper and I.R. Epstein, J. Am. Chem. Soc., 104 (1982) 504.
13. J. Boissnade and P. De Kepper, J. Phys. Chem., 84 (1980) 501.
14. F. Ali and M. Menzinger, J. Phys. Chem., 95 (1991) 6408.
15. A. Pacault, P. Hanusse, P. De Depper, C. Vidal and J. Biossonade, Acc. Chem. Res., 9 (1976) 438.
16. R.M. Noyes and S.D. Furrow, J. Am. Chem. Soc., 104 (1982) 85.
17. S.D. Furrow and R.M. Noyes, J. Am. Chem. Soc., 104 (1982) 38.
18. P. De Kepper, I.R. Epstein and K. Kustin, J. Am. Chem. Soc., 103 (1982) 6121.
19. G.A. Papsin, A. Hanna and K. Showalter, J. Phys. Chem., 85 (1981) 2575.
20. R.P. Rastogi, I. Das, M.K. Verma and A.R. Singh, J. Non-equilib. Thermodyn., 8 (1983) 255.
21. (a) R.P. Rastogi, A.R. Singh and H. Malchow, J. Non-equilib. Thermodyn, 10 (1985) 193. (b) R.P. Rastogi, Ishwar Das and A.R. Singh, J. Phys. Chem., 88 (1984) 5132.
22. P. De Kepper, I.R. Epstein and K. Kustin, J. Am. Chem. Soc., 103 (1981) 2133.
23. A.K. Dutt and S.C. Muller, J. Phys. Chem., 97 (1993) 10059.
24. Q. Oyang, Z. Noszticzius and H.L. Swiney, J. Phys. Chem., 96 (1992) 6773.
25. P. De Kepper, V. Castes, E. Dulos and J. Boissonade, J. Phys. D, 49 (1991) 161.
26. M. Giordano, M. Marteneli, L. Pardia and S. Saitucci, Phy. Rev. Lett., 59 (1987) 327.
27. E. Anbay and D. Gourier, J. Phys. Chem., 96 (1992) 5513.
28. C. Vigreux, L. Binet and D. Gourier, J. Phys. Chem. B, 102 (1998) 1176.
29. T. Teorell, J. Gen. Phys., 42 (1959) 831.
30. T. Teorell, J. Gen. Phys., 42 (1959) 847.
31. P. Meares and K.R. Page, Philos. Trans. R. Soc. London, A 272 (1972) 1.
32. P. Meares and K.R. Page, Proc. R. Soc. London, A 339 (1974) 513.
33. Y. Kobatake and H. Fujita, J. Chem. Phys., 40 (1964) 2212.
34. Y. Kobatake and H. Fujita, J. Chem. Phys., 40 (1964) 2219.
35. R.P. Rastogi, R.C. Srivastava and S.N. Singh, Chem. Rev., 93 (1993) 1945.
36. R.P. Rastogi, G. Srinivas, R.C. Srivasatva, R.C. Pandey, A.P. Mishra and A.R. Singh, J Coll. Interf. Sci., 175 (1995) 262.
37. R.P. Rastogi, G.P. Mishra, P.C. Pandey, Kanchan Bala, K. Kumar; J. Coll. Interface Sci. 217 (1999) 275–287.
38. R.P. Rastogi, Kehar Singh, Raj Kumar and S.A. Khan, J. Phys. Chem., 81 (1977) 2114.
39. Y. Kobatake, Physica, 48 (1970) 301.
40. U.F. Franek, Ber. Bunsenges Phys. Chem., 67 (1963) 657.
41. I. Prigogine, in Non-eqilibrium Thermodynamics – Variational Techniques and Stability, R.J. Donelly, R. Herman and I. Prigogine (eds), University of Chicago Press, Chicago, 1953.
42. P. Glansdorff, Mol. Phys., 3 (1960) 277.
43. P. Glansdorff and I. Prigogine, Physica, 30 (1964) 351.
44. P. Glansdorff and I. Prigogine, Hays D Phys. Fluids, 5 (1962) 144.
45. M.K. Jain, The Biomolecular Lipid Membrane, Van Nostrand Reinhold, New York, 1972, 333.
46. R. Stampfli, Helv. Physiol. Acta, 16 (1958) 127.
47. I.R. Segal, Nature, 182 (1959) 1370.
48. J.W. Moore, Nature, 183 (1959) 265.
49. N. Shinor, W. Hohmann, M. Kraus, J. Miller, J. Miinster and F.W. Schneider, Phys. Chem., 1 (1999) 827.

Chapter 9

TIME ORDER - CHEMICAL OSCILLATIONS

9.1. Introduction

Frontiers of non-equilibrium physical chemistry are continuously advancing on account of increasing interest in the study of far equilibrium region in real systems such as living systems, socio-political system, finance and market economy. In the earlier part, we had discussed the development and nature of steady states, close to equilibrium and their mode of transition to multiple steady states.

Equilibrium structures are maintained without any exchange of energy and/or matter. A crystal is the prototype of an equilibrium structure. Dissipative structures, on the other hand, are maintained through exchange of energy (and in some cases also exchange of matter) with the outside world. Since entropy is dissipated in open systems under specific circumstances, emerging structures in such systems are called dissipative structures.

In this chapter, it is intended to discuss more complex phenomena as we further move away from equilibrium. Even through $dS > 0$ for such open systems, on account of export of entropy to the surrounding, order can be established with respect to time and space as observed. Further, in more complicated reaction–diffusion system, exotic dissipative structure such as Fractals can appear.

Oscillatory reactions are a typical class of phenomena, which display unusual features. After the discovery of Belousov–Zhabotinskii (B–Z) reaction, there has been a tremendous flurry of activity [1] and a large number of such reactions have been discovered during recent years. Biochemical reactions [2–10] such as glycolytic oscillations and peroxidase catalysed oxidation of nicotinamide adenosine deoxyhydrogenase (NADH) have also generated considerable interest. The interest in such reactions is still sustained in view of their importance in understanding cardiac and neuronal oscillations. In the case of many oscillatory chemical reactions [1], detailed reaction mechanisms have been postulated and verified with the help of numerical computation. This has also been particularly so for B–Z reaction where Field–Koros–Noyes (FKN) mechanism [11] has been invoked.

9.2. Isothermal chemical dissipative structures

There are two fundamental requisites for the experimental study of isothermal chemical dissipative structures: (i) The reaction must be maintained in constant non-equilibrium condition; and (ii) The reacting medium must be kept homogenous in order reduce the set of dynamical variables to a unique concentration variable per intermediate species. These conditions are easily satisfied in the continuous stirred tank reactor (CSTR). This device has been extensively used during last several decades. Experimental studies [1] have led to

(a) observation of oscillations over an extended period of time;
(b) evidence of different types of state multiplicities;
(c) observation of different types of complex oscillations including chaos;
(d) characterization of several types of chemical chaos;
(e) evidence for different routes to chaos; and
(f) discovery of large number of different types of oscillatory reactions.

These are sensitive to changes in concentration, nature of reactions and temperature etc.

Hopf bifurcation and Jug Handle (JH) type bifurcation are most commonly observed in chemical systems. In the former case, at the bifurcation point, time period remains finite but amplitude tends to zero. However, in the latter case both the time period and amplitude have finite values.

Bifurcation features and deformation of limit cycles for NaH_2PO_4 + cycle hexanone/acetone + Mn^{2+} + BrO_3^- + H_2SO_4 oscillatory reaction system have been reported [12]. When acetone is used as the organic substrate, the reaction system yields unperturbed limit cycles. However, in the case of cyclohexanone system, the limit cycles are deformed and knot is obtained. At the upper limit, Hopf bifurcation is obtained in the case of acetone while JH type bifurcation is obtained in the case of cyclohexanone.

9.3. Chemical oscillators

Oscillatory chemical reactions have received considerable attention during the latter half of the last century, in view of the importance of such exotic non-equilibrium phenomena in Physics, Chemistry, Biology, Physiology, Engineering and Finance. The discovery of the first bromate driven oscillatory reaction by Belousov in 1958 and reinvestigated by Zhabotinskii in 1964 and subsequent mechanistic simulation through the discovery of Brusselator [13] in 1968, triggered considerable activity in the field of chemical oscillations. Since then, a large number of new oscillatory reactions have been discovered. Good accounts of these are available in several monographs [1, 4, 12].

Chemical oscillators can be divided in following groups:

(a) homogeneous reaction system;
(b) heterogeneous reaction system; and
(c) interface-controlled oscillations.

9.3.1. Homogeneous reaction systems

Number of oscillators have been discovered which belong to different types such as (i) Bromate-driven oscillators; (ii) Iodate-driven oscillators; (iii) chlorate-driven oscillators; (iv) p-H driven oscillators; and (v) Miscellaneous oscillators. Some of these are depicted [1] in Figs. 9.1–9.4.

Different variants of the B–Z oscilltor (bromate + Ce^{4+} + malonic acid + H_2SO_4) are known. Some of these are:

(a) minimum B–Z oscillator (without organic substrate). A typical example is $BrO_3^- + Br^- + Mn^{2+}/Ce^{4+} + H_2SO_4$ system;
(b) uncatalysed B–Z oscillator (Phenol + BrO_3^- + H_2SO_4) [14] and catalysed B–Z oscillator;
(c) non-Br^- controlled oscillators e.g. free-radical controlled oscillators (Glucose + Oxalic acid + Ce^{4+} + BrO_3^- + H_2SO_4)

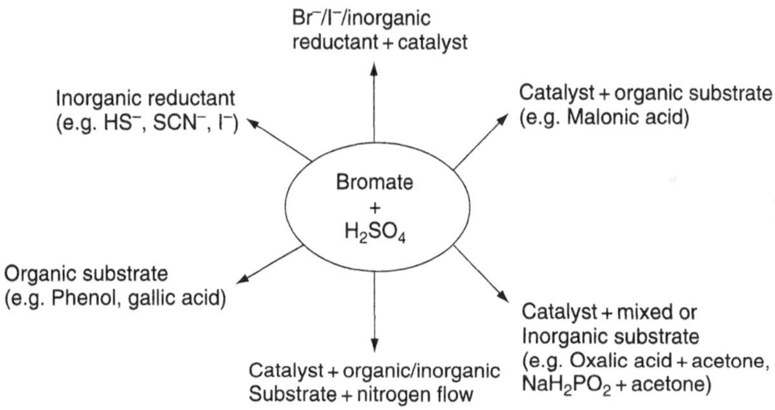

Figure 9.1. Bromate-driven oscillators.

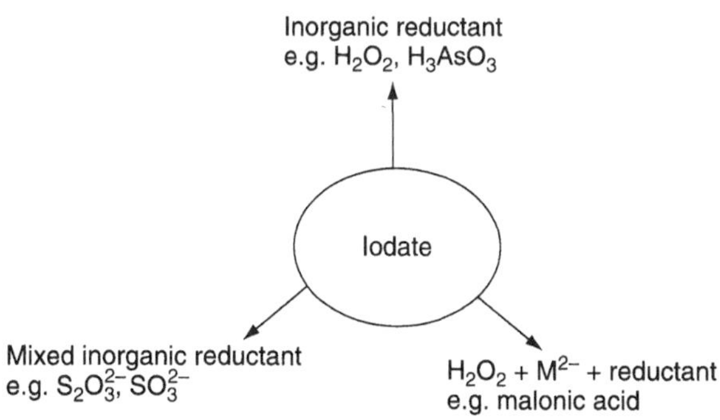

Figure 9.2. Iodate-driven odcillators.

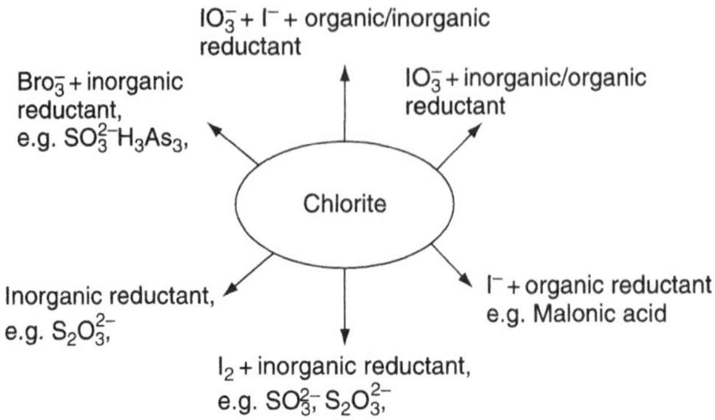

Figure 9.3. Chlorate-driven oscillators.

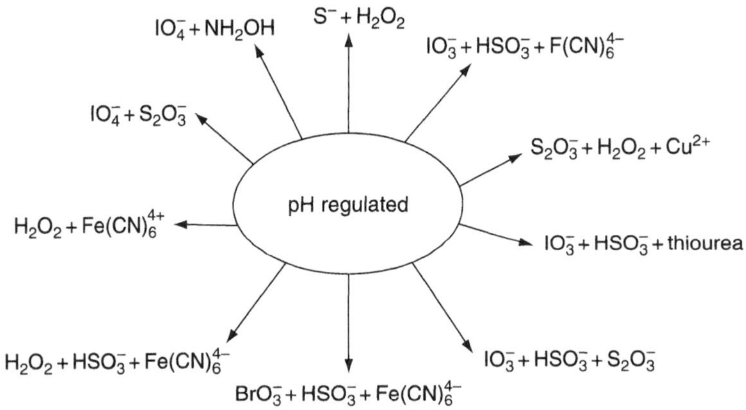

Figure 9.4. pH-driven oscillators.

9.3.2. Oscillations in heterogeneous reaction systems

Experimental observations clearly indicate that rate of several oxidation reactions oscillate continuously and never attain a steady state [15]. Following are some of the examples of such reactions:

(a) oxidation of hydrogen by oxygen on nickel;
(b) oxidation of hydrogen by oxygen on supported Pt-catalysts;
(c) oxidation of CO on supported catalysts; and
(d) oxidation of CO on Pt-surface.

For investigating oscillatory characteristics, following techniques are adopted by monitoring:

(a) contract potential difference (CPD) by the vibrating capacitor technique;
(b) surface temperature oscillation; and
(c) dependence of the activation energy of the rate constant on the surface coverage

Shift between multiple stead states due to slow adsorption–desorption of an inert species.

9.3.3. A simple demonstration experiment

Aerial oxidation of Aldehydes: Chemical oscillations have been noted in a system consisting of benzaldehyde, acetaldehyde, sodium bromide, cobalt (II) acetate and nitric acid. When air is passed through the reaction mixture at 70°C, colour of the solution oscillates between pink and dark brown. The system is quite suitable for classroom demonstration. If a platinum electrode is inserted in the medium and is coupled to a calomel electrode, changes in redox potential are observed which is synchronized with the colour changes [16, 17] (Figs. 9.5 and 9.6).

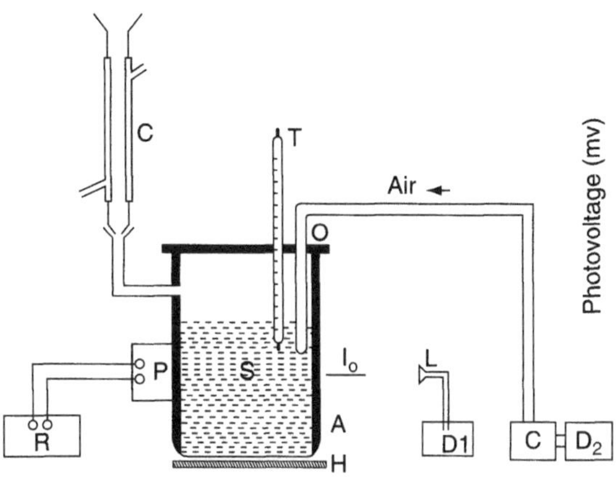

Figure 9.5. Reaction cell used for monitoring oscillations in colour during oxidation of benzaldehyde; P = Photocell, C_1 = Condenser, T = Thermometer, S = Oscillatory solution, A = Corning glass cell, L = Light source, D_1, D_2 = Dimmerstat, C_2 = Compressor, H = Hot plate, R = Recorder or Digital multimeter.

This is a good classroom demonstration experiment. Hydroquinone stops the oscillations indicating the important role played by the free radicals. In such reactions, auto catalytic production C_6H_5CHO or CH_3CHO has been envisaged as it happens in gas

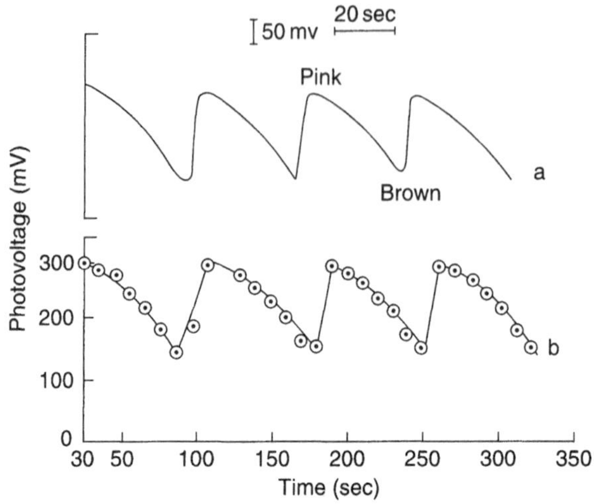

Figure 9.6. Change in photovoltage with time during oxidation of benzaldehyde; Curve (a) and (b) show the result obtained simultaneously by using X – t recorder and digital multimeter respectively. Conditions: [Cobalt acetate] = 20 mM. [Sodium bromide] = 4 mM, [Benzaldehyde] = 750 mM, Temperature = 72 ± 1 °C.

phase oxidation of aldehydes [18]. The positive and negative feedback (auto-catalytic and inhibitory reaction) for benzaldehyde can be postulated as follows:

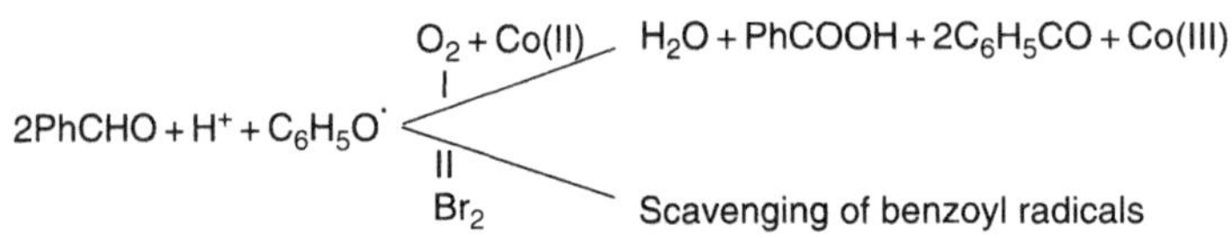

involving autocatalysis of the free radical $C_6H_5O^{\bullet}$ and inhibition of the free radical by Br_2.

9.3.4. Thermodynamics and thermochemistry of oscillatory reactions

Let us examine the occurrence of oscillatory reactions (which involve a number of elementary step) from the angle of thermodynamics. Obviously, the free energy change ΔG of the net reaction is responsible for driving the oscillations. For a reaction to occur in a in closed system, the necessary condition is $\Delta G < 0$. It is little difficult to conceive how this can happen when the concentrations of certain species in the reaction system are periodically increasing or decreasing. The explanation becomes easy if we consider

$$(\Delta G)_{\text{Total}} = \sum G_{\text{Product}} - \sum G_{\text{Reactants}}$$

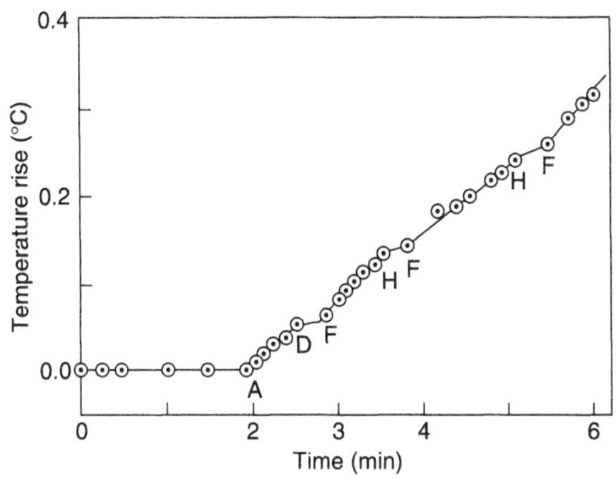

Figure 9.7. A typical run for the oscillatory B–Z reaction with malonic acid as organic substrate; AD, DF and FH are different phases.

What we required is that under all circumstances $\sum G_{\text{Product}}$ has to be less than $\sum G_{\text{Reactants}}$. It does not matter whether individual reactants or products increase or decrease during the course of reaction.

The component reactions of oscillatory reaction can be exothermic or endothermic. This aspect is expected to influence oscillatory characteristics. In this case of B–Z reaction involving malonic acid as substrate, it has been found that the rate of temperature rise oscillates rather than the temperature of the reaction mixture [19]. All the component reactions are found to be exothermic. Typical results are given in Fig. 9.7. Temperature begins to fall after 1½ hr.

All the component reactions investigated are found to be exothermic. Initial temperature rise of bromide + bromate reaction was found to be the highest (0.55°C/min) while that of cerous + bromate + malonic acid was to be found to be quite low (0.085°C/min). Thus in the first stage, when the reaction was mixed, the latter reaction involving autocatalysis predominates and the temperature rise is very slow. On the other hand, when $Br^- + BrO_3^-$ reaction involving inhibition reaction becomes dominant, there is a sharp rise in temperature. The thermochemical behaviour is thus in conformity with the FKN mechanism (Br^--control mechanism).

9.3.5. Complexity in reaction systems

In order to understand the complexity in oscillatory reactions, it would be worthwhile to examine their relationship with different types of chemical reactions [10], which have been summarized in Fig. 9.8 in increasing order of complexity viz., irreversible reactions → reversible reactions → parallel reaction → consecutive reaction → autocatalytic reactions → damped oscillations → aperiodic oscillations → spatio-temporal oscillations → chaotic oscillations. Further, Fig. 9.8 shows the concentration

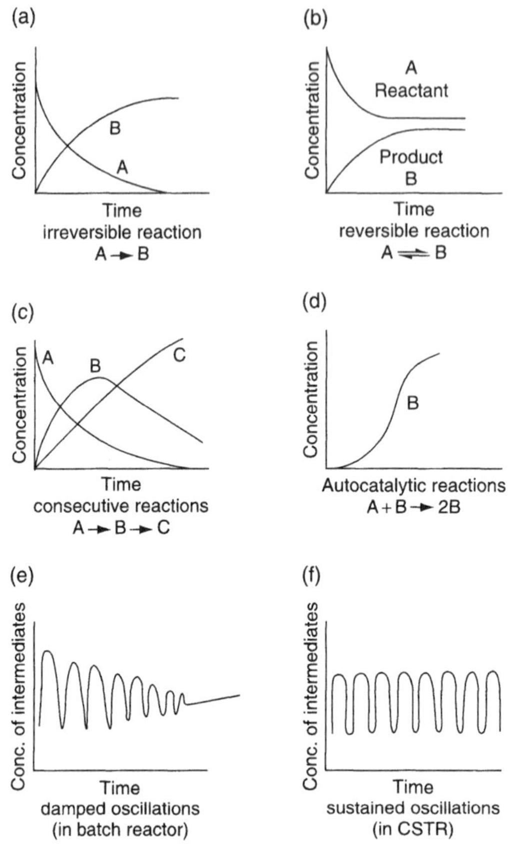

Figure 9.8. Graphical representation of time evolution of concentration of reactants/products in different types of reactions.

of reactant/product/intermediate changes with time in each case. Equations of the type $\mathrm{d}x/\mathrm{d}t = f(x)$ are linear. On the other hand when,

$$\mathrm{d}x/\mathrm{d}t = f(x, y, \text{and other variables})$$

the differential equation is non-linear since the rate equation depends on x, y and other variables. The rate equation depends on x, y and other variables. The rate equations for different types of chemical reactions are summarized in Table 9.1, which shows that such equations are non-linear in the case of consecutive, autocatalytic, autoinhibitory and oscillatory reactions.

It is evident that in case of autocatalytic reaction (v) in Table 9.1, the product accelerates its own production. The process is also called *positive feedback* whereas in case of auto inhibitory reaction [Table 9.1 (vi)], the product retards its own production; such processes are called *negative feedback.*

Table 9.1. Rate equations for simple and complex reactions.

Reactions		Rate law	Nature	Dependence
(i)	First order $A \xrightarrow{k} B$	$-\frac{d[A]}{dt} = k[A]$	Linear	One variable
(ii)	*n*th order $nA \xrightarrow{k} B$	$-\frac{d[A]}{dt} = k[A]^n$	Linear	One variable
(iii)	Reversible reaction $A \underset{k_{-1}}{\overset{k_{-1}}{=}} B$	$\frac{d}{dt} = k_1[A] - k_1[B]$	Linear	One variable
(iv)	Consecutive Reaction $A \overset{k_1}{—} B \overset{k_1}{—} C$	$-\frac{d[A]}{dt} = k_1[A]$	Linear	One variable
		$-\frac{d[B]}{dt} = -k_1[A] + k_2[B]$	Non-linear	Two variables
		$-\frac{d[C]}{dt} = k_2[C]$	Linear	One variable
(v)	Autocatalytic reaction $A + X \xrightarrow{k} 2X$	$-\frac{d[X]}{dt} = k[A][X]$	Non-linear	Two variable
(vi)	Autoinhibitory Reaction $A + B \xrightarrow{k} C$	$\frac{d[C]}{dt} = \frac{k[A][B]}{[C]}$	Non-linear	Three variables
(vii)	Chain reaction $A \xrightarrow{k_1} R$			
	$R + A \xrightarrow{k_2} P + R$	$\frac{d[P]}{dt} = k_1[A] + k_2[R][A] - k_3[R]$	Non-linear	Two variables
	$R \xrightarrow{k_3}$ destruction			
(viii)	Oscillatory reaction $A_0 \xrightarrow{k_1} A$	$dA/dt = k_1 A_0$	Linear	One variable
	$A + B \xrightarrow{k_2} 2B$	$dB/dt = k_2[A][B] - k_3[B]^{1/2}$	Non-linear	Two variables
	$B \xrightarrow{k_3} C$	$^*dC/dt = k_3[B]^{1/2}$	Linear	One variable
	*half-order kinetics			
(ix)	Reaction–diffusion equation			
	Autocatalytic reaction	$\frac{dB}{dt} = D_B \frac{\partial^2 B}{\partial x^2} + f(A, B, \ldots)$	Non-linear	
	Diffusion of species	Fick's law diffusion term Reaction term		

Table 9.2. Feedbacks in reaction systems.

Type of control	Example
(a) Negative feedback	$A \rightarrow B \rightarrow C \rightarrow D$ $-\uparrow$________$\lrcorner$
(b) Positive feedback	$A \rightarrow B \rightarrow C \rightarrow D$ $+\uparrow$________$\lrcorner$
(c) Negative feedforward	$A \rightarrow B \rightarrow C \rightarrow D$ $\llcorner$______$\uparrow -$
(d) Positive feedforward	$A \rightarrow B \rightarrow C \rightarrow D$ $\llcorner$______$\uparrow +$

$\llcorner\_\_\uparrow^{-}$ Inhibition of the reaction

$\llcorner\_\_\uparrow^{+}$ Acceleration of the reaction

One can see that in case of Fig. 9.8 (c), where two consecutive reactions are involved, the intermediate B does not oscillate. Only its concentration increases and then falls down. In case of Fig. 9.8 (d), oscillations are not possible, unless the concentration of B is brought down by some step and the cycle repeats over and over again. Similar argument would hold good in case of autoinhibitory reaction where some additional step for enhancing the concentration would be required.

In Table 9.2, a simple case of series of consecutive reaction evolving three steps has been considered in order to understand how negative feedback occurs. In case of (a), the product C inhibits the production of B. Hence, the process is called negative feedback. On the other hand, in case of (b), the product C accelerates the production of B. The process is, therefore, called positive feedback. In contrast to this, in case of (c), intermediate B inhibits the production of D which is a case of negative feedforward, while in case of (d), B accelerates the production of D which is obviously the case of positive feedforward.

9.4. Modelling of oscillatory reactions

Before we discuss the mechanism of the B–Z reaction, it would be worthwhile to review the earlier hypothetical reaction schemes, which generate oscillations by taking recourse to computer solutions of differential equations. These were pioneering efforts since earlier it was difficult to conceive how a reaction system could generate oscillations. In fact, the principles and paradigms emanating from such analyses paved the way for formulating the reaction mechanism, as we shall see in the next section.

The first attempt in understanding the chemistry of isothermal oscillatory reaction was made by Lotka [20, 21] who proposed a model based on preypredator model. i.e.

$$
\begin{aligned}
A + X &\rightleftharpoons 2X \\
X + Y &\rightleftharpoons 2Y \\
Y &\rightleftharpoons E
\end{aligned}
$$

The major weakness of the above model was that both amplitude and time period depend on the initial values of the oscillating species X and Y. Hence, it could generate infinite number of trajectories in X–Y plan depending on initial conditions.

Next major attempt was made by Lefever and Prigogine [22] which results in a simple model called "Brusselator" as given below

$$\begin{aligned} A &\longrightarrow X \\ B + X &\longrightarrow Y + D \\ Y + 2X &\longrightarrow 3X \\ X &\longrightarrow E \end{aligned}$$

This model yields sustained oscillations even if back reactions are considered in the above scheme, provided some assumptions are made regarding the values of the rate constants.

It has also been shown [23] that the principle of detailed balance is not violated. However, there are still some difficulties with the model. Nevertheless, one of the most impressive triumphs of the model was that it predicted limit cycle, which does not depend on initial values of X and Y which oscillate. Scott [24] has considered a simplified version of the above model as indicated below and examined its characteristics in great detail.

$$\begin{aligned} P &\longrightarrow A \\ A &\longrightarrow B \\ A + 2B &\longrightarrow 3B \\ B &\longrightarrow C \end{aligned}$$

9.5. Mechanism of B–Z reaction; positive and negative feedback

In this section, we will comment on the broad features of the mechanism of B–Z reaction. As we have seen in an oscillatory reaction system, number of components, by-products and intermediates (including free radicals) are involved which can yield numerous possible reaction steps. Hence, deciphering of mechanism is quite complicated. For example, in Gyorgyi, Tura'nyi, Feild (GTF) model, the number of postulated reactions steps is 80, involving 26 reactants [25]. However, all the steps involve electron-transfer and free-radical reactions.

In the case of oscillatory reaction under discussion, reactions are ionic in nature and oscillating species are ions. The oscillating species Br^- and Ce^{4+}/Ce^{3+} are detected by bromide and platinum sensitive electrodes in conjunction with standard calomel electrode. The essential challenging task of developing a reaction mechanism is to postulate how the concentration of Ce^{4+} and Br^- builds-up in the course of time and how it is periodically inhibited. In the light of Brusselator model discovered by

Prigogine and co-workers in 1968, build-up is supposed to be due to autocatalysis of some species produced in the reaction and decay is related to destruction of the growth of autocatalysed species by another reaction. Hence, the first step needed for suggesting a plausible mechanism involves identifying (i) the autocatalytic step and the species being autocatalysed; (ii) the inhibitory step; (iii) steps involved in the generation of species which initiate the autocatalytic reactions; and (iv) step involved in the removal inhibitor.

A simple example of an autocatalytic step is

$$A + X \longrightarrow 2X$$

It is easy to visualize it in terms of population dynamics; in case of chemical reactions, it is somewhat difficult to conceive but it can be visualized that it involves a combination of steps of the following type:

$$\begin{array}{r} A + X \longrightarrow Y \\ Y \longrightarrow 2X \\ \hline A + Y \longrightarrow 2X \\ \hline \end{array}$$

The real problem arises in identifying these two steps.

For proceeding further, it is important to have an idea of stoichiometry, intermediate species formed in the reaction and kinetics of intermediate reactions. This requires detection of intermediates, study of kinetics and assignment of rate constants.

In the B–Z reaction, BrO_3^- can generate the following species: Br^-, HOBr, $HBrO_2$, $BrO_2^\bullet$, Br_2, and Br_2O_4. Similarly, organic reactants can give a variety of products including free radicals and bromo-derivatives along with CO_2. The interconversion between Ce^{3+} and Ce^{4+} takes place and in this process total cerium remains constant. H_2SO_4, in excess, only provides hydrogen ion and controls standard redox potential E_0 of ionic redox reactions. In the overall oscillatory reaction, there can be reactions among organic species and reactions between organic and inorganic species [26].

The electron-transfer reaction involving oxybromine species have been identified using thermodynamic and kinetic considerations. The direction of reaction can easily be understood by the following redox potential (volt) diagram:

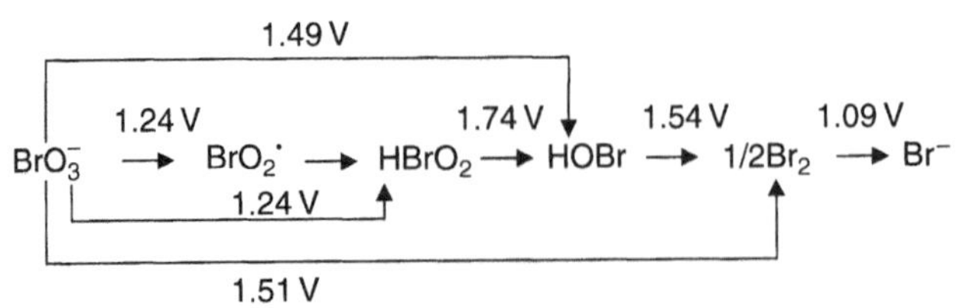

In addition to inorganic free radicals, organic free radicals are also produced in the system, which obey general principles of free radical mechanism during elementary reactions.

Taking a simplistic view in the early stages, Field, Koros and Noyes postulated the following mechanism (usually known os Oregonator) [27] based on the following sequence of reactions:

$$A + Y \xrightarrow{K_1} X \qquad \text{(generation of autocatalysing species)}$$

$$X + Y \xrightarrow{K_2} 2P \qquad \text{(inhibitory reaction)}$$

$$A + X \xrightarrow{K_3} 2X + Z \qquad \text{(autocatalytic step)}$$

$$2X \xrightarrow{K_4} P + A \qquad \text{(destruction of autocatalytic species)}$$

$$Z \xrightarrow{K_5} fY \qquad \text{(generation of inhibiting species)}$$

where $X = HBrO_2$, $Y = Br^-$, $Z = 2Ce^{4+}$, $A = BrO_3^-$, $P = HOBr$ and f is the stoichiometric coefficient.

When appropriate values of the rate constants were inserted in the non-linear differential equations involving X and Y, numerical computations showed that oscillations are generated.

Subsequently, the model was modified [28] and reversible steps were taken into consideration. Again computer calculations predicted oscillations.

In case of Racz's oscillator [29] or fructose oscillator [30], FKN mechanism still holds but a different type of control mechanism had to be postulated. The following autocatalytic reaction has been proposed:

$$HBrO_2 + H^+ + BrO_3^- \rightleftharpoons Br_2O_4 + H_2O$$

$$Br_2O_4 \rightleftharpoons 2BrO_2$$

$$Ce^{3+} + BrO_2^{\cdot} + H^+ \rightleftharpoons Ce^{4+} + HBrO_2$$

$$Ce^{3+} + BrO_3^{2-} + BrO_2^{\cdot} + 2H^+ \rightleftharpoons Ce^{4+} + 2BrO_2 + H_2O$$

The inhibitory step involves interaction of $BrO_2^{\bullet}$ with free radical generated from malonic acid or fructose. In case of B–Z reactions [31–33] involving mixed organic substrate, the hard core of FKN mechanism still remains intact. Only difference in mechanistic scheme arises on account of Br^- and Br_2 removal mechanism. In case of mixed substrate, whereas although the acid reduces Ce^{4+}, acetone gets brominated and thus removes bromine. In other versions of B–Z reactions, Br_2 is removed by physical methods such as nitrogen bubbling [34] or by partitioning with the addition of benzene [33]. The case of ferroin-bromate reaction is a little complicated where it is difficult to comprehend the mechanism in view of the fact that the oxidized form of $Fe(phen)_3^{2+}$ are inert to substitution and have reduction potential of about 1.0–1.3V.

Sequential oscillations in case of ascorbic acid and acetone mixed substrate have been interpreted satisfactorily in terms of FKN mechanism [35].

It needs to be emphasized that the modified version of FKN mechanism in most of the cases has been able to satisfy tests based on computer studies. In the same manner, the mechanism of such oscillation where some characteristic features of oscillations are perturbed by addition of Ag^+ ions [36, 37] could also be explained satisfactorily on the basis of FKN mechanism.

On tinkering the hard core of oscillatory mechanism by (i) induction of some additional steps; or (ii) by acceleration and deceleration of some steps, one can change the oscillatory features and the reaction mechanism.

9.5.1. Test of mechanism

In fact, Oregonator (reaction mechanism for B–Z reaction proposed for the first time) provides only a skeleton mechanism which was improved and modified at later stages. Although the subject has advanced considerably, one is still interested in predicting and understanding other features of oscillatory reaction such as bistability, multiperiodicity and chaos including its generation and control.

As stated earlier in the section, for a comprehensive investigation of mechanism of oscillatory reactions, detailed study of kinetics (determination of rate constants) and mechanism of component reactions is also needed as a supporting study to provide information relevant for computer modelling of modified FKN mechanism.

As in other cases of study of elementary or slightly complex reactions, it is also necessary to confirm the existence of the intermediate in oscillatory reactions by using various techniques such as UV, IR Mass, ESR, Spectra and HPCL. An acceptable test of the level of understanding of a mechanism is by computer simulation. Such a test is by no means unequivocal, but can often reveal gross deficiencies in assumption about a mechanism [38]. More rigorous test of the mechanism would be its capability to predict other characteristic features such as the bifurcation point, aperiodicity and chaos.

The B–Z reaction is quite complex. However, Noyes and co-workers were successful in identifying (i) the species viz., $HBrO_2$ which is produced autocatalytically; (ii) species viz., Br^{2-} ions which destroy $HBrO_2$ and prevents accumulation; and (iii) removal mechanism of Br^{2-}/Br_2. The detailed reaction are given below:

$$
\begin{array}{lrcl}
\text{(i)} & Br^- + BrO_3^- + 2H^+ & \rightleftharpoons & HOBr + HBrO_2 \\
 & H^+ + HBrO_2 + Br^2 & \rightleftharpoons & 2HOBr \\
 & Br_2 + H_2O & \rightleftharpoons & HOBr + Br^- + H^+ \\
\text{(ii)} & HBrO_2 + BrO_3^- + H^+ & \rightleftharpoons & Br_2O_4 + H_2O \\
 & Br_2O_4 & \rightleftharpoons & 2BrO_2^{\bullet} \\
 & [Ce^{3+} + BrO_2^{\bullet} + H^+ & \rightleftharpoons & Ce^{4+} + HBrO_2] \times 2 \\
\hline
 & HBrO_2 + 2Ce^{3+} + BrO_3^- + 3H^+ & \rightleftharpoons & Ce^{4+} + 2HBrO_2 + H_2O \\
\hline
\end{array}
$$

(iii) 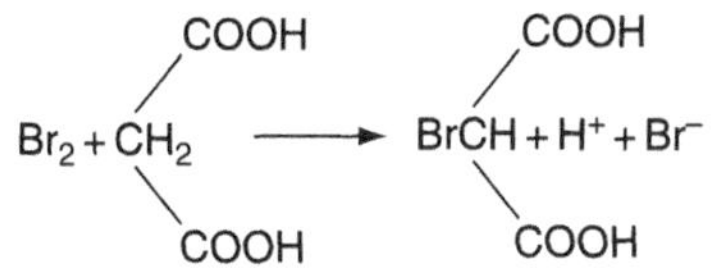

(iv) 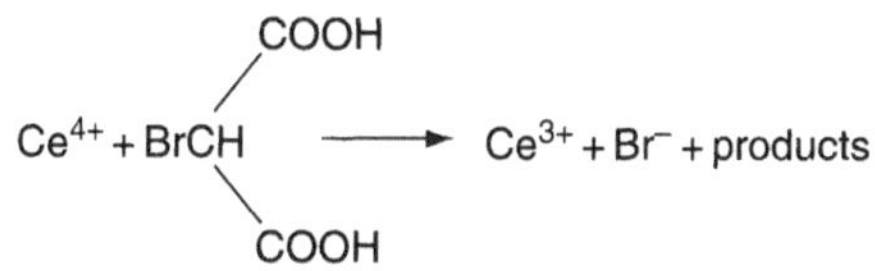

An understanding of the detailed mechanism of each step is not vital for explaining the specific features of oscillations. However, in certain suggested classroom experiments, attention has been focused on the bromination mechanism of methylene group of malonic acid, which proceeds by enolization mechanism.

Several variants of B–Z reactions have been quoted by Noyes [39] which are illustrated in Fig. 9.1–9.4. The hard core of the mechanism remains intact. The modifications only arise on account of (i) species being produced autocatalytically; (ii) autocatalitic reaction; and (iii) bromine removal mechanism.

However, for a class of oscillations which do not require any metal ion catalyst [40–42] but have derivatives of phenols or aniline as an organic substrate, the following autocatalytic reaction has been postulated (for the case of phenol):

$$BrO_3^- + HBrO_2 + C_6H_4(OH)_2 + H^+ \rightleftharpoons 2HBrO_2 + C_6H_4O_2 + H_2O$$

where $C_6H_4(OH)_2$ is generated by a series of reactions from $C_6H_4O_2$ which is produced by the following reaction:

$$6C_6H_4(OH)_2 + BrO_3^- \rightleftharpoons 6C_6H_4O_2 + Br^- + H_2O$$

9.6. Alternate control mechanism

The FKN [1, 11] mechanism for the B–Z reaction or its modified version has proved to be remarkably successful in explaining and simulating the detailed behaviour of oscillatory features which is generally not expected of any oscillatory mechanism. That is why it has not only stimulated interest in the mathematical, computational and experimental studies of the B–Z reaction, it has also reinforced tremendous interest in the study of non-linear phenomena in reaction systems. However, FKN mechanism involves Br^--control mechanism. Successful attempts have been made to explore alternative control mechanism.

Nosticzius *et al.* [29, 43] have suggested that for Racz's systems (0.6 M malonic acid in $3NH_2SO_4$), it is the autocatalytic production of $BrO_2^{\bullet}$ which is important and malonyl radical is the control intermediate, whereas in the original FKN mechanism, the autocatalytic production of $HBrO_2$ and the control intermediate is the bromide ion. The autocatalytic steps and two types of control mechanisms depend on the circumstances.

Gyorgyi, Turanyi and Field [25] modified the FKN mechanism and suggested an extended mechanism involving 80 elementary reactions and 2,6 reactant species. Subsequently, these workers [26] suggested a further modified version of the above mechanism involving 42 reactions and 22 species which was found to be in agreement with the experimental results. The model contains two negative feedback loops viz., the Br^--controlled and the organic free-radical controlled.

There is much greater possibility of more than one control mechanism. One can expect multiple control mechanisms to be operative in socio-political and socio-economic systems.

9.7. Dual control mechanism

Normally, if saccharides are substituted for malonic acid in classical B–Z system, oscillations are observed since Br^--control mechanism operates. This has been ascribed to accumulation of Br^- in the system such that $[Br^-]$ exceeds the critical Br concentration. Critical $[Br^-]$ is necessary for oscillations to occur [1]. Hence, in order to reduce the prevailing $[Br^-]$ to the level of $[Br_2]$ critical following steps can be undertaken:

(a) addition of an additional organic substrate which may get easily brominated [44]; and
(b) bubbling Br_2 out of the system using an inert gas [45].

Comprehensive studies of the oscillatory features of B–Z reaction involving double substrate such as (i) glucose + acetone; (ii) fructose + acetone; and (iii) sucrose + acetone have been made. One of the interesting features is that the systems display the lower and upper critical limits of [acetone] between which oscillations occurred on increasing [F]. Another significant observation was that beyond a certain [F], oscillations occurred even when no acetone was present in the system. Obviously, at this stage Br^--control mechanism is not in operation [46] and an alternative free-radical control

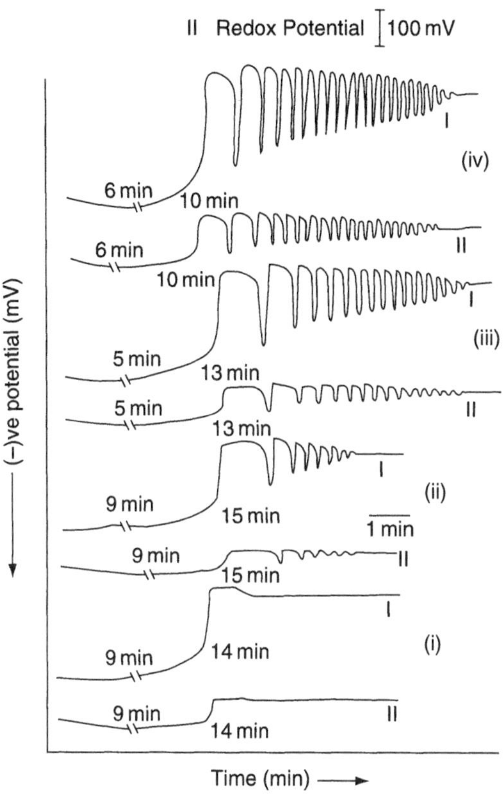

Figure 9.9. Oscillations in bromide ion potential for the system BrO_3^-(0.06 M) + Ce^{4+}(1.45 10^{-3} × M) + H_2SO_4(1.5 M) + [Fructose]; (a) 0.02 M (b) 0.03 M (c) 0.035 M (d) 0.04 M (e) 0.05 M (f) 0.06 M (g) 0.07 M (h) 0.075 M.

mechanism has been suggested [47]. Further studies with B–Z oscillator having (i) fructose + tartaric acid [48]; (ii) fructose + oxalic acid [49]; (iii) xylose + oxalic acid [50]; and (iv) glucose + oxalic acid [51] were performed.

Oscillations in fructose + BrO_3^- + Ce^{4+} + H_2SO_4 without any bromine scavenger have been observed in the range 0.035–0.7 M fructose in a batch reactor as well as CSTR (Fig. 9.9) [46]. Tartaric acid is found to promote oscillations when added to non-oscillation fructose system although no oscillations are observed in tartaric acid + BrO_3^- + Ce^{4+} + H_2SO_4 system of similar concentration. Within a certain range of fructose concentration, damped complex oscillations are observed in the batch reactor. In CSTR, aperiodic oscillations are observed at higher [F] with decreasing flow rated although sustained periodic oscillations are observed at low (Fig. 9.10) [Fructose].

In a particular concentration range, two types of oscillations in Fructose [F] + acetone oscillator separated by time pause have also been observed in the batch reactor. Oscillations that occur before time pause (first type oscillations are insensitive to addition of

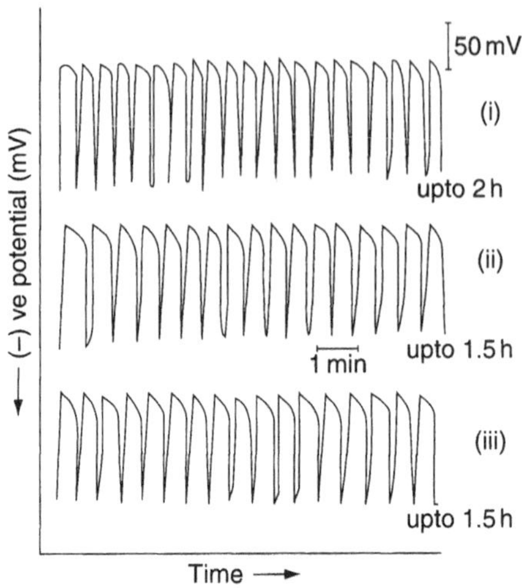

Figure 9.10. Oscillations in bromide ion potential in CSTR for the system; Fructose (0.075 M) + BrO_3^- (0.06 M) + Ce^{4+} ($1.45 \times 10^{-3} \times$ M) + H_2SO_4 (1.5 M) at flow rate equal to (a) 0.08 ml/min (b) 2.0 ml/min (c) 2.50 ml/min.

Br^- up to 10^{-2} M and are F + OA similar to that observed in the corresponding fructose oscillator without acetone. On the other hand, oscillations that occur after time pause (second type oscillations) are stopped when $[Br^-] \approx 10^{-2}$ M. This gives an indication that the first type is non-bromide ion-controlled while the second is Br^--controlled. Computer simulation studies confirm that both control mechanisms are operative in the system [46].

Experimental studies on B–Z oscillator having fructose + tartaric acid as double substrate have been undertaken by keeping either of the two organic substrates at a fixed concentration and varying the concentration of the other [48]. A modified mechanism by including following additional steps (7) to (13) in the FKN mechanism has been used to explain the observations.

$$Br^- + BrO_3^- + 2H^+ \rightleftharpoons HOBr + HBrO_2 \quad (1)$$

$$Br^- + HBrO_2 + H^+ \rightleftharpoons 2HOBr \quad (2)$$

$$Br^- + HOBr + H^+ \rightleftharpoons Br_2 + H_2O \quad (3)$$

$$2HBrO_2 \rightleftharpoons HOBr + BrO_3^- + H^+ \quad (4)$$

$$HBrO_2 + BrO_3^- + H^+ \rightleftharpoons 2BrO_2 + H_2O \quad (5)$$

$$Ce^{3+} + BrO_2^* + H^+ \rightleftharpoons Ce^{4+} + HBrO_2 \quad (6)$$

$$Ce^{4+} + TA \rightleftharpoons Ce^{3+} + TA + H^+ \quad (7)$$

(1)–(7): FKN mechanism

$$TA + BrO_2^* \rightleftharpoons Product \quad (8)$$

$$F + BrO_3^- + H^+ \longleftarrow BrO_2^* + Product \quad (9)$$

$$F + BrO_2^* \longrightarrow Product + Br^* \quad (10)$$

$$Br^* + Br^* \longrightarrow Br_2 \quad (11)$$

$$F + Ce^{4+} \longrightarrow F + Ce^{3+} + H^+ \quad (12)$$

$$F + BrO_2^* \longrightarrow Product \quad (13)$$

It may be noted that F, G and OA are not Bromine scavengers. However, acetone acts as a Bromine scavenger. Examination of the above scheme shows that

(a) Combination of $(5) + 2 \times (6)$ yields the autocatalytic reaction for $HBrO_2$. The inhibiting reactions are (2) and (3). If a proper balance can be maintained between the two, oscillations can occur and the oscillatory reaction would be Br-controlled.
(b) If combination of (5) and (6) is favoured, autocatalysis of $BrO_2^{\bullet}$ would occur and the corresponding inhibiting reactions would be (7), (8), (12) and (13). Here again if the rates are adjustable, oscillations would occur which will be free-radical controlled.

Oscillatory features of a similar type of B–Z oscillator, Fructose (F) + oxalic acid $[OA] + Ce^{4+} + BrO_3^- + H_2SO_4$ have also been investigated recently [49]. The induction time is found to be usually small or negligible. Both single frequency oscillations and two oscillatory states separated by a time pause are observed. Oscillations occur between two critical limits of [F] and [OA]. Computer simulation correctly predicts some of the oscillatory features such as (i) time of inhibition; (ii) critical limits of [OA]; and (iii) stoppage of oscillation by higher $[Br^-]$, which conforms the proposed mechanism involving the primary role of Br^--control mechanism. The tentative mechanism involves both Br^--control and free-radical control mechanism.

Experimental and computational studies on B–Z oscillator with fructose alone as substrate [52] have been recently studied. The phenomena of multiple control mechanism is possible in specific systems. If in a system, number of sets of autocatalytic and inhibitory reaction are present, one can have larger number of control mechanism. To this belongs another example of the B–Z oscillator having double substrate. Thus, oscillations in a B–Z system having oxalic acid (OA) and glucose (G) have been investigated where none of the substrate acts as Br scavenger [51]. Studies have been performed for (i) varying concentration of G while keeping the [OA] fixed; and (ii) varying [OA] but keeping [G] fixed in a batch reactor. In both cases, upper and lower critical limits occur, between which oscillations are observed. Both single and double frequency oscillations have been observed in a wide range of concentration of G as well as of OA. The induction period in most of the cases was < 1 min.

When [G] is fixed and [OA] is varied, the time pause between the sequential oscillations increases with an increase in [OA]. On the other hand, when [OA] is fixed and G is varied, the time pause increases with an increase in [G]. The first-type of oscillations is Br^--controlled, where the second is non-Br^- controlled. The order of addition of G and OA in the last has no influence on the induction period. However, it influences the oscillatory characteristics. A tentative dual control mechanism has been suggested involving autocatalysis of $HBrO_2$ and $BrO_2^\bullet$.

The detailed mechanism is as follows:

$$Br^- + BrO_3^- + 2H^+ \rightleftharpoons HOBr + HBrO_2 \quad (1)$$

$$Br^- + HBrO_2 + H^+ \rightleftharpoons 2HOBr \quad (2)$$

$$Br^- + HOBr + H^+ \rightleftharpoons Br_2 + H_2O \quad (3)$$

$$2HBrO_2 \rightleftharpoons HOBr + BrO_3^- + H^+ \quad (4)$$

$$HBrO_2 + BrO_3^- + H^+ \rightleftharpoons Br_2O_4 + H_2O \quad (5)$$

$$Br_2O_4 \rightleftharpoons 2BrO_2 \quad (6)$$

$$Ce^{3+} + BrO_2^* + H^+ \rightleftharpoons Ce^{4+} + HBrO_2 \quad (7)$$

which are important steps of FKN mechanism.

The network of reactions around G involves a number of reactions including oxidative products of G i.e.

$$G + BrO_3^- + H^+ \longrightarrow Q\ (\text{gluconic acid}) + HBrO_2 \quad (8)$$

$$G + HOBr \longrightarrow Q + H^+ + Br^- \quad (9)$$

$$Q + BrO_3^- + H^+ \longrightarrow P\ (\text{gluconic acid}) + HOBr + H_2O \quad (10)$$

$$G + Br_2 + H_2O \longrightarrow Q\ (\text{gluconic acid}) + 2Br^- + 2H^+ \quad (11)$$

$$G + Ce^{4+} \longrightarrow G + Ce^{3+} + H^+ \quad (12)$$

$$G + Ce^{4+} + H_2O \longrightarrow P\ (\text{gluconic acid}) + Ce^{3+} + H^+ \quad (13)$$

The acids P and Q generate free radicals by a one-electron transfer reaction with Ce^{4+} as follows:

$$Q + Ce^{4+} \longrightarrow Q + Ce^{3+} + H^+ \quad (14)$$

$$P + Ce^{4+} \longrightarrow P + Ce^{3+} + H^+ \quad (15)$$

The free radical Q and P can reach with $BrO_2^\cdot$ free radical as follows:

$$P + BrO_2^* \longrightarrow \text{Product} + Br^- + H^+ \quad (16)$$

$$Q + BrO_2^* \longrightarrow P + HOBr \quad (17)$$

Oxalic acid on the other hand can also undergo the following sequence of reactions:

$$Ce^{4+} + OA \longrightarrow Ce^{3+} + COOH + CO_2 + H^+ \quad (18)$$

$$Ce^{4+} + COOH \longrightarrow Ce^{3+} + H^+ + CO_2 \quad (19)$$

$$BrO_2^{\bullet} + COOH \longrightarrow HBrO_2 + CO_2 \quad (20)$$

$$OA + BrO_3^- + H^+ \longrightarrow HBrO_2 + 2CO_2 + H_2O \quad (21)$$

One can also have following type of free-radical recombination reaction:

$$2P \longrightarrow \text{product} \quad (22)$$

$$P + Q \longrightarrow \text{product} \quad (23)$$

$$PQ \longrightarrow \text{product} \quad (24)$$

Two types of auto catalytic reactions are involved in the above reaction scheme:

(a) On combing (5), (6) and twice the step of (7), we get

$$HBrO_2 + BrO_3^- + Ce^{3+} + 3H^+ \rightleftharpoons 2Ce^{4+} + 2HBrO_2 + H_2O \quad (25)$$

which leads to autocatalysis of $HBrO_2^{\bullet}$ on the other hand.

(b) On adding (5), (6) and (7), we get

$$BrO_2^{*} + BrO_3^- + Ce^{3+} + 2H^+ \rightleftharpoons Ce^{4+} + 2BrO_2 + H_2O \quad (26)$$

which leads to autocatalysis of $BrO_2^{\bullet}$.

Thus, in the reaction system, there are two processes autocatalysis of (i) $HBrO_2$; and (ii) $BrO_2^{\bullet}$. Any species reacting with them can inhibit the whole autocatalytic process. The normal control mechanism in B–Z reaction essentially involves step (25) which is inhibited by Br^- according to step (2). This is called Br^--control mechanism

The second type of control mechanism involves (i) autocatalysis of $BrO_2^{\bullet}$ step (26); and (ii) inhibitory reactions (16) and (17). This type of mechanism is called free radical control mechanism.

The first one controls the type I oscillations, while the second controls type II as indicated by experiments as well as by computer simulation.

Thus, both the G + OA as well as F + OA oscillators provide examples of dual control mechanism. Multiple-control mechanisms are a distinct possibility.

A simple explanation of dual control mechanism can be easily provided. In the case of type I oscillations, initially in the presence of oxalic acid, a sufficient amount of $HBrO_2$ is produced, and in turn, by a sequence of reactions Br^- is in greater excess than critically needed for balance of positive and negative feedback. However, when G is in excess, free radicals are generated from acid produced by oxidation of G, which reduced the concentration of $HBrO_2$ and Br^-. Thus, eventually a situation is reached when the autocatalytic production of $HBrO_2$ can be balanced by the inhibition through Br^-. Under such circumstances, the oscillations are Br^--controlled

In the case of type II, after a time pause sufficient amount of free radicals are produced via interaction of Ce^{4+} with acids produced from the oxidation of glucose. At a certain stage, autocatalytic production of $BrO_2^{\bullet}$ can be balanced by the negative feedback, involving interaction of $BrO_2^{\bullet}$ and P and Q. In course of time, the production of P and Q goes on increasing and when this exceeds the critical concentration necessary for countering the autocatalysis production of $BrO_2^{\bullet}$, the type II oscillation stops. Thus, two type II oscillations are free-radical controlled (non-Br^- controlled).

9.8. Coupled oscillators

The concept of coupled oscillators is important from the viewpoint of understanding oscillatory phenomena in biochemical systems where the basic question relates to the type of dynamic behaviour when two are more such systems are coupled together. The concept of coupling is quite relevant in the context of spatial dissipative structures in cases where chemical oscillators are coupled through diffusion.

On the basis of physical properties, the coupled oscillators can be divided in three categories:

(1) physically coupled oscillator;
(2) chemically coupled oscillator; and
(3) physically and chemically coupled oscillator.

A single oscillator may contain more than one set of reactions capable of oscillations. However, a coupled system exhibits more complex dynamical behaviour. The following types of complexities can occur in coupled oscillators:

(a) *Entrainment of frequency locking*: In coupled oscillators, the oscillators co-ordinate motion so that one oscillator goes through n cycles in exactly the time required for the other to go through m cycles where both m and n are integers. Such behaviour is known as entrainment or frequency locking.
(b) *Rhythmogenesis*: It is the onset of oscillations when the two steady states are coupled.
(c) *Phase death*: It is a type of steady state produced by coupling two or more oscillators.
(d) Quasi-periodicity and complex periodic oscillations where multiple periodicity can occur.
(e) Chaos.

9.8.1. Physically coupled oscillators

In case (1), coupled oscillators operate under different constraints such as input concentration to a flow reactor (CSTR). The coupling is achieved by limited contact between the separate cells. Physically coupled oscillators can be coupled (i) electrically; (ii) by mass transport; or (iii) by coupling between individual oscillators [53]. At low coupling rate, two CSTRs oscillate at their own frequency and amplitude. However, at very high coupling rates, the oscillations resemble that of a CSTR under average conditions. In some medium, a region is found where two oscillators at a steady state unexpectedly stop oscillating.

9.8.2. Electrically coupled oscillators [54]

For electrical coupling, large-area platinum electrodes are placed in each of the reactors to be coupled, and their circuit is completed by a wire between the electrodes and by an ion-bridge between the two reactors. Several phenomena such as entrainment, quasi-periodicity and chaos have been observed in such oscillations [55].

9.8.3. Mass transport coupled oscillators

In this case, the driving force for the coupling is the concentration difference between the cells, when a pair of chemical oscillators with a common species is coupled. Crowley and Epstein studied such oscillations using B–Z oscillator with acetyl acetone as substrate. Entrainment and phase death were observed depending on mass coupling strength. Citri and Epstein [56] have proposed a model for such oscillators. Numerical simulation gives good agreement with the observed exotic behaviour.

9.8.4. Chemically coupled oscillators

Chemically coupled oscillators can be of the following types:

(a) oscillation where coupling takes place between two distinctly different oscillators in a common pool;
(b) coupling takes place between the main oscillator and oscillator having the decay product as the main substrate;
(c) when two or more subsystems each of which is capable of oscillating independently and oscillating systems have one or more species in common. Thus, if we have three systems, A, B and C mixed in a single reactor, A+B and B+C in pairs give rise to oscillations.
    Maselko for the first time [57] investigated chemical coupling in CSTR involving B–Z reactions with mixed substrates such as malonic acid–citric acid and malonic acid-oxalo-acetic acid.

Chemically coupled oscillation in batch reactor involving the use of two or more organic substrates in case of B–Z oscillator and Briggs–Rauscher [58] have been studied. Both systems exhibited a bust of oscillations, characteristic of one substrate followed by a quiescent period and then a bust of oscillations characteristic of the other substrate.

Citri and Epstein have reported a mechanistic study of bromate-chlorate-Iodide and a bromate-oscillator coupled via a common iodine containing species. It may be noted that bromide-ion is a key intermediate A mechanism for chlorite-iodide and bromate-iodide oscillators [59, 60].

Coupled systems exhibit a variety of dynamical phenomena such as birhythmicity, chaos and various degree of complexity which are not observed in uncoupled systems. It has been claimed that suggested mechanism explains the behaviour of the coupled system as well as the component oscillators in many respects.

9.8.5. Oscillators coupled both physically and chemically

Such oscillators are of great interest in Physiology since they give rise to complex behaviour. Oscillatory behaviour of pancreatic islet cells is a typical example when interaction between glycolytic oscillations and membrane transport of glucose is involved [61].

Another example is the typical cellular oscillator which is located in the surface of the membrane or in the cytoplasm. These two oscillators can very well interact if they have components in common, such as calcium. From a biological angle, such systems where there is an interaction between membrane transports are of considerable interest.

9.9. Logic function

Non-linear system may be used to process information. The implementation of logic functions and computations has been reviewed by Helmfelt and Ress [62]. As a typical example, the logical network built of cyclic enzymatic reactions which are coupled with enzymatic reactions via common reagents has been investigated [63].

References

1. R.J. Field and M. Burger (eds) Oscillations and Chemical Waves in Chemical Systems, Wiley, New York, 1985.
2. H. Degn, L.F. Olsen and J.W. Perram, Ann. NY Acad. Sci., 316 (1979) 622.
3. A. Goldbeter and R. Caplan, Am. Rev. Biophys. Bioeng., 5 (1976) 449.
4. A. Goldbeter, Biochemical Oscillations and Cellular Rhythms, The Molecular Phases of Periodic and Chaotic Behaviour, Cambridge University Press, London 1996.

5. B. Chance, E.K. Pye, A.K. Ghosh and B. Hess, Biological and Biochemical Oscillatorys, Academic Press, London and New York, 1973.
6. T. Geest, C.G. Steinmetz, R. Larter and L.F. Olsen, J. Chem. Phys., 96 (1992) 5678.
7. M.S. Sample, Y-F. Hung and J. Ross, J. Phys., 96 (1992) 7338.
8. I. Schneider, Y-F. Itang and J. Ross, J. Phys., 100 (1994\6) 8556.
9. A. Forster, T. Hauck and F.W. Schneider, J. Chem. Phys., 98 (1994) 184.
10. A. Lekebush, A. Forster and F.W. Schneider, J. Chem. Phys., 95 (1995) 681.
11. R.J. Field, E. Koros and R.M. Noyes, J. Am. Chem. Soc., 94 (1972) 8649.
12. (a) R.P. Rastogi and G.P. Mishra, Indian J. Chem., 29A (1990) 941. (b) R.P. Rastogi, M.M. Husain, R. Khare and P. Chand, Chem. Edu. Rev., 15 (1999) 5.
13. G. Nicolis and I. Prigogine, Self-Organization in Non-Equilibrium Systems, John Wiley and Sons, New York, 1977, p. 93.
14. R.P. Rastogi and P.K. Srivastava, J. Non-equilib. Thermodyn., 12 (1987) 123.
15. M. Eisiwirth, P. Molar, K. Wetzl, R. Imbihl and G. Ertl, J. Chem. Phys., 90 (1989) 510.
16. R.P. Rastogi and I. Das, Indian J. Chem., 23A (1984) 363.
17. R.P. Rastogi, I. Das, V. Pandey and K. Jaiswal, Chem. Educ., Oct.–Dec. (1990) 46.
18. C.H. Yang and B.F. Gray, J. Phys. Chem., 73 (1969) 3395.
19. R.P. Rastogi, K.D.S. Yadava and A. Kumar, Indian J. Chem., 12 (1974) 1280.
20. A.J. Lotka, J. Am. Chem. Soc., 42 (1920) 1595.
21. M.V. Volkenstein, Biophysics, Mir Puliishes, Moscow, 1982, p. 515.
22. R. Lefever and I. Prigogine, J. Chem. Phys., 48 (1968) 1995.
23. P. Gary, S.K. Scott and J.H.J. Merkin, Chem. Soc. Faraday Trans., 84 (1988) 993.
24. S.K. Scott, Acc. Chem. Res., 20 (1987) 186.
25. L. Gyorgyi, T. Turanyi and R.J. Field, J. Phys. Chem., 96 (1990) 7162.
26. T. Turanyi, L. Gyorgyi and R.J. Field, J. Phys. Chem., 97 (1993) 1931.
27. R.J. Field and E. Koros, R.M. Noyas, J. Chem., Phys., 60 (1974) 1877.
28. R.J. Field, J. Chem. Phys., 63 (1975) 2284.
29. H.D. Forsterling, S. Muranyi and Z. Noszticzius, J. Phys. Chem., 94 (1990) 2915.
30. R.P. Rastogi, R. Khare, G.P. Mishra and S. Srivastava, Indian J. Chem., 36A (1997) 19.
31. R.P. Rastogi, P. Rastogi and K.D.S. Yadav, Indian J. Chem., 15A (1977) 338.
32. Z. Noszticzius and J. Bodiss, J. Am. Chem. Soc., 101 (1979) 3177.
33. R.P. Rastogi and M.K. Verma, Indian J. Chem., 22A (1983) 827.
34. M.T. Beck, G. Bazsa and K. Hauck, Ber. Bunsen. Phys. Chem., 84 (1980) 408.
35. R.P. Rastogi, I. Das and A. Sharma, J. Chem Soc. Faraday Trans. I, 85 (5) (1989) 2011.
36. R.P. Rastogi and K. Mani, Chem. Phys. Letts., 164 (1989) 545.
37. R.P. Rastogi and G.P. Mishra, J. Phys. Chem., 96 (1992) 4426.
38. S.D. Furrow and R.M. Noyes, J. Am. Chem. Soc., 104 (1982) 38.
39. R.M. Noyes, J. Am. Chem. Soc., 102 (1980) 4644.
40. M. Orban, E. Koros and R.M. Noyes, J. Phys. Chem., 83 (1979) 3056.
41. M. Orban and E. Koros, J. Phys. Chem., 82 (1978) 1972.
42. R.P. Rastogi and P.K. Srivastava, J. Non-equib. Thermodyn., 12 (1987) 123.
43. H.D. Forsterling and Z. Nostiezius, J. Phys. Chem., 93 (1989) 2740.
44. R.P. Rastogi, R.D. Yadav, S. Singh and A. Sharma, Indian J. Chem., 24A (1985) 43.
45. P. Seveik and L.J. Adamcikova, J. Phys. Chem., 89 (1985) 5178.
46. R.P. Rastogi, R. Khare, G.P. Misra and S. Srivastava , Indian J. Chem., 36A (1997) 19.

47. R.P. Rastogi, M.M. Husian, P. Chand, G.P. Misra and M. Das, Chem. Phys. Letts., 353 (2002) 40.
48. R.P. Rastogi, M.M. Husain and P. Chand, Indian J. Chem., 19A (2000) 679–684.
49. R.P. Rastogi, and P. Chand, Chem. Phys. Letts., 369 (2003) 434.
50. R.P. Rastogi, S.N. Singh and P. Chand, Chem. Phys. Letts., 385 (2004) 403–408.
51. R.P. Rastogi, P. Chand, M.K. Pandey and M. Das, J. Phys. Chem. A, 109 (2005) 4562–4567.
52. R.P. Rastogi, M.M. Husain, P. Chand, G.P. Misra and M. Das, Chem. Phys. Letts., 353 (2002) 40–48.
53. K. Bar-Eli, in Non-equilibrium Phenomena in Chemical Dynamics, C. Vidal and A. Pacault, (eds), Springer-Verlag, Berlin 1981.
54. M.F. Crowley and R.J. Field, J. Phys. Chem., 90 (1986) 1907.
55. M.F. Crowey and I.R. Epstein, J. Phys. Chem., 93 (1989) 2496.
56. O. Citri and I.R. Epstein, J. Phys. Chem., 929 (1988) 381.
57. M. Maselko, J. Chem. Phys., 78 (1983) 381.
58. D.O. Cooke, Int. J. Chem. Kinet., 14 (1982) 1047.
59. O. Citri and I.R. Epstein , J. Am. Chem. Soc., 108 (1986) 357.
60. I.R. Epstein, Chem. Eng. News, 13 (1987) 65.
61. E.K. Mathews and M.D.L.O. Conner, J. Expt. Biol., 81 (1979) 75.
62. Helmfelt and Ress, Physica D, 84 (1995) 180.
63. M. Okomoto, Y. Mako, T. Sakiguchi and S. Yoshida, Physica D, 84 (1995) 194.

Chapter 10

CHEMICAL WAVES AND STATIONARY PATTERNS

10.1. Introduction

In the earlier chapter, we have discussed the emergence of time-order in chemical reactions in continuously stirred tank reactor (CSTR) and have discussed the concept of negative and positive feedback for occurrence of oscillatory reactions. In this respect, experimental studies of oscillatory reactions in batch reactors have been investigated in great depth which has provided convincing evidence for the important role of auto-catalytic and inhibitory reactions in oscillatory reactions. Rate of internal production is controlled by the influx of reactants from external source.

Increase in entropy is associated with (i) decrease in order, (ii) increase in disorder and (iii) loss of information as predicted by Statistical Mechanics. Everything tends towards disorder. Any process that converts from one energy to another must lose heat. The universe is one-way street. Entropy must always increase in the universe and in any hypothetical system, within it. Although d_iS is greater than zero d_eS can be less than zero under certain circumstances, thereby generating a particular type of ordered structure which is called dissipative structure.

Nature forms patterns. Some are orderly in space. Some patterns are fractals, exhibiting self-similar in scale. Others give rise to steady states or oscillating ones. We will discuss fractals in Chapter 13.

In case of space order and spatio-temporal oscillations, diffusion of the reactants control the internal production of entropy so that dissipative structure occurs. Diffusion contributes to the dissipation of entropy.

10.2. Chemical waves and stationary patterns

Reacting systems involving non-linear kinetics (and systems) involving interplay of reaction and diffusion (R–D systems) exhibit a number of exotic phenomena. Several reviews (and symposium volumes) have been addressed to these issues [1–9]. In general, we come across the following types of systems:

Temporally oscillating systems: Reaction kinetics of such systems is governed by non-linear differential equations. Under certain conditions, their solution leads to limit cycle; under such circumstances, we have relaxation oscillations [10] of concentrations of some intermediate species with time. In these systems under specific circumstances, we come across the phenomena of bistability and chaos in the dynamic system.

Travelling waves: These may be classified as (i) phase or kinematic waves, (ii) trigger waves and (iii) self-accelerating meta-waves, a general account of which is given below.

Phase or kinematic waves [11–16]: These are generated due to the phase gradient in the system undergoing bulk temporal oscillations taking place not synchronously at various points in the system. It has been reported that if a temporarily oscillating system is left undisturbed, there can be non-synchronous oscillations. This generates a phase gradient due to a temperature or concentration gradient prevailing in the system [16]. Such a system will show a wave propagation though nothing moves along the passage of such wave, in the system. Hence, it can pass through impenetrable barrier. The gradual change of the phase of the system at its various point gives the impression of a moving wave. The waves cannot be stopped, by placing a barrier in its path.

Trigger waves [11–20]: These are generated due to interaction of non-linear kinetics and diffusion. The non–linear kinetics involves an autocatalytic and inhibitory reaction steps. The essential requirement of such a wave is that though the system may be at steady state, it should be excitable by a small change of concentration of certain species. On account of amplification of concentration of that particular intermediate, diffusion will set in resulting in the generation of trigger waves in the ongoing regions. Since the triggering of the waves is diffusion-dependent, a barrier in their path will obstruct their passage. Trigger waves have importance in the context of biological excitable tissues, marine and ecological system and multiplayer semiconductor devices [21]. Such waves in nature occur as wind-driven forest fire and epidemics.

Self-accelerating or meta-waves [12]: These waves propagate with increasing velocity and are caused by additional feedbacks like thermo-kinetic processes in addition to diffusion.

10.3. One-dimensional chemical waves

A simple experiment for generation of spatio-temporal oscillations has been performed as follows [22]: 10 ml each of malonic acid (1.1538 M), cerous sulphate (0.0042 M) and potassium bromate (0.2994 M in 1.5 M H_2SO_4) were homogenized in a boiling tube at 30 ± 0.05°C. After 4 h, 5 ml of the solution was transferred to a corning

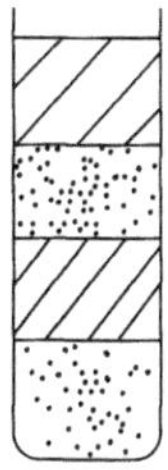

Figure 10.1. Chemical waves in a test tube experiment.

glass tube (dia. 0.5 cm, length 40 cm), 10 drops of ferroin indicator added and thoroughly mixed by inverting the tubes several times. Blue-coloured reaction waves started from the bottom of the tube and propagated through the solution. The appearance of chemical waves at the bottom was periodic. If the system is disturbed, by shaking, wave formation started from the bottom and the cycle was repeated, but period of appearance of the waves increased. After an initial fluctuation in velocity of propagation, it finally attained a steady value (Fig. 10.1).

Spatio-temporal oscillations or chemical waves [23] are a special class of phenomena, which have been extensively investigated during the last few decades. Chemical waves involving B–Z system are present (i) in homogeneous aqueous solutions [1–10], (ii) in gel systems [11], (iii) on adsorbing surface such as hydrated collodoin matrix [12], porous porcelain [13], millipore filters [14] and (iv) on ion exchange beads [15]. The subject has been reviewed in a few monographs [16–18].

10.4. Mechanism of wave propagation

The wave movement has been explained in the following way: The concentration of Br^- ions decreases at the beginning and the oxidation stage takes place after sometime; the leading edge of the wave (sharp blue boundary), with high concentration of $HBrO_2$, Fe^{3+} and low concentration of Br^- is formed. The reducing stage takes place behind the boundary and the reaction medium becomes red again due to the oxidation reaction:

$$4Fe^{3+} + Br.CH(COOH)_2 + 2H_2O \longrightarrow Br^- + 4Fe^{2+} + HCOOH + 2CO_2 + 5H^+$$

In a region with a high $[Br^-]$, a new wave cannot be initiated. $HBrO_2$ and Br^- diffuse in opposite directions through the boundary of the moving edge as indicated in Fig. 10.2.

Their mutual reaction

$$HBrO_2 + Br^- + H^+ \longrightarrow 2HOBr$$

$$HOBr + CH_2(COOH)_2 \longrightarrow Br.CH(COOH)_2 + H_2O$$

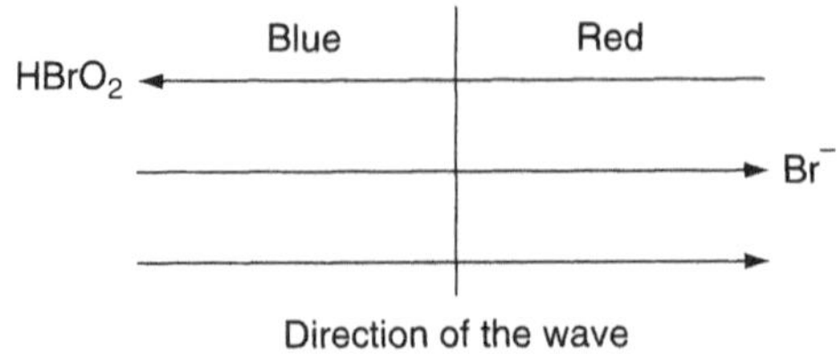

Figure 10.2. Diffusion of Br^- and $HBrO_2$ in opposite directions during wave propagation.

causes a decrease of the Br^- in the front of the leading edge and thus a new start of the oxidation process begins. Further, the decrease of $[Br^-]$ enables the start of the autocatalytic step and the formation of the leading boundary of the wave, characterized by a high concentration of Fe^{3+} and $HBrO_2$. High concentration of Br^- close behind the leading wave boundary is sharply increased. The Br^- ions are then again consumed by the reaction step until the value of $[Br^-]$ is reached when the autocatalytic step can start again.

A quantitative theory of wave propagation based on the above ideas has been developed by Field and Noyes [18] using Fick's second law, FKN mechanism of B–Z reaction and non-linear kinetics.

Basic steps of FKN mechanism which are used in the formulation are as follows:

$$\text{(R1) } HOBr + Br^- + H^+ = Br_2 + H_2O \qquad k_1$$

$$\text{(R2) } HBrO_2 + Br^- + H^+ = 2HOBr \qquad k_2$$

$$\text{(R3) } BrO_3^- + Br^- + 2H^+ = HBrO_2 + HOBr \qquad k_3$$

$$\text{(R4) } HOBr = BrO_3^- + HOBr + H^+ \qquad k_4$$

$$\text{(R5) } BrO_3^- + H^+ + HBrO_2 = 2BrO_2^{\cdot} + H_2O \qquad k_5$$

$$BrO_2^{\cdot} + Ce^{3+}/Fe(II) + H^+ = HBrO_2 + Ce^{4+}/Fe(III)$$

$$\text{(R6) } BrO_2^{\cdot} + Fe(II) + H^+ = HBrO_2 + Fe(III)$$

It is assumed that the velocity of band propagation is controlled by the concentration of Br^- (B) and bromous acid (A) in the band front and that trailing band does not affect the velocity of band propagation. Wave is due to (i) autocatalytic production and destruction of $HBrO_2$ and (ii) diffusion of $HBrO_2$.

Both chemical reaction and diffusion are involved in concentration changes at any position X and time t. The resulting reaction–diffusion equation can be written as follows:

$$(\partial C_A/\partial t)_X = D_A(\partial^2 C_A/\partial X^2)_t + k_3[H^+][BrO_3^-]C_B - k_2[H^+]C_A C_B + k_5[H^+][BrO_3^-]C_A - k_4 C_A^2 \tag{10.1}$$

$$(\partial C_B/\partial t)_X = D_B(\partial^2 C_B/\partial X^2)_t - k_3[H^+][BrO_3^-]C_B - k_4 C_A^2 \tag{10.2}$$

Equations (10.1) and (10.2) cannot be simultaneously solved. Hence, we have to adopt an approximate solution. Putting

$$V = -(\mathrm{d}X/\mathrm{d}t)_{CA,CB};\ (\mathrm{d}C_A/\mathrm{d}t)_X = V(\mathrm{d}C_A/\mathrm{d}X)_t$$

and ignoring the last three terms of Eq. (10.1), the approximate form of equation would be

$$D_A(\mathrm{d}^2C_A/\mathrm{d}X^2) - V(\mathrm{d}\,C_A/\mathrm{d}X) + k_5[H^+][BrO_3^-]C_A = 0$$

The stationary solution of the above equation yields

$$V = \surd(4D_A k_5[H^+][BrO_3^-]) \tag{10.3}$$

10.4.1. Experimental test

Experimental results on velocity of trigger waves in acidic bromate oxidation of ferroin have been reported by Field and Noyes [18] and Showalter [24]. A semi-quantitative comparison with experimental results shows that the observed velocity is proportional to square root of the product of $[H^+]$ and $[BrO_3^-]$ as theoretically predicted. However, deviations have been observed depending on experimental conditions. Thus, in the case of malonic acid + bromate + manganous sulphate + ferroin in H_2SO_4 medium, the wave velocity is found [25] to be proportional to $[BrO_3^-]^{1/2}[H^+]^{1/2}$. Similar results are obtained for citric acid + BrO_3^- + Mn^{2+} + H_2SO_4 system [26].

It is significant that chemical waves originate from the point where specks are first formed. Bands appear at heterogeneous centres or dust particles or at glass surfaces. Nucleation of chemical wave and nucleation of a new phase have a formal similarity [27].

10.5. Wave formation on membranes

Formation of two-dimensional patterns on porcelain membrane has also been reported [23]. Wave velocity in (i) ferroin + malonic acid + bromate + H_2SO_4, (ii) ferroin + malonic acid + acetone + bromate + H_2SO_4, (iii) ferroin + Ce^{4+} + malonic acid + bromate + H_2SO_4, (iv) ferroin + Ce^{4+} + malonic acid + bromate + acetone systems has been investigated with and without porous unglazed porcelain membrane. Wave velocity is found to be higher on membrane than in solution.

Wave equation is satisfied in case (i), while in other cases modified form of the equation is found to be valid. Typical results for the various systems with and without membranes are summarized in Table 10.1.

In Table 10.1, k_1, k_2, k_3 and k_4 are the rate constants for the wave velocities in the solution while k'_1, k'_2, k'_3 and k'_4 are the rate constants for the wave velocity on

Table 10.1. Wave velocity dependence on $[H^+]$ and $[BrO_3^-]$ in solution and on membrane.

System	Solution	Membrane
Ferroin + malonic acid + BrO_3^- + H_2SO_4	$V = k_1[H^+]^{1/2}[BrO_3^-]^{1/2}$	$V = k_1'[H^+]^{1/2}[BrO_3^-]^{1/2}$
Acetone + ferroin + malonic acid + BrO_3^- + H_2SO_4	$V = k_2[H^+]^{1/2}[BrO_3^-]^{1/2} + b_1$	$V = k_2'[H^+]^{1/2}[BrO_3^-]^{1/2} + b_1$
Ferroin + Ce^{4+} + malonic acid + BrO_3^- + H_2SO_4	$V = k_3[H^+]^{-1}[BrO_3^-]^{1/2}$	$V = k_3'[H^+]^{-1}[BrO_3^-]^{1/2}$
Acetone + ferroin + Ce^{4+} malonic acid + BrO_3^- + H_2SO_4	$V = k_4[H^+]^{-1}[BrO_3^-]^{1/2} + b_2$	$V = k_4'[H^+]^{-1}[BrO_3^-]^{1/2} + b_2$

the membrane for the similar type of solutions. b_1 and b_2 are constants for ferroin system and ferroin + Ce^{4+} systems.

For first and third systems, when no acetone is present, the classical equation (10.3) of wave velocity is obeyed both for solution as well as for membrane. However, when acetone is present in second and fourth systems as an extra reactant, complications appear on account of which an additional term b_1 or b_2 appears in the wave equation.

10.6. Wave structures

Different types of wave structures have been reported. Some of these are discussed below:

10.6.1. Mosaic structure

During one-dimensional study of uncatalysed oscillatory reaction systems, a specific type of structures called mosaic structures have been observed in phenol–bromate–H_2SO_4 [28] and ferroin–bromate–H_2SO_4 systems [29]. Analytical studies showed that the structure in the latter case is simply an agglomeration of fine grains of solid solutions of the reaction products, tri-bromophenol and *p*-benzoquinone formed during periodic nucleation [30]. Mosaic structures were not formed in a tube which had been freshly cleaned by steaming, indicating the occurrence of heterogeneous nucleation.

10.6.2. Stationary structure (Turing structure)

Castets *et al.* [31] have defined Turing structure as follows: "These are stationary patterns originating with Sole coupling of reaction and diffusion processes. They correspond to stable stationary solutions of a set of reaction diffusion equations". However, Turing [32] used the word stationary to denote the situation when a spatial pattern does not move.

10.7. Turing instability

In a reaction–diffusion (R–D) system with two intermediates X and Y where X is an activator and Y is an inhibitor, criterion for Turing symmetry breaking instability is that

$$D_X << D_Y$$

where D_X and D_Y denote the diffusion coefficients of X and Y, respectively. In the B–Z system, this condition is not satisfied. The diffusion coefficient of the activator $HBrO_2$ is greater than the diffusion coefficient of inhibitor Br^-. Hence, B–Z system cannot be used for obtaining Turing pattern. However, diffusion coefficient of activator can be reduced by complexation of the activator X [33]. Lengyel and Epstein have used this concept to obtain Turing patterns in Chlorite, Iodide and Malonic acid (CIMA) reaction system.

For developing a theoretical formalism, under such circumstances, the modified form of reaction–diffusion equation would be as follows:

$$\partial x/\partial t = f(x, y, p) + D_X\, \partial^2 x/\partial Z^2 \tag{10.4}$$

$$\partial y/\partial t = f(x, y, p) + D_Y\, \partial^2 y/\partial Z^2 \tag{10.5}$$

where x and y denote the concentration of X and Y, Z denotes the distance travelled by the band and p is a parameter depending on the kinetics of complex formation between X and species S such that

$$S + X \rightleftharpoons SX$$

When [S] is large, it is likely that diffusion coefficient D_{SX} may be very low, nearly equal to zero. The complex formation modifies the stability of steady-state of the system without changing the steady-state composition. In the absence of S, a system may have a unique unstable steady state and hence oscillating. On the other hand, with complex formation, the homogeneous steady state may be stabilized at the same set of parameter values. Introduction of S broadens the range of steady state making it possible for the onset of diffusion-induced instability in the regions of parametric space inside the region held by oscillating state in complex free system.

Primary function of diffusion in the Turing instability is thus to spatially disengage the counteracting species.

10.7.1. Convection and chemical waves

Convective effects [34] depend on density gradient in the system which can arise from (i) enthalpy change ΔH of the reactants and (ii) volume change. It has been suggested that when sign of ΔH and ΔV are same, simple convection will occur. If sign changes of ΔH and ΔV are not same, simple convection will not occur and complication would arise on account of ΔH, ΔV and the direction of the propagation of the wave. Such a situation has been investigated analytically by Epstein and co-workers [35, 36].

In the case of B–Z reaction, it has been shown that [34]

(a) The magnitude of convective flow in a horizontal layer increases with increase of layer thickness.
(b) In a vertical tube, wave can generate density gradient which may make it unstable.
(c) In tubes having dia. <0.2 cm, convection effect would diminish.
(d) For wider tubes, the downward convection seriously disturbs the advancing wave.

In the case of IO_3–arsenite system [35] convection enhances the velocity of wave front [33]. The density is decreased on account of exothermicity of reaction as well as positive V. This effect is similar to the observed effect.

More complex behaviour during wave propagation has also been observed. Thus, Bazsa and Epstein [37] in the case of Fe-HNO_3 system have found anisotropic wave velocity in the system as follows:

Descending wave front > velocity of ascending wave front > velocity of wave front due to R–D process. They [38] have reported similar observations in chlonite–thiosulphate reaction system at lower concentrations but a reversal of behaviour at higher concentrations.

10.7.2. Experimental techniques for obtaining Turing pattern

A CSTR provides an effective method of obtaining temporal dissipative structures under sustained and controlled conditions, since it is difficult to obtain stationary structures using a batch reactor. Control of convection is desirable, since the stationary structures are vitiated by convective effects [39].

In this context, couvette reactor (CR) provides a technique which eliminates the problem of (i) varying reactant concentration, (ii) diffusivity constraints and (iii) convective interferences.

The Couvette reactor is made up of two horizontally mounted coaxial glass cylinders. Reactions occur in the annular space between the two coaxial cylinders. The annular space is connected to two CSTRs, one being connected at each end. Each CSTR has an inlet and outlet pump. During the experiments, the rates of input and output are carefully balanced so that there is no net flow through the annular reactor. Transport only occurs on account of turbulent diffusion. The whole system is immersed in thermostat. Each CSTR is provided with a Pt electrode to monitor the electrochemical potential. The patterns in the annular space are indicated by adding appropriate indicator and are recorded by a video camera.

The outer cylinder is fixed and inner cylinder can be rotated at varying speed. When the system is rotated at large Reynold's number, the flow inside the CR becomes turbulent enough and the fluid is well-mixed.

CIMA reaction:
$MA + I_2 \rightarrow IMA + I^- + H^+$
$ClO_2^- + I^- \rightarrow ClO_2^- + {}^1\!/_2 I_2$
$ClO_2^- + 4H^+ + 4I^- \rightarrow Cl^- + 2I_2 + 2H_2O$

In this reaction, ClO_2^- is self-inhibitor and I^- is autocatalyst. Kepper *et al.* [40] investigated the pattern formation in CIMA reaction by using the above type of reactor. For the CIMA system, the composition of CSTR was kept constant whereas that of other CSTRs was varied stepwise. Different sequences of spatial and spatio-temporal patterns were observed. At large ClO_2^- flow rates, stationary patterns were observed. In the CIMA reaction in CR, the concentrations of I_2, MA and H^+ were same in both the CSTRs but spatial gradient in ClO_2^- concentration may be imposed by applying different feed concentrations in the CSTR. Thus $[ClO_2^-]$ was allowed to vary with distance along the reactor but not with time.

Detailed mechanistic interpretation of patterns in the case of CIMA reaction in the Couvette reactor has been proposed by using modified forms of R–D equations [40].

Depending on the experimental conditions different types of patterns are observed. One such example is flow-distributed oscillatory (FDO) pattern.

Flow-distributed oscillatory patterns: FDO patterns were first predicted by Kuznetsov *et al.* [41] and later developed by Andersen and co-workers [42] using Brusselator model. Such patterns were first experimentally observed by Kaern and Menzinger [43] in the B–Z reaction system. The experimental results were found to be in agreement with computer simulation using a modified theoretical framework using Zhabotinskii–Rovinsky [44] model for B–Z reaction. The reactants are fed into to CSTR through two inlet tubes through pumps. The mixed reactants are then fed in the tubular reactor (TR) packed with spherical glass beads in a vertically mounted cylindrical glass tube.

The vertical tubular reactor (TR) is essentially a I-D plug flow reactor, through which the reactant mixture flows with a uniform velocity.

The reaction–diffusion equation would include (i) a diffusion term, (ii) kinetic terms depending upon the reaction mechanism, (iii) a flow term.

Diffusion flow-induced chemical instability: Since one of the criterion for obtaining Turing structure is D activator $<<$ D inhibitor, it essentially means a device to spatially disengage the counteracting species. The key species may by disengaged by an alternative mechanism, i.e. by differential transport. It involves the destabilization of steady state by flows of activator and inhibitor at different rates (irrespective of one which is faster) leading to pattern formation. Bulk differential flow is easier to control than Turing's condition of diffusibility at molecular level. The mechanism of such a flow-driven pattern formation has been developed by Rovinsky and Menzinger [45, 46] which is called differential flow-induced chemical instability (DFICI).

B–Z reaction using ferroin as a catalyst has been used for pattern formation studies for such a system. According to FKN mechanism, $HBrO_2$ is the autocatalytic species. On commencement of autocatalysis the $[Br^-]$ is depleted to a value lower than the critical value. The autocatalytic production of $HBrO_2$ is accompanied by oxidation of ferroin to firrin $Fe(phen_3)^{3+}$. Br^- formation occurs due to reaction of $Fe(phen_3)^{3+}$ with bromo-malonic acid. Thus, a differential flow between $HBrO_2$ and Br^- can be achieved.

In order to generate a differential flow of activator and inhibitor, the ferroin is immobilized on a cation-exchange resin that is packed in a vertical tubular reactor, while the rest of the reactants are forced to flow through the tube. The cation-exchange resin beads were loaded with ferroin solution and filled in tubular reactor. The rest of the reactants were allowed to flow through the loaded beads under suitable varying pressures.

10.7.3. Self-organization – varying type of Turing pattern in chemical and biological systems

The spatial patterns produced in the above and similar cases are widely known as spiral and target patterns, which are actually wave structures. There is an increasing number of (quasi) two-dimensional chemical and biological systems, i.e. thin layer of surfaces, for which the propagation of circular and spiral-shaped wave-fronts of excitation has been reported. Wave-propagation in the classical non-equilibrium systems has been studied mostly in thin layers of ferroin-catalysed reaction. In recent years, cerium and ruthenium have been frequently used as catalyst and indicator. Spiral wave dynamics has been critically reviewed by Muller and Plesser [47, 48], which gives an overview of experimental and numerical investigations on self-organization. The quantitative analysis of spiral yields a detailed picture of concentration distribution inside and outside the spiral core. This information can be used to clarify the details in the mechanistic steps of the reaction coupled with diffusion.

Electric field influences the spiral patterns. It has been found that in a small glass capillary, which can roughly be considered as a one-dimensional system, applying an electric field can cause a chemical wave propagating towards the anode to accelerate. Changing the polarity may stop or split the wave, depending on the strength of the field. The external perturbation forces the spiral cores to drift towards the anode accompanied by a deformation, characteristic for moving wave sources.

We discuss below some of the specific type of Turing patterns which depend on (i) reactor geometry, (ii) concentration of the reactants and (iii) ratio of diffusion coefficient of inhibitor and activator [49, 50]:

(1) Three-dimensional pattern consisting of parallel lines and periodic spots (body-centred cubic structure) [50]. The pattern is influenced by temperature which controls the reactionh rates.
(2) Quasi-two-dimensional structure in CIMA reaction (chlorite + iodide + malonic acid + starch). Initially transient yellow circles emerge and start to grow in a blue surrounding. These structures dissolve and break into dot patterns demonstrating a wide distribution of sizes. The dots evolve to a stationary structure consisting of yellow hexagons of different orientations [51, 52]. Similar structures have been observed in petri dish experiments with PA-MBO reaction [53].
(3) Hexagon stripes have been observed in PA-MBO system in which case gel is removed from the petri dish [53]. Similar observations have been reported in CIMA reaction with disc reactor [52] and also PA-MBO system with petri dish [53].

Black-eye pattern: A more complex structure appears well beyond the primary instability of a hexagonal array through increase in the malonic acid concentration [52].

Transient Turing-like patterns in PA-MBO reaction (Polyacrylamide–methylene blue–sulphide-oxygen) system are obtained. Non-Turing stationary patterns in FIS reaction (Hexacyanoferrate (II)–iodate–sulphite system) are obtained using an experimental technique similar to that used in CIMA reaction. These patterns develop through propagation of chemical fronts from the initial perturbation [54].

Dividing blobs, chemical flowers and patterned islands: Different modes of propagation of the patterned state have been observed in CDIMA reaction (chlorine dioxide–iodine-malonic acid – PVA in a one-sided-fed reactor) in an absolutely unstable uniform state, among which are a spot division and finger printing mode [55].

More complicated structures have been reported in the literature. Coupling of flows with reaction–diffusion yields interesting travelling structures. The addition of flow to chemical reaction–diffusion system provides a robust pattern forming mechanism which is expected to occur in a wide variety of natural and artificial systems [56]. Typical examples are (i) Ca^{2+} waves in the slime mould dictyostelium discoideum, (ii) heterogeneous catalysis; CO oxidation on Pt, (iii) oxidation wave in B–Z reaction.

10.8. Logic functions

Efforts have been made to develop chemical diode utilizing the concept of wave propagation. Propagating pulses of concentration in an excitable system may carry information because regions of high concentration of a selected reagent may be associated with logical reactors constructed of active (excitable) and passive areas in which diffusion of reagents can be applied to transform such information.

Response of simple reactors to regular trains of pulses has been investigated in the case of few reactors [57] which can act as chemical diode (a cross-junction switch). It has been demonstrated that response of such a signal processing device may depend on input frequency.

Logic gates based on chemical wave propagation in geometrically constrained excitable media have been investigated in a Belousor–Zhabatbuskit membrane system [58].

10.9. Precipitation patterns

10.9.1. Introduction

Inter-diffusion of one electrolyte into another reacting electrolyte may lead to a rhythmic deposition of precipitate, which is known commonly as Liesegang phenomenon. Liesegang ring-type patterns are observed both in living and non-living systems including bacterial and fungal growth. When a solution of large concentration of $K_2Cr_2O_7$ is allowed to diffuse into a less concentrated solution of lead nitrate containing a lyophilic gel (agar-agar) medium, periodic deposition of precipitate appears after several hours.

Figure 10.3 shows Liesegang ring-type patterns observed during crystallization of ascorbic acid from its alcoholic solution, periodic precipitation of copper chromate in agar-agar gel, aggregate bands, growth of the bacteria *Pr. vulgaris* on agar plate and the fungi *Necteria cinnabarina*. Rings of this kind are also observed in nature in minerals such as malachite, limonite, variscite and wings of certain colourful butterflies [59].

The Liesegang ring formation may be considered as possible example of Turing phenomena.

10.9.2. Liesegang patterns in flow reactors (CFUR, DPL, open ring reactor)

Two-dimensional experiments [59] on periodic precipitation of $PbCrO_4$, $CuCrO_4$ and HgI_2 have been carried out with a simple and inexpensive Das–Pushkarna–Lall (DPL) flow reactor [60].

The reactor is open with respect to the flow of one reactant as the other electrolyte was taken in the gel. In this reactor, the concentration and level of the electrolyte in the empty space were always kept constant. The DPL reactor was further modified by Das and co-workers to make it open with respect to the flow of both the reactants [61].

In the modified open gel-ring reactor, a gel ring (Fig. 10.4b) separates the two electrolyte solutions, which are fed continuously at the same flow rates. These reactors

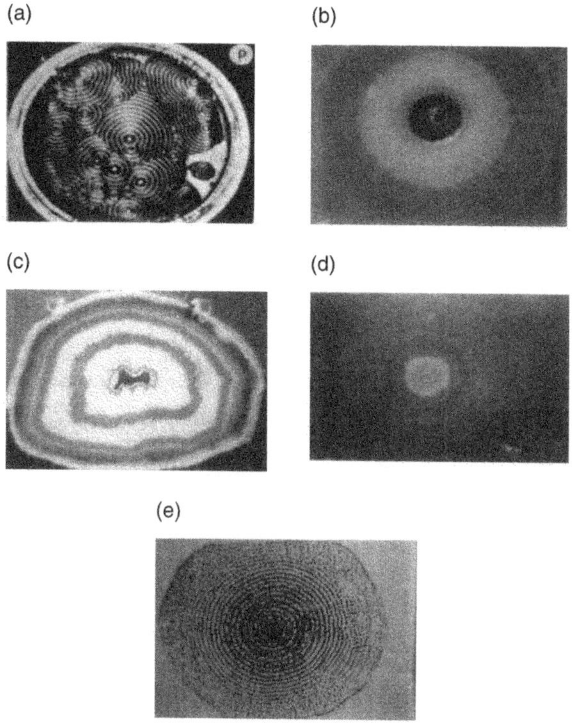

Figure 10.3. Periodic patterns in living and non-living systems: (a) concentric rings of ascorbic acid crystallized from methanolic solution, (b) copper chromate rings, (c) agate thin slab, (d) *Proteous vulgaris*, (e) *Necteria cinnabarina* [59]. See also Colour Plate Section.

have many advantages over the closed or batch reactors during slower diffusion and chemical reaction process.

The kinetics of propagation of the precipitation front has been investigated and it satisfies the relation $d = ct^m$ where d is the distance traversed and t denotes time while c and m are constants.

10.9.3. Mechanism

Three different approaches have been used involving three models:

(1) Super saturation theory [62]
(2) Competitive particle growth models [63]
(3) Sol coagulation model [64, 65]

Super saturation theory: According to model (1), the two reactants A and B diffuse from opposite direction and co-exist in the gel until the solubility product reaches a critical value above which the nucleation occurs according to the reaction

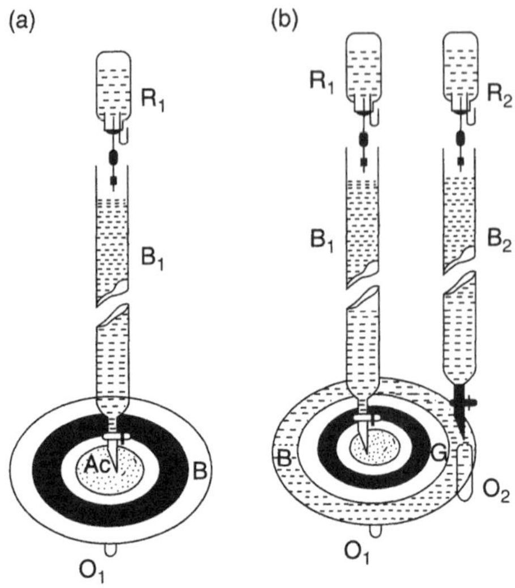

Figure 10.4. Experimental setup for Leisegang ring in (a) DPL Continuous flow reactor and (b) gel-ring reactor. The reactor (a) contains 1% agar-agar gel and a reactant at lower concentration, Reservoir R_1 contains another electrolyte of higher concentration. In gel-ring reactor (b), 1% aqueous agar-agar gel was used to form gel-ring and reservoir R_1 and R_2 contain aqueous solutions of reactants. O_1, O_2 represents outer nozzles and B_1, B_2 are Corning burets.

$$A + B \longrightarrow AB \text{ (solid) (model I)}$$

Once the nucleation has started, a depletion of concentration of A and B occurs in the surroundings. Hence, a stage comes when the formation of precipitate stops and a clear space is formed. However, diffusion of ions persists and after a certain stage (beyond the super saturation limits) spontaneous nucleation takes place and another band is formed in a similar manner. The process is repeated until the supply of one of the electrolyte is exhausted.

In a more recent version of the theory [66], the following modified picture emerges. The two species A and B are assumed to react to produce a new species C, which also diffuses into the gel subsequently; nuclei are formed followed by crystallization according to the following scheme:

$$A + B \longrightarrow C$$

$$\text{C Nuclei} \longrightarrow$$

$$\text{Nuclei} \longrightarrow \text{Crystal}$$

As a result, separate bands are obtained.

10.9.4. Light-dependent liesegang phenomena

There are several cases in which the periodicity occurs in darkness. Exposure to light, however, can induce the onset of periodicity or cause some disturbance. Such an example is found in the case of silver chromate [67] in an illuminated condition.

Lead chromate is another photosensitive material which invoked many workers to carry out detailed investigation. Hatschek [68] and Das *et al.* [69] reported results of the light-induced periodic precipitation of lead chromate in agar-agar gel medium. In order to study the effect of light on pattern formation, experiments were done both in the presence of sunlight and in complete darkness. Homogeneous precipitation was observed when the experiment was performed in complete darkness while rhythmicity was observed in an illuminated condition.

Das *et al.* later on observed that red and yellow forms of mercuric iodide precipitate when mercuric chloride reacts with potassium iodide in the agar-agar gel under different experimental conditions. The colour of the precipitate was yellow when KI of relatively higher concentration diffused into a less concentrated solution of $HgCl_2$ containing 1.5% agar-agar in complete darkness and precipitated downward in a corning tube. Red form of the HgI_2 precipitated when KI solution of relatively higher concentration diffused into the gel containing $HgCl_2$ of lower concentration in complete darkness. The chemical reactions may be written as

$$\mathrm{2KI}\ (C_1) + \mathrm{HgCl_2}\ (C_2) \longrightarrow \mathrm{2KCl} + \mathrm{HgI_2}\ \text{(yellow)}$$

$$\mathrm{HgCl_2}\ (C_1) + \mathrm{2KI}\ (C_2) \longrightarrow \mathrm{2KCl} + \mathrm{HgI_2}\ \text{(red)}$$

$$(C_1 > C_2)$$

The kinetics of yellow and red wave propagation was studied that satisfied the relation $d^2 = kt$, where d is the extent of propagation from the initial junction; k and t are the rate constant and time, respectively. The energy of activation for yellow wave propagation was found to be 9.2 kcal mol^{-1}. Potential and $[H^+]$ during the propagation of the advancing front were also monitored as a function of time. Results lead to the conclusion that the phenomena involved the transition from one state to another.

Das *et al.* [70] earlier reported that

Yellow wave was produced when aqueous [KI] > $[Hg^{2+}]$ in gel and
Red wave is produced when aqueous $[Hg^{2+}]$ > [KI] in gel in complete darkness.

Later on, Das *et al.* reported that a single red band bifurcates into several revert spaced bands in the presence of natural light of wavelength $\lambda < 600$ nm. However, the yellow band did not bifurcate under these conditions. Other remarkable observations were:

The light induced spatial bifurcation of the yellow wave into a number of alternate red and yellow bands at low pH of the gel containing Hg^{2+} (pH 2.72);
dependence of the precipitation pattern on the orientation of tubes; and
influence of electric field on the kinetics of yellow and red wave propagation.

It was observed that the yellow wave propagated downward in the presence or absence of natural light at pH 4.68 of lower content. However, a remarkable change in its characteristics could be noticed when the pH was reduced by adding a small amount of acetic acid. At pH$<$2.72, the yellow wave bifurcates into many alternate red and yellow bands at the lower end of the tube in the presence of natural light.

Experiments on the yellow wave propagation at different tube orientations were carried out by Das *et al.* [71]. Results revealed that the location of bands in the horizontal tube (H) was approximately half-way between those in the $g^{\uparrow}$ and $g^{\downarrow}$ directions.

10.9.5. Mechanism of light-induced propagation

The situations in illuminated systems are slightly different. The interaction of light radiation with one of the species becomes an important and necessary step. Nitzan *et al.* [72] analysed the general illuminated reacting–diffusing system. In the lead system the band formation does not depend upon the dimensions of the system. Therefore, it is postulated, in formulating the mechanism, that this system uses intrinsic length scaling and involves more than two reacting species. The light absorption by one of the species, nucleation of colloidal particles, and colloidal growth which is found to be an autocatalytic step are important steps in the mechanism. Lead chromate is almost insoluble in water, therefore the degree of ionization is very small. It is proposed therefore that the degree of ionization increases in the illuminated system.

For an illuminated system, it was postulated that chemical instability would occur due to light absorption by one of the species. Das *et al.* [71] observed that light of wavelength 400–550 nm was most effective for periodic precipitation of $PbCrO_4$ in agar-agar. For a complete reacting–diffusing system of illuminated lead chromate, the following sequence of events was proposed:

Absorption of light by one of the species;
nucleation of energized particles and colloidal growth; and
distribution of colloids and band formation.

Results showed that equilibrium state was very unstable under illumination.

10.9.6. Light-induced zonation in fungal growth [73]

Zonation in the formation of fruiting structure in concentric ring on an agar surface is a common phenomenon. Zonation of reproductive structures in some fungi is neither dependent upon light nor upon temperature fluctuation; presumably nutritional factors are involved. The occurrence of light-induced zonation is known. A simple explanation is difficult because two extreme types of response to an alternation of light and dark periods are known, viz.:

The fungus sporulates well in darkness, much less or not at all during the light period. Examples are *Cephalothecium roseum* and *Sclerotinia fructicola.*

Certain organisms sporulate poorly in the dark and form a zone of spores after even a brief exposure to light, e.g. neither *Fusarium spp.*

Temperature alternations also induce the formation of zones of reproductive structures; it appears especially from the data that the immediate stimulus for sporulation is a check in growth.

Zonation of some fungi is dependent neither upon light nor on temperature fluctuations. Presumably, nutritional factors are involved here but they are not understood. pH of the medium also has an effect.

Any hypothesis to explain the action of light on sporulation will have to take into account several facts, viz.: (i) the indifference of many fungi to light stimuli, (ii) the high specificity of the response and (iii) whether the light simulates or inhibits sporulation.

10.9.7. Influence of electric field on band propagation

Influence of external electric field on the propagation of yellow wave has been studied in agar-agar gel at different field intensities employing the experimental setup as shown in Fig. 10.5.

Non-linear plots of location of bands (d) as a function of time were obtained. Plots of d^2 versus t yield straight line indicating that the relation $d^2 = kt + c$ was obeyed. In addition to these observations the following conclusions could be drawn:

(a) In the absence of an electric field the velocity of propagation was relatively faster than those observed in the presence of electric field.

(b) In the absence of an electric field I^- ions diffuse into the gel on account of concentration gradient. When an electric field was applied Hg^{2+} ions tried to move with greater velocity and I^- diffusion would be reduced. As a result the velocity of propagation decreased on applying the voltage.

At a fixed field intensity, velocity of propagation increased with an increase in $[I^-]$. Plots of d^2 versus t yielded a straight line obeying the relation $d^2 = kt + c$. A similar

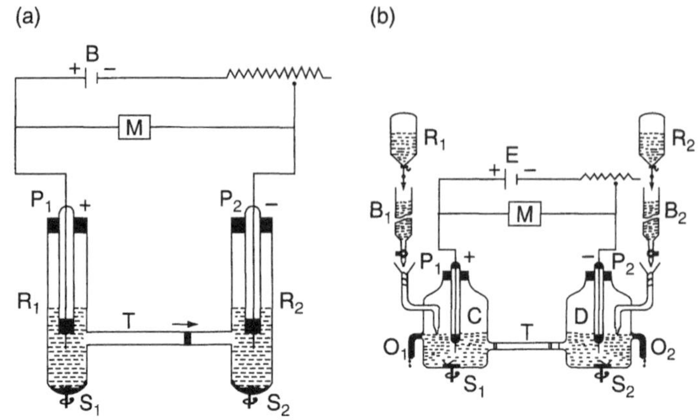

Figure 10.5. Experimental setup for investigating the electric field influence experiments on one-dimensional pattern formation. (a) batch and (b) flow reactor. P_1, P_2: platinum electrodes. R_1, R_2: Reservoirs containing aqueous copper nitrate (0.01 M) and potassium chromate (0.1 M) solutions, respectively. B: DC voltage source, M: digital multimeter; T: tubular reactor S_1, S_2: magnetic stirrers.

trend for the dependence of $[Hg^{2+}]$ on the velocity of red wave propagation has been observed in the presence of electric field.

Another interesting feature of the system was the propagation of continuous precipitation of red HgI at low [KI] in the presence of an electric field.

Interaction of chemical waves with an applied electric field is an important feature of electrochemical systems [74]. The influence of an external electric field on the precipitation of yellow and red mercuric iodide in one-dimensional gel media was studied by Das *et al.* [71] using an experimental setup, shown in Fig. 10.5a. In the absence of an electric field, the colour of the precipitate remained unchanged. In the presence of electric field, transitions from yellow to red (e, g) and red to yellow (f, h) HgI_2 were obtained at the field intensity $0.567\ V\ cm^{-1}$.

In the absence of an electric field, I diffuses into the gel due to concentration gradient and a yellow precipitate is formed. When an electric field is imposed, I moves towards the anode but at the same time Hg^{2+} would also have the tendency to move towards the cathode. As a result of counter diffusion, the movement of I ions would be restricted, resulting in the reduced I ion concentration near the anode in the gel and Hg^{2+} ion concentration would comparatively be increased resulting in the precipitation of red HgI_2. A similar explanation was given by Das *et al.* [66] for a red to yellow transition. A schematic representation showing ion movements during precipitation of (i) yellow and (ii) red HgI_2 in the presence and absence of an electric field is given in Fig. 10.6.

In a separate experiment, the field intensity was also varied in the range $0.0–0.567\ V\ cm^{-1}$ during the precipitation of yellow HgI_2. KI (0.2 M) diffuses into 1%

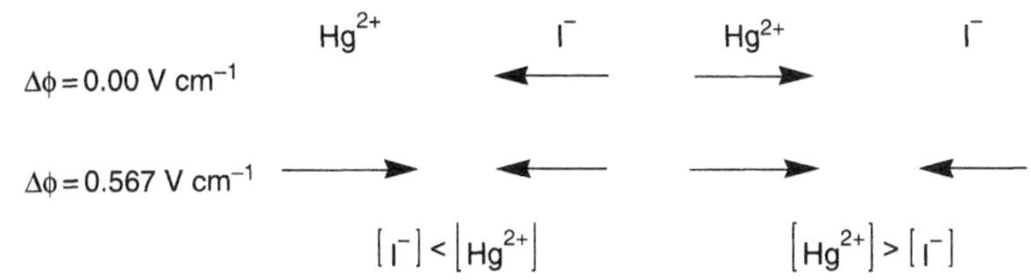

Figure 10.6. Ion movement in the presence and absence of electric field.

agar-agar containing $HgCl_2$ (0.01 M). It was observed that a critical field intensity V_c exists at 0.283 V cm^{-1} at which the transition from yellow to red wave took place.

The influence of external electric field on the propagation of yellow wave has been studied in agar-agar gel. Field intensity was varied in the range 0–0.383 V cm^{-1} employing the experimental setup shown in Fig. 10.5a. Non-linear plots of location of bands as a function of time at various field intensities were obtained. Plots of d^2 versus t yield straight line indicating that the relation $d^2 = kt + c$ was obeyed. Results also indicated that in the absence of an electric field the velocity of propagation was relatively faster than those observed in the presence of electric field.

In addition to these observations, the following conclusions could also be drawn:

In the absence of an electric field, I^- ions diffuse into the gel on account of concentration gradient. When an electric field was applied Hg^{2+} ions tried to move with greater velocity and I^- diffusion would be reduced. As a result the velocity of propagation decreased on applying the voltage.

At a fixed field intensity, velocity of propagation increased with an increase in $[I^-]$. Plots of d^2 vs t yielded straight line obeying the relation $d^2 = \mathrm{k}t + c$. A similar trend for the dependence of $[Hg^{2+}]$ on the velocity of red wave propagation has been observed in the presence of electric field.

Another interesting feature of the system was the propagation of continuous precipitation of red HgI at low [KI] in the presence of an electric field intensity greater than 0.084 V cm^{-1} and the revert spaced bands at field intensity in the range 0–0.84 V/cm^{-1}.

10.9.8. Light-induced spatial bifurcation of HgI_2

In the earlier communication, Das *et al.* [67] reported results of the investigation on the one-dimensional propagation of yellow/red mercuric iodide. In a later communication, Das *et al.* [68] reported new results on the two-dimensional propagation of a red/yellow wave of mercuric iodide, in gel media in batch, DPL and gel-ring reactors. Salient features of the investigation are as follows: Precipitation was carried out in a DPL continuous flow reactor, as described earlier, in which the concentration and level of the entering reagent in the empty space were always kept constant. Experiments were performed in the DPL reactor for the propagation of yellow and red waves of HgI_2. The location of the precipitate front from the initial function was measured as a function of time. Plots of d^2 versus t yielded straight lines obeying the equation

$$d^2 = kt + c$$

The kinetics of yellow wave propagation at different [KI] studied in the DPL reactor results showed that the velocity of propagation increased with increase in [KI].

The velocity of yellow wave propagation also depended on temperature. The velocity increased with temperature.

Das *et al.* [71] also observed the light-induced spatial bifurcation of a yellow wave into alternate red and yellow bands at low pH in the DPL reactor.

Due to interest in the growth of complex structures under non-equilibrium conditions, pattern formation of HgI_2 on thin film of agar-agar was studied. Results showed that (i) rhythmic or complex pattern of HgI_2 are formed depending on experimental conditions; (ii) rhythmicity was observed during crystallization of red mercuric iodide at low [KI] (0.001 M), whereas at relatively high [KI] (0.05 M) rhythmicity disappeared.

References

1. P. Glansdroff and I. Prigogine, Thermodynamic Theory of Structures, Stability and Fluctuations, Wiley, New York, 1971.
2. G. Nicolis and I. Prigogine, Self Organisation in Non-equilibrium Systems, Wiley, New York, 1977.
3. G. Nicolis and I. Portnow, Chem. Rev., 73(4) (1973) 365.
4. R.J. Field and M. Burger, Oscillations and Travelling Waves in Chemical Systems, Wiley, New York, 1985.
5. J. Phys. Chem., 93(7) (1989) dedicated in honour of Prof. Richard M. Noyes on his 70th birth anniversary.
6. P. Grey and S.K. Scott, Chemical Oscillations and Instabilities in Non-linear Chemical Kinetics, Clarendon, Oxford, 1994.
7. S.K. Scott, Oscillations, Waves and Chaos, Oxford University Press, Oxford, 1994.
8. R. Kapral and K. Showalter, Chemical Waves and Patterns, Kluwer, Dordrecht, 1994.
9. I.R. Epstien and J.A. Pojman, An Introduction to Non-linear Chemical Dynamics, Oxford University Press, Oxford, 1998.
10. M. Minorsky, Non-linear Oscillations, D. Van Nostrand Co. Inc., Princeton, NJ, Affiliated East-West Press Ltd, New Delhi, 1968.
11. E.J. Reusser and R.J. Field, J. Am. Chem. Soc., 101(5) (1979) 1063.
12. L. Kuhnert, H.J. King and L. Pohlmann, J. Phys. Chem., 899 (1985) 2022.
13. (a) A.T. Winfree, Science, 175 (1972) 634. (b) A.T. Winfree, Science, 181 (1973) 937. (c) A.T. Winfree, Sci. Am., 230 (1974) 82.
14. M.T. Beck and Z.B. Vardi, Nat. Phys. Sci., 253 (1973) 15.
15. D. Thoenes, Nat. Phys. Sci., 243 (1973) 18.

16. N. Kopell and L.N. Howard, Science, 180 (1973) 1171.
17. R.J. Field and R.M. Noyes, Nature, 237 (1972) 390.
18. R.J. Field and R.M. Noyes, J. Am. Chem. Soc., 96 (1974) 2001.
19. W. Jahnke, W.E. Skaggs and A.T. Winfree, J. Phys. Chem., 93 (1989) 740.
20. F. Sagues and I.R. Epstein, Roy. Soc. Chem. Dalton Trans., (2003) 1201.
21. A.E. Bugarin, A.M. Zhabotinskii and I.R. Epstein, Phys. Rev. Lett., 75 (1995) 1206.
22. R.P. Rastogi and K.D.S. Yadava, Nature, London, 240 (1972) 19.
23. I. Das, V. Pandey, R.P. Rastogi and G.P. Mishra, Indian J. Chem., 39A (2000) 618.
24. K. Showalter, J. Phys. Chem, 85 (1981) 2001.
25. R.P. Rastogi, K.D.S. Yadava and K. Prasad, Indian J. Chem., 13 (1975) 352.
26. R.P. Rastogi, P. Rastogi and R.B. Rai, Indian J. Chem., 15A (1977) 295.
27. R.P. Rastogi, K. Singh, P. Rastogi and R.B. Rai, Indian J. Chem., 15A (1975) 295.
28. M. Orban, J. Am. Chem. Soc., 103 (1980) 4311.
29. K. Showalter, J. Chem. Phys., 73 (1980) 3735.
30. R.P. Rastogi and P.K. Srivastava, Indian J. Chem., 24A (1985) 999.
31. V. Castets, E. Dulos, J. Boissonade and P. De Kepper, Phys. Rev. Lett., 64 (1990) 2953.
32. A.M. Turing, Philos. Trans. Roy. Soc., London, B237 (1952) 137.
33. I. Lengyel and I.R. Epstein, Proc. Natl. Acad. Sci, USA, 89 (1992) 3977.
34. J.A. Pojman and I.R. Epstein, J. Phys. Chem., 94 (1990) 4966.
35. J.A. Pojman, I.R. Epstein T.J. McMamanus and K. Showalter J. Phys. Chem., 95 (1991) 1299.
36. J.A. Pojman, I.P. Nagy and I.R. Epstein, J. Phys. Chem., 95 (1991) 1306.
37. G. Bazsa and I.R. Epstein, J. Phys. Chem., 89 (1985) 3050.
38. I. Nagyapal, G. Bazsa and I.R. Epstein, J. Am. Chem. Soc., 108 (1986) 3635.
39. J.C. Michean, M. Grimenez, P. Borckman and G. Dewal, Nature, 305 (1983) 43.
40. Q. Qyang, V. Castets, J. Boissouade, J.C. Roux, P. De Kepper and H.L. Swinney, J. Chem. Phys., 95 (1991) 351.
41. S.P. Kuznetsov, E. Mosekilde, G. Dewel and P. Borckmans, J. Chem. Phys., 106 (1997) 7609.
42. P. Andersen, M. Bache, E. Mosekilde, G. Dewel and P. Borckmans, Phys. Rev. E, 60 (1999) 297.
43. M. Kaern and M. Menzinger, Phys. Rev. E, 60 (1999) 3471.
44. A.M. Zhabotinskii and A.B. Rovinsky, J. Phys. Chem., 94 (1990) 8001.
45. A.B. Rovinsky and M. Menzinger, Phys. Rev. Lett., 69 (1992) 1193.
46. A.B. Rovinsky and M. Menzinger, Phys. Rev. Lett., 70 (1993) 778.
47. S.C. Muller and T. Plesser, 'Chemical Waves and Patterns', R. Kapral and K. Showalter (eds), Kluwer, Dordrecht, 1995, Ch. 2.
48. S. Grill, V.S. Zykov and S.C. Muller, J. Phys. Chem., 100 (1996) 19082.
49. P.K. Maini, K.J. Painter and H.N. Phong Chau J. Chem. Soc., Faraday Trans., 93 (1997) 3601.
50. P. De Kepper, V. Castets, E. Dulos, and J. Boissonade, Physica D, 49 (1991) 161.
51. Q. Ouyang and H.L. Swinney, Nature (London), 352 (1991) 61.
52. G.H. Gunraine, Q. Ouyang and H.L. Swinney, Phys. Rev. E, 50 (1994) 2802.
53. M. Watzel and A.F. Munster, Chem. Phys. Lett., 242 (1995) 2802.
54. K. Lee, W.D. McCormc, Q. Ouyang and H.L. Swinney, Science, 261 (1993) 192.

55. P.W. Davies, P. Blanchedeau, E. Dulos, and P. De Kepper, J. Phys. Chem. A, 102 (1998) 8236.
56. A. Taylor, Advances in Complex Systems, Vol. 6, No. 1 (2003) p. 155.
57. J. Sielewiesluk and J. Gorecki, Phys. Chem. Chem. Phys., 4 (2002) 1326.
58. O. Steinbock, P. Kettumen and K. Showalter J. Phys. Chem., 100 (1996) 18970.
59. I. Das and A. Bajpai, J. Sci. Ind. Res., 60 (2001) 10.
60. I. Das, A. Pushkarna and R.S. Lall, J. Cryst. Growth, 82 (1987) 36.
61. I. Das, A. Pushkarna and S. Chand, J. Colloid Interface Sci., 130(1) (1989) 176.
62. S. Prager, J. Chem. Phys., 259 (1956) 279.
63. M. Flicker and J. Ross, J. Chem. Phys., 60(9) (11974) 3458.
64. N.R. Dhar and A.C. Chaterjee, Kolloid Z., 39 (1925) 2.
65. A.C. Chaterjee and M.C. Rastogi, J. Indian Chem. Soc., 28 (1951) 138.
66. E.C. Zimmermann, M. Shell and J. Ross, J. Chem. Phys., 81 (1984) 1327.
67. H.K. Henisch, Crystals in Gels and Liesegang Rings, Cambridge University Press, Cambridge, 1988.
68. E. Hatschek, Proc. Roy. Soc., 99(4) (1921) 496.
69. I. Das and A. Pushkarna, J. Non-equilib. Thermodyn., 13(3) (1988) 209.
70. I. Das, A. Pushkarna and N.R. Agarwal, J. Phys. Chem., 93 (1989) 7265.
71. I. Das, A. Pushkarna and A. Bhattacharjee, J. Phys. Chem., 94 (1990) 896.
72. A. Nitzan, P. Orteleva and J. Ross, J. Chem. Phys., 60 (1974) 3134.
73. V.W. Cochrane Physiology of Fungi, Wiley, London, 1958.
74. N. Kanniah, F.D. Gnanam and P. Ramasamy, J. Colloid and Interface Sci., 80 (1981) 377.

Part Three

COMPLEX NON-EQUILIBRIUM PHENOMENA FAR FROM EQUILIBRIUM

Chapter 11

DYNAMIC INSTABILITY AT INTERFACES

11.1. Introduction

Dynamic instability in complex systems [1–6] is of considerable current interest in view of its applications in the study of physico-chemical, biological, physiological, geophysical systems, dynamics of social and economic change including behavioural sciences. Dynamics of weather changes, population and ecology have also stimulated good deal of interest. Out of all these, most intensive and comprehensive studies have been undertaken for chemically reactive systems as elaborated in previous chapters.

Non-linear phenomena such as temporal oscillations and chemical waves in the case of chemical reactions are governed by the autocatalytic reaction (positive feedback) and reaction where the product of autocatalysis is destroyed by some other species (negative feedback). Of course in the case of chemical waves (spatio-temporal oscillations), diffusion does play a role, and the concept of reaction diffusion equation is evoked to predict the dependence of velocity of chemical waves on different parameters. In this chapter, we propose to discuss electrical potential oscillations generated due to coupling of volume flux, solute flux and electric current through solid–liquid interface (membrane systems), liquid–liquid interface, solid–liquid–liquid interface (density oscillator) and liquid–liquid–vapour interface.

Oscillatory transport phenomena under discussion are quite complex where departures from distribution, ionic and adsorption equilibrium occur. Role of Helmholtz electrical double layer becomes important in such cases.

Two types of cases generally occur in the above type of oscillatory phenomena:

(1) Where external stimulus is needed. In this category, we can have further two types:
   (i) Teorell type of oscillators, where volume flux is involved. In this case, hydrodynamic flux is opposed by electro-osmotic flux.
   (ii) where ion-flux is involved, which disturbs the diffusion potential $(\Delta\phi/\Delta\mu)_{I=0}$. Pant and Rosenberg [2] have reported sustained and coupled electrical and mechanical oscillations in BLM systems in the presence of inorganic ions. In their experiments, eclectic potential oscillations were obtained only on application of external stimulus.

(2) Self-excited oscillations have been reported to occur in L–B films of dioleyl lecithin (Y-films and Z-films). The films separated equimolar solutions of KCl and NaCl solutions, and also in polytetrafluoroethylene membrane doped with triolein and monolein [3, 4]. No satisfactory theory has been proposed so far to explain self-excited oscillations in such systems. However, it has been postulated that such a situation can occur on account of any one of the following factors [5]
   (i) when different electrolytes are taken on the two sides of the membrane
   (ii) change of permeability due to (i) periodic sol–gel transformation of the membrane and (ii) periodic openings of specific channels.

Self-excited oscillations are quite relevant in the context of biological oscillator. Typical self-excited oscillations observed in the case of (i) density oscillator, (ii) liquid–liquid interface oscillator and (iii) liquid–vapour interface oscillator, the mechanism of which is much clearer.

The above phenomena have a bearing on various physiological processes, e.g. membrane transport, oscillatory phenomena in cells and living organism, models for sense of taste and smell and flow system associated with cardiac rhythms.

11.2. Dynamic instability at solid–liquid interface

11.2.1. Electro-kinetic oscillators

Electro-kinetic phenomena have proved to be useful non-equilibrium phenomena as compared to other phenomena (e.g. thermo-osmotic pressure, thermal diffusion and Soret effect), since the former can be experimentally investigated in the entire non-equilibrium range, from linear to far from equilibrium by controlling the variables. The results can be analysed using paradigms of non-equilibrium thermodynamics and non-linear dynamics. Besides, such phenomena are also interesting from physiological angle in connection with the ion transport through nerve cell and rhythmic action potential [6, 7].

11.2.2. Experimental studies

The pioneering experiments in this area are due to Teorell [6]. In this setup, aqueous solutions of electrolytes of different concentrations were kept in the two compartments (NaCl, KCl, LiCl) separated by the membrane, and a current of fixed magnitude was passed. Synchronous oscillations in electric potentials, resistance and pressure differences were observed. Teorell observed that (i) there exists a "threshold value" of current density below which one obtains highly damped or moderately damped oscillations, (ii) in the undamped case the oscillations go on for hours and start to die away when current becomes too low or Ag/AgCl electrodes gets exhausted, (iii) oscillations are observed in narrow concentration range 0.1–0.09 N NaCl in one compartment and 0.012–0.008 N in the other compartment and (iv) results of experiments with KCl and NaCl and LiCl were similar. Meares and Page [8, 9] using a similar setup obtained similar results using a millipore filter membrane. In the experiments of Teorell, pyrex membrane was used, in which case relaxation oscillations were obtained.

The subject has been reviewed by Larter [7] and Rastogi et al[10, 11]:. An earlier review is due to Mikulecky and Caplan [12].

Recent experimental studies are due to Rastogi and co-workers [13, 14]. A schematic diagram of their experimental setup is shown in Fig. 11.1, which is similar to that of Teorell with some modifications.

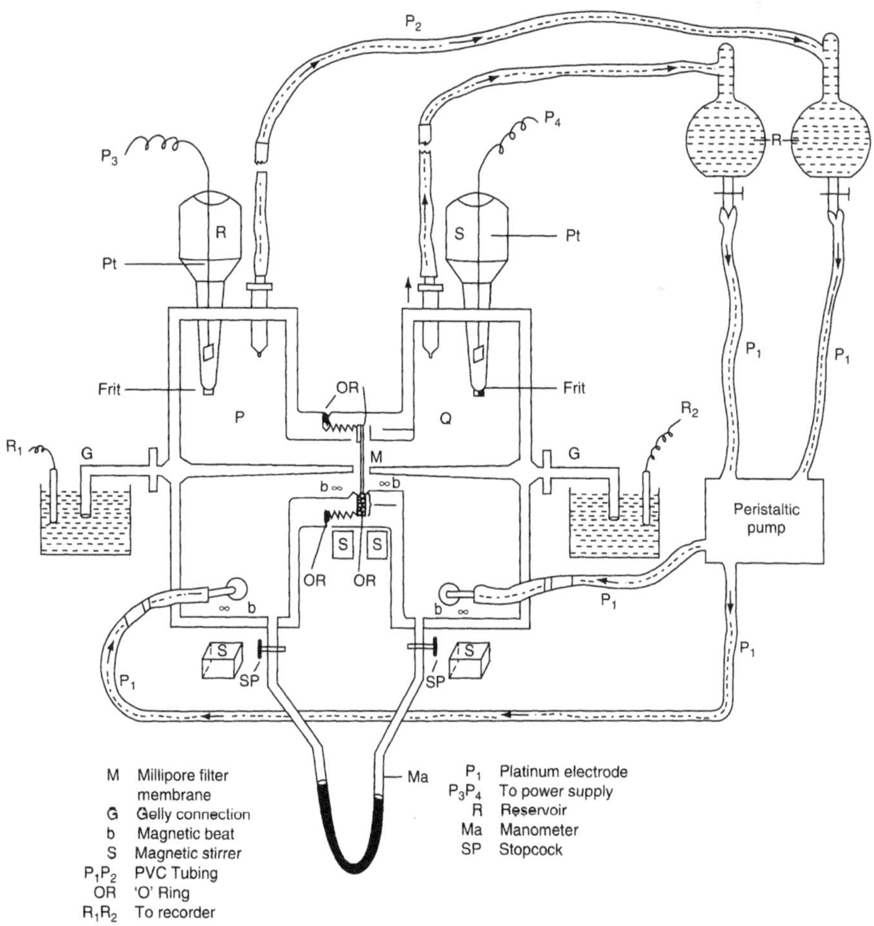

Figure 11.1. Experimental setup for monitoring oscillations in trans-membrane potential across the membrane.

A single capillary oscillator has also been used for oscillation in case of non-electrolytes [15]. Interesting results have been obtained with four types of membranes, viz. Pyrex, millipore filter, silver coated and inorganic membranes for 0.1 N NaCl/KCl||0.01 N NaCl/KCl systems. Relaxation oscillations are obtained for Pyrex membrane, whereas high frequency sinusoidal oscillations are observed for the other membranes. Bistability is also observed in some cases as discussed in earlier Chapter 8. Typical results on oscillatory features are recorded in Fig. 11.2.

The current is the bifurcation parameter as indicated by amplitude of oscillation versus current plot. Figure 11.3. The bifurcation point occurs at <1 mA for millipore filter.

Similar oscillation using a single capillary in place of membrane for the first time has also been reported recently using solutions of non-electrolytes [15]. Current is the bifurcation parameter as in the previous case. Multi-periodic oscillations are observed in case of solution of urea. In the above case, the electro-osmotic flow in one direction and hydrodynamic flow in the other direction are responsible for oscillations. Non-linearity is imposed by concentration difference across the membrane. This point will be discussed in greater depth in the subsequent section.

In a slightly different type of oscillator, the oscillations occur on account of balancing of streaming current and ohmic current flowing in opposite direction. Recent experimental studies on doped membranes conducted by Aisawa and Misawa [16] and Kim and Larter belong to this type of oscillators. Schematic diagram of their experimental setup [17] is described in Fig. 11.4. Same concentration of electrolyte KCI was kept on both sides of the membranes. Oscillations were observed on simultaneous application of pressure gradient (~ 30 mmHg) and current of magnitude 5×10^{-8}–1×10^{-6} A.

Similarly experiments reported by Arisawa and Furukawa [18] combine the features of both the above type of oscillators. They have reported that when a DOPH-doped filter is placed between salt solutions of different concentrations and a pressure gradient (between 10 and 50 mmHg) is applied, oscillations in membrane potential are observed. Here solute flow from one compartment to the other due to concentration difference across the membrane is opposed by the solute flow in hydrodynamic flux in addition to balancing of streaming current and ohmic current flowing in opposite direction.

A periodic and chaotic membrane oscillations have been observed in case of membrane doped by DOPH [19, 20]. After 30 min, a thick opaque di-oleyl phosphate (DOPH/oleyl alcohol/water (D/O/W) emulsion appears on the low-pressure surface of the membrane. The growth of this gel-like layer triggers a dramatic rise in the electrical resistance of the membrane and consequently oscillation in membrane potential. Chaotic behaviour appears as a part of periodic-chaotic sequence, also quite commonly associated with chaos. However fair amount of noise appears during membrane oscillations (Fig. 11.5) creating doubt regarding the existence of deterministic chaos. It has been shown that aperiodic behaviour in the system by way of a period-doubling sequence of bifurcation [19].

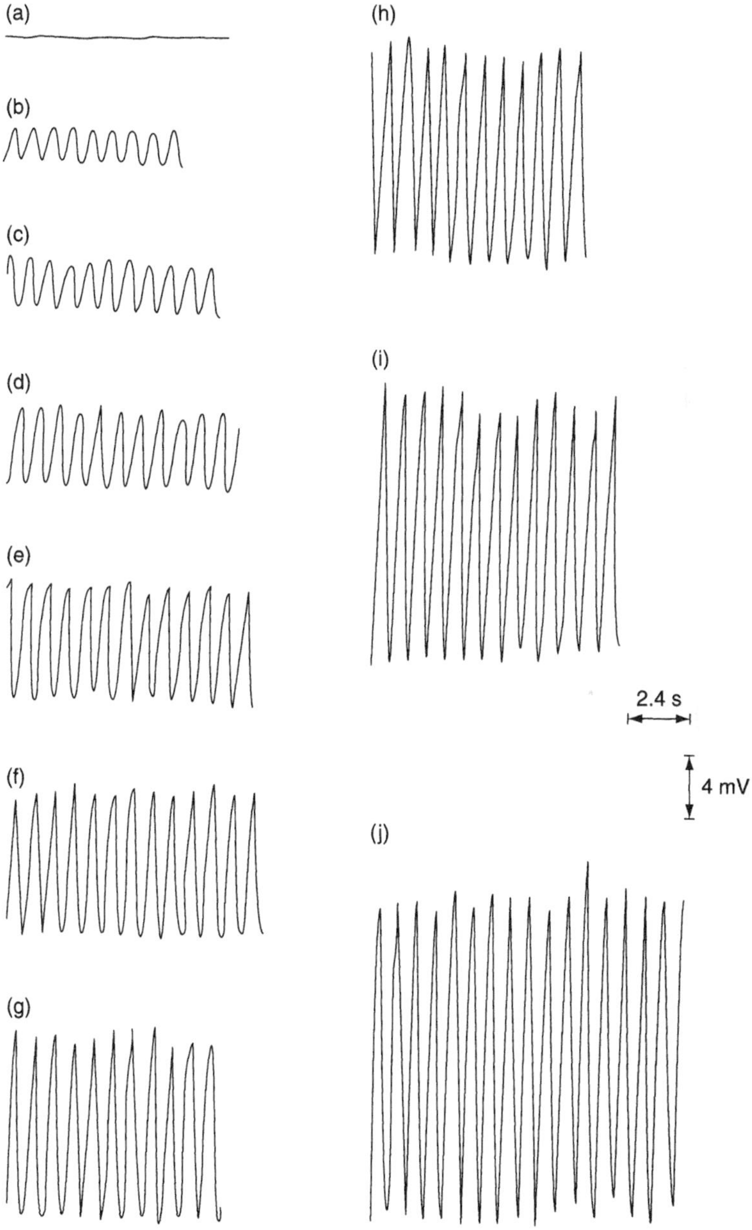

Figure 11.2. Typical electrical potential oscillations using millipore filter membrane (pore size 0.88 μm) in a system 0.1 M KCl/millipore filter membrane/0.01 M KCl. Current strength: (a) 0.0 mA, (b) 2.4 mA, (c) 3.4 mA, (d) 5.4 mA, (e) 7.3 mA, (f) 9.3 mA, (g) 11.4 mA, (h) 13.3 mA, (i) 16.2 mA, (j) 20.3 mA.

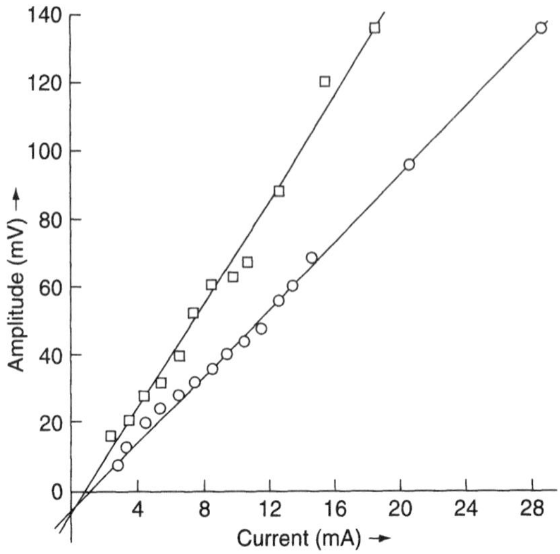

Figure 11.3. Dependence of amplitude on current for (a) Whatman filter membrane/Nacl system, pore size 14 μm [0], (b) KCl system pore size 14 μm [].

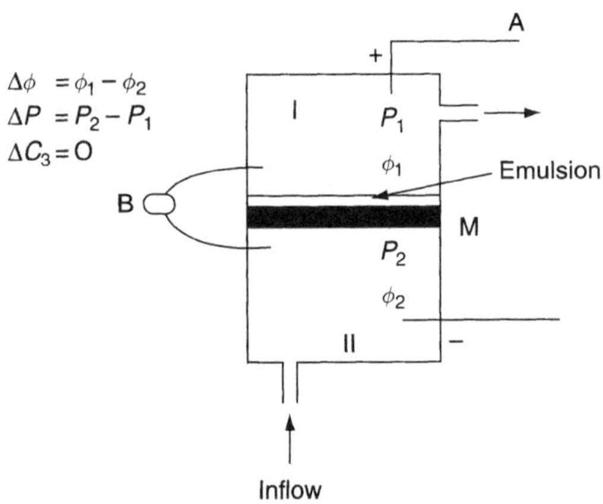

Figure 11.4. Schematic diagram of Kim and Larter's oscillator: (A) current supply, (B) voltage measuring device, (M) membrane, (I) chamber I, (II) chamber II.

Such oscillations in a lipid-alcohol-doped milipore filter for the existence of deterministic chaos can be studied using standard methods from non-linear dynamic theory, inspite of the fact that reliable Lyapunov exponent could be obtained from the data using low values of the evolution time. A possible source of the noise in the system could be spatial non-uniformities or erratic variations in thickness of the gel, which is not easy to study.

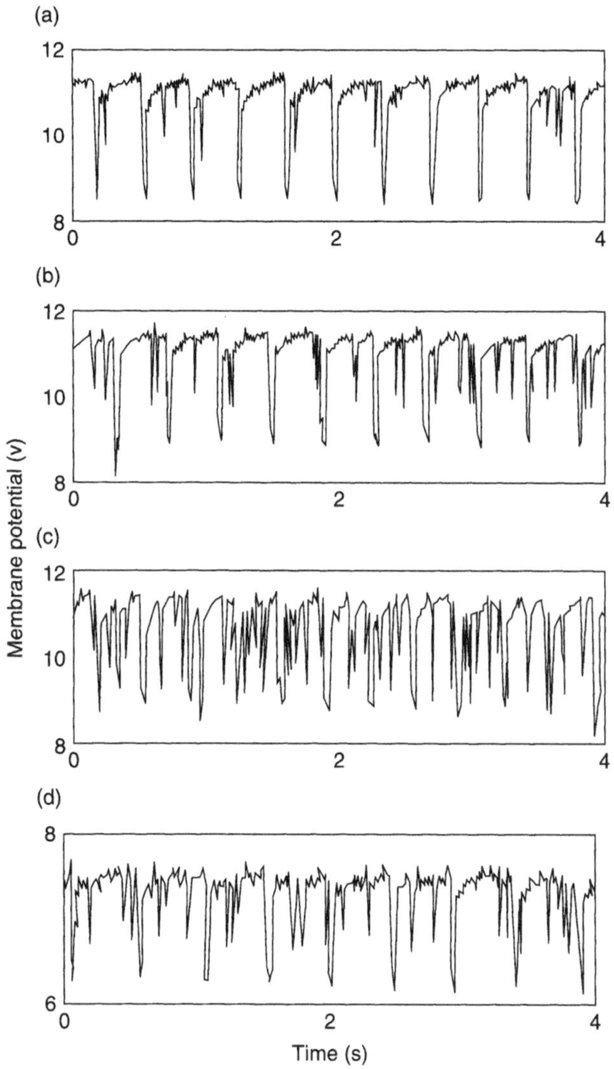

Figure 11.5. Time series of membrane potential: $I = 0.5\,\mu A$ and $\Delta P = 30$–40 mmHg. Different oscillatory states a, b, c and d were recorded as the gel expanded.

11.2.3. Mechanism and theoretical considerations

First attempt in formulation theory of electro-kinetic oscillations was reported by Teorell [6]. Alternative theories were first suggested by Kobatake and Fujita [20, 21] and subsequently by Mears and Page [8]. The former belong to the class of macroscopic theory, while the latter are concerned with the microscopic details of the phenomena.

The governing equation used by Kobatake and Fujita [21] is the simple equation for motion of fluids using a simple model which incidentally does not reproduce the conditions of Teorell's experiments. In this model, it is assumed that mass flow inside the membrane is slow enough and in a steady state at every instant. Furthermore, the density of the solution is assumed to be equal to that of water. The theoretical formulation ignores the electro-osmotic effect which is quite important. In addition, the theory cannot be quantitatively checked on account of various difficulties as pointed by Mikulecky and Caplan [12]. Similar problem arises with respect to the formalism developed by Arnow [22]. Mears and Page developed a theory [8] based on the following force-balance equation valid at any instant:

$$F_{\mathrm{i}} = (P_{\mathrm{e}} - F_{\eta}) + P_{\varepsilon} = 0 \tag{11.1}$$

where P_{ε} is the hydrostatic pressure difference, P_{e} the electro-osmotic pressure difference, F_{η} the viscous drag force and F_{i} the inertial force.

Using several assumptions, they obtained an equation of Van der Pol type. The theory is a differential theory in contrast with phenomenological theory of Teorell. But it is a pseudo-steady-state theory similar to that of Kobatake and Fujita.

The oscillatory phenomena can be visualized as follows on the basis of experimental observation under a variety of conditions. When current flows through dilute solution in compartment II to compartment I containing concentrated solution, electro-osmotic flow occurs in the same direction. Consequently, dilute solution flows in the same direction so that electro-osmotic pressure increases in chamber I. Both membrane resistance R and potential difference $\Delta\phi$ increase in this process. This flow is resisted by hydrodynamic flow due to pressure difference ΔP in the reverse direction. When this happens, concentrated solution enters from compartment II, thereby decreasing the membrane resistance. Normally, when concentration difference is zero, a steady state is attained, but due to finite ΔC, instability is caused and under appropriate conditions rhythmic changes in ΔP and $\Delta\phi$ are observed. However, when the concentration difference is of lower magnitude, either damped oscillation or no oscillation is observed.

Teorell started with phenomenological relation for volume flow J_{v} and I given by

$$J_{\mathrm{v}} = L_{11}\Delta P + L_{12}\Delta\phi \tag{11.2}$$

$$I = L_{22}\Delta\phi \tag{11.3}$$

where L_{ik} $(i, k = 1, 2)$ are phenomenological coefficients. However, the streaming potential term $L_{21}\Delta P$ term was ignored. It is not unjustified, since as reported by Rastogi and associates [13, 14] in case of pyrex membrane, it is of the order of 10^{-7} A which is fairly low as compared to ohmic current. In order to construct a two-variable time-dependent differential equation, Eq. (11.3) was replaced by an empirical equation:

$$\mathrm{d}R/\mathrm{d}t = -k(R - R_{\mathrm{v}}\ldots) \tag{11.4}$$

where k is a constant and R_{v} the resistance when flow $= J_{\mathrm{v}}$.

Furthermore, assuming that

$$R_v \ldots = r' J_v + R_o \ldots, \tag{11.5}$$

Eq. (11.4) can be rewritten as

$$\mathrm{d}R/\mathrm{d}t = -k(R - (r' J_v + R_o)) \tag{11.6}$$

where r' is a constant and R_o the initial resitance.

Using these assumptions, it has been shown [6, 13] that van der Pol type equation can be deduced. Theoretical formulations of Teorell as well as that of Kobatake and Fujita have been compared, and it has been shown [13–15] that both involve certain assumptions but both yield van der Pol type equation. It also turns out that there is no relationship between bistability and oscillations. The validity of the two-variable model has been further examined by comparing the experimental results with computer simulation based on the use of experimentally determined parameters [14], which did not yield satisfactory results.

Further modifications in the theory have been made, which also lead to van der Pol type equation. Computer simulation yields satisfactory results [23]. In the case of coupling of the two forces J_v and I, two types of steady states can occur; when $J_v = 0$ or when $I = 0$. In the case of Teorell oscillator as depicted in Fig. 11.1, the steady state corresponds to the first situation. In order to keep the system far from equilibrium and have oscillations, ΔC has to be sufficiently large, of the order of 0.1 M. The magnitude of streaming current is quite low as compared to ohmic current. Since

$$J_v = q\,\mathrm{d}(\Delta P)/\mathrm{d}t \quad \text{and} \quad IR = \Delta\phi \tag{11.7}$$

we can write the equation for J_v, as

$$J_v = q\,\mathrm{d}(\Delta P)/\mathrm{d}t = -L_{11}\Delta P + L_{12}IR \tag{11.8}$$

where $q = n\pi r^2/\rho g$, n is the number of capillaries, r the radius of the capillaries, ρ the density, g the acceleration due to gravity and R the resistance of the membrane. Instead of expressing $\mathrm{d}R/\mathrm{d}t$ as a function of J_v, a vectorial quantity as suggested by Teorell, it has been suggested [23] that it is more appropriate to use the relation

$$\mathrm{d}R/\mathrm{d}t = -k(\Delta P) \tag{11.9}$$

since R would depend on the inflow and outflow of the fluid in the membrane which would affect the concentration and hence the resistance of the membrane. This is supported by the earlier experimental work that oscillations in electrical potential and pressure difference are synchronous [6, 8, 9, 13, 14]. Differentiating Eq. (11.8) with respect to time t and making use of Eqs. (11.7) and (11.9), we get

$$q\,\mathrm{d}^2(\Delta P)/\mathrm{d}t + L_{11}\mathrm{d}(\Delta P)/\mathrm{d}t + L_{12}Ik\;\Delta P = 0 \tag{11.10}$$

The above equation corresponds to the van der Pol type of equation and predicts oscillation. Using the above equations and the experimental data [23], computer simulation predicts synchronous oscillations in both ΔP and R.

11.2.4. Streaming current-driven oscillations at fixed pressure difference

In the former case, we had considered electro-osmotic-driven oscillations at fixed value of current while ΔP was allowed to vary, on account of which membrane resistance could vary. An interesting case of electro-kinetic oscillations also occurs when ΔP is fixed, and the resistance varies due to progressive thickness of the emulsion on the membrane surface and the oscillations occur around the steady state corresponding to $I = 0$. Such electric potential oscillations, in a number of cases using DOPH-doped membranes have been reported [16, 19]. Chaotic oscillations in membrane oscillator have been reported and analysed from the angle of non-linear dynamics.

Experimental studies

The experimental studies reported by Kim and Larter [17, 19] are well defined where formation of emulsion has been reported. Schematic diagram of their experimental setup is given in Fig. 11.4. A DOPH membrane, along with dioleyl alcohol, was used as membrane, where DOPH is $[CH_2\{(CH)_2\}7CH = CH\{(CH)_2\}_7]_2 \cdot POH$ and oleyl alcohol is $[CH_2\{(CH_2)_7\} \cdot CH = CH\{(CH_2)_7\}_2] \cdot OH$. The special features of the oscillator are the following:

(a) Concentration difference across the membrane is zero.
(b) The applied pressure difference ΔP (~30–40 mmHg) is kept constant.
(c) The typical range of applied constant current was ~0.05–0.01 μA.
(d) Formation of D/O/W emulsion was found necessary to generate oscillations.
(e) Current and pressure differences act as bifurcation parameters.
(f) Transition from periodic oscillations to chaos is observed with increasing current stimulus.

Mechanistic and theoretical studies

Attempts have been made to elucidate the mechanism of generation of such oscillations. Kim and Larter suggested a mechanism based on (i) balancing of hydrodynamic and electro-osmotic flux and (b) influence of these fluxes on the anion concentration gradient, which may influence the microscopic structure of the emulsion. However, the above tentative explanation is not tenable since hydrodynamic flux will have a fixed value on account of fixed pressure difference across the membrane. Another complication is that on account of fixed ΔP, streaming current will have a fixed value along with ohmic current. A workable mechanism has recently been suggested [23], by considering the oscillator as an L–C–R type of circuit. If emulsion can be considered as coil (self-inductance) having inductance L and membrane channels can be considered as a capacitor C on account of Helmholtz double layer, back emf can be produced by change in the self-inductance itself, so that Lenz's law is obeyed where the inductance L of the coil is constant. $\Delta\phi$ is the applied voltage corresponding to imposed current I; we can write for the oscillator as L-C-R circuit [24]:

$$L\,\mathrm{d}I/\mathrm{d}t + Q/C + RI = 0 \tag{11.11}$$

$$L\,\mathrm{d}^2Q/\mathrm{d}t^2 + R\,\mathrm{d}Q/\mathrm{d}t + Q/C = 0 \tag{11.12}$$

since according to Lenz's law,

$$\text{potential difference across inductance} = L\mathrm{d}t/\mathrm{d}t.$$

Furthermore, potential difference due to streaming current would be equal to RI, while potential difference across the capacitor would be equal to Q/C where Q is the charge and $C = \varepsilon A/d$, where ε is the permittivity, A is the surface area and d is the distance between parallel plates.

Capacitance would be constant in case of the present oscillator since pressure difference and concentration difference of the electrolyte remain constant during oscillation. Using the above relationships Eq. (11.12) can be transformed as follows:

$$C\,L\,\mathrm{d}^2(\Delta\phi)/\mathrm{d}t^2 + CR\,\mathrm{d}(\Delta\phi)/\mathrm{d}t + \Delta\phi = 0 \tag{11.13}$$

The above equation is of van der Pol type which predicts oscillations under certain conditions. Erratic and abnormal oscillations as observed [17, 19] are expected on the basis of the above model if the emulsion were destabilized due to either low or high pressure difference. A similar type of instability may also arise when the current is too high or too low. These types of instabilities belong to system instabilities [25].

11.3. Dynamic instability at liquid–liquid interface along with solid–liquid interface

11.3.1. Introduction

Instability at the liquid–liquid interface can also give rise to exotic phenomena. We will discuss in the following sections the generation of electric potential oscillations due to such instabilities at the interface (i) between nitrobenzene containing picric acid and water containing CTAB and (ii) between liquids, e.g. aqueous solutions of electrolytes or solutions of non-electrolytes of higher density and lower density.

The initial important experiments in this area are due to Yoshikawa and Matsubara [26]. The interface involves the upper layer containing 2-nitropropane/1-nitropropane and picric acid while the lower layer contains CTAB (hexadecyl trimethyl ammonium bromide). Oscillatory variations of pH in the aqueous phase and electric potential oscillations between organic and aqueous phases were observed.

Yoshikawa *et al.* suggested a mathematical formalism by taking into account (i) reactions in both phases and (ii) different equations for the ascending and descending portions of oscillations. The mechanism is complicated and too involved. An alternative simpler mechanism has recently been suggested [27]. The essential features of the suggested mechanism are the following:

(a) Initially picric acid (HP) would distribute between the two phases at a certain rate.
(b) Subsequently, the picrate ion (P^-) would exchange with Br^- of CTAB and the ion pair would be formed and would accumulate at the interface.
(c) Ion pairs would be transferred to the non-aqueous phase.
(d) Reverse micelles would be formed in the non-aqueous phase due to the lipophilic character of CTA^+P^- when a critical concentration is reached. The time involved would depend on the series of reactions [28]:

$$\text{Monomer} \rightarrow \text{dimmer} \rightarrow \text{trimer} \rightarrow \text{------} \rightarrow \gamma - \text{mer}$$

It may be noted that due to reduction of P^- in the aqueous medium, more HP will be transferred to the aqueous phase from the non-aqueous phase. The dynamics at the interface would be quite complicated due to fusion-scission kinetics of CTAB and CTAP and the possibility of formation of liposomes and non-spherical lamellar micelles. The sequence of events involved in the mechanism has already been discussed by Rastogi *et al.* [27].

CAT^+ would vary periodically in both the phases along with the ratio $r = [CAT^+]\text{org}/[CAT^+]\text{aq}$, which would be responsible for electric potential oscillations. The formation of reverse micelles in the non-aqueous phase would be responsible for the decrease in concentration of $[CAT^+]_{org}$ in the non-aqueous phase. On the other hand, it would be compensated by the transfer from the aqueous phase. But this would require more H^+ in the aqueous phase to interact with CTAB to generate CTA^+. The required amount of H^+ would be produced by transfer of HP from the non-aqueous phase and the cycle would be repeated over and over again, giving rise to oscillations in electric potential and PH. Using relevant kinetic equations [27] and linear stability analyses, mathematical formalism has been developed.

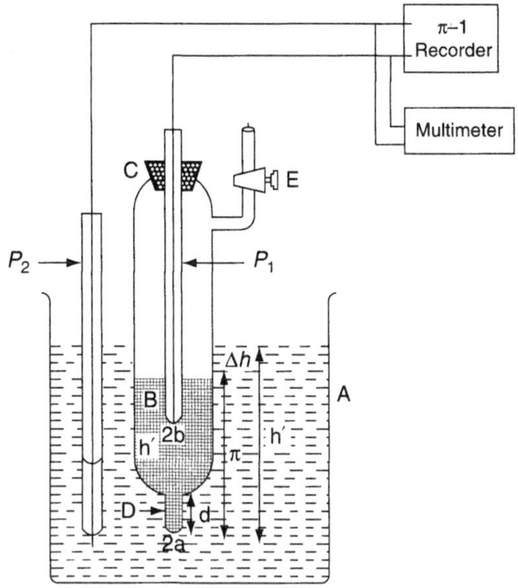

Figure 11.6. Schematic diagram of the experimental setup. A, outer vessel; B, inner vessel; C, glass joint; D, capillary; E, stop cock.

11.3.2. Density oscillator

Salt-water oscillator which is also called density oscillator, maintained quite far away from equilibrium, is another example of oscillatory transport phenomena. A typical setup for observing oscillations in such system is illustrated in Fig. 11.6.

Outer vessel contains liquid of lower density while the inner vessel contains liquid of higher density. The system is intrinsically unstable and is naturally maintained far away from equilibrium. Up and down flow of liquid occurs through the capillary, which is reflected by the oscillatory movement of the fluid in the inner vessel. When the electrodes in the two chambers are connected to a voltage-measuring device, oscillations in electric potentials are also observed for the cases when aqueous solutions of electrolyte–water system or aqueous solution of polar non-electrolytes are used in the system. This type of hydrodynamic instability is different from Benard instability or Taylor instability [29].

Experimental studies

The phenomenon was discovered by Martin [30] in 1970, who observed rhythmic oscillations of water flow generated in a vertical hypodermic syringe (plunger removed) which is filled with salt water and submerged in a beaker containing pure water. Allfredson and Lagersted [31] used a similar setup in which a small hole connected the two fluids through which these could be interchanged with distinct pulses. These were experimentally measured, and attempts were made to identify the relevant parameters governing the pulse time using dimensional analyses. The analysis showed that the pulse time depends on density difference, viscosity, gravitational force and the diameter of discharge orifice. Furthermore, pulse time was found to be insensitive to water depth. Phenomenon is apparently related to the classical problem of hydrodynamics concerned with the stability of the interface between the two fluids of different densities.

These developments were followed by the discovery of salt-water oscillator by Yoshikawa and co-workers, who emphasized the non-linear dynamical character of the oscillator which exhibits various non-linear phenomena such as limit cycle, bifurcation and entrainment [32, 33].

Typical results for H_2O–D_2O oscillators are given in Fig. 11.7. Oscillatory features of some typical systems are recorded in Table 11.1.

Typical results for (a) urea–water, (b) glucose–water and (c) sucrose–water system are given in Fig. 11.8.

The above table shows that there are two types of systems. In one type (non-polar–non-polar system), only oscillatory flow occurs, while in the other type involving aqueous solutions of electrolytes or polar solvents, both oscillatory flow and electric potential oscillations are observed.

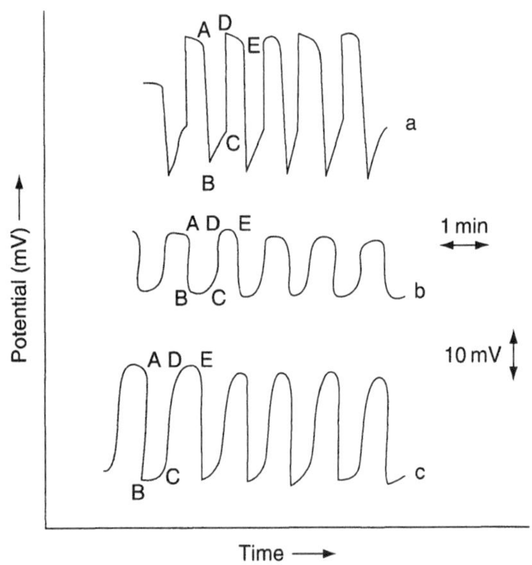

Figure 11.7. Electrical potential oscillations using D_2O–H_2O in hydrodynamic oscillator D_2O in the inner tube (a) 5 mole dm^{-3}, (b) 4.4 mole dm^{-3}, (c) 2.8 mole dm^{-3}; A, B, C represent downward flow and D, E upward flow (R.P. Rastogi, G. Srinivas, H.P. Maity and S. Kumar, Ind. J. Chem., 40A (2001) 119, Fig. 4.

Table 11.1. Oscillatory features in some typical systems.

System potential	Whether oscillatory flow observed	Whether electric oscillation observed
	Non-polar/non-polar	
Benzene–toluene (density difference Not adequate	No	No
CS_2/hexane	Yes	No
CS_2/octane	Yes	No
	Polar/polar non-electrolyte	
D_2O/H_2O	Yes	Yes
Glycerol/water	Yes	Yes
Formamide/Water	Yes	Yes
Ethylene glycol/water	Yes	Yes
Water/acetonitrile	Yes	Yes
Water/acetone	Yes	Yes
Water/demethyl formamide	Yes	Yes
	Aqueous solution of non-electrolyte/water	
Urea–water	Yes	Yes
Glucose–water	Yes	Yes
Sucrose–water	Yes	Yes
	Aqueous solution of electrolyte/water	
Nacl	Yes	Yes
KCl	Yes	Yes
$BaCl_2$	Yes	Yes
$AlCl_3$	Yes	Yes
H_2SO_4	Yes	Yes

For qualitative understanding of the origin of oscillatory fluid flow, one has to examine the role of (i) density gradient, (ii) osmotic pressure and (iii) interfacial tension, if any.

It may be noted that for non-polar CS_2/hexane system [34], the development of osmotic pressure is out of question. Even for aqueous solutions of electrolyte–water system, experiments [35] with a U-tube system rule out the possibility of generation of oscillations due to osmotic pressure. It has also been reported [35] that experiments with aqueous solutions of electrolyte/aqueous solution of surfactant cetyl pyridinium chloride system (the latter kept in the outer vessel) and the system without surfactant show that the interfacial effects are also not significant since the amplitude of oscillations is unaffected.

Mechanism and theoretical considerations

On the basis of experimental studies it turns out that viscosity and density difference of the two fluids play a major role in controlling oscillations. Following simple picture

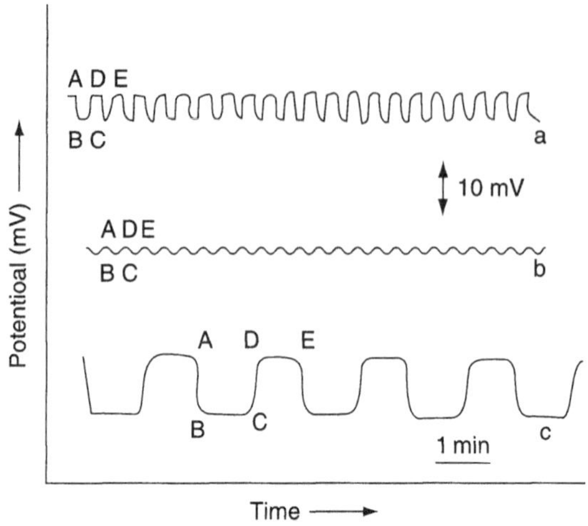

Figure 11.8. Electric potential oscillations (P_t–P_t) for system [(a) 3.0 mole dm^{-3} urea–water; (b) 1.0 mole dm^{-3} glucose–water; (c) 1.0 mole dm^{-3} sucrose–water; A, B, C represent downward flow and C, D, E represent upward flow; temperrature = 30 ± 1°C].

can be envisaged. When the density of the fluid in the inner vessel is higher than that in the outer vessel, the fluid from the inner vessel flows downwards on account of hydrostatic pressure difference. But the flow is resisted by viscosity and buoyancy. The latter would be more significant when the density difference is large. After a certain stage, the downward flow stops, and then upward flow of the fluid from the outer vessel to the inner one occurs, which is driven by buoyancy. The upward flow also stops after sometime and downward flow starts again and the cycle repeats again and again. Based on this picture, the oscillatory flow can be represented by the following force balance equation which is valid at any instant:

Force due to acceleration = viscous force + force due to pressure gradient + buoyancy

Mathematical formalism has been developed using semi-empirical considerations [36, 37]. Computer simulation studies show that resulting equation predicts oscillations. Attempt has been made to provide justification on the bases of Navier–Stokes equation but it is open to question. Dimensional analysis has recently been employed for investigating the phenomena [31]. Flow dynamics and stability in a density oscillator have been examined by Steinbock and co-workers [38]. They have related it to Rayleigh–Taylor instability of two different dense viscous liquids. A theoretical description has been presented which is based on a one-fluid model and a steady state approximation for a two-dimensional flow using Navier–Stokes equation. However, the treatment is quite complex and cannot explain the generation of electric potential oscillations.

A simple mechanism has been proposed by Rastogi *et al.* [27], which provides acceptable explanation of the origin of simultaneous oscillatory flow and electric potential oscillations. Salient features of experimental observation are summarized below [10, 23].

(a) Electric potential builds up in the direction of upward/downward flow. It may be noted that the electrode in the inner chamber is connected to terminals of recorder as well as digital voltmeter. Positive charges move in the direction of the flow.
(b) The time period depends on the density difference, gravitational force and diameter or discharge orifice.
(c) Oscillations are observed in the case of KCl, NH_4Cl also indicating that no diffusion potential is involved.
(d) Oscillatory flow and electric potential oscillations are synchronous.
(e) The fluid flow gives rise to streaming current and subsequently streaming potential.
(f) Oscillations occur in a definite concentration range of the inner liquid. In the case of glycerol–water system, oscillations occur in the range 0.68–13.6 mol dm^{-3} for glucose.

Let us consider the schematic diagram of the experimental setup (Fig. 11.6). Let P_1 denote the pressure in the inner chamber and P_2 denote the pressure in the outer chamber containing a liquid of lower density. Fluid flow across the capillary is mainly responsible for generating oscillations. The upward flow is opposed by the viscosity of heavier fluid in the inner vessel, while the downward flow is opposed by buoyancy. The buoyancy depends on the density difference $(P_1 - P_2)$ where P_1 and P_2 are the densities of heavier liquid and lighter liquid, respectively. This is so since the magnitude of the buoyancy force would depend on the difference of vertical components of fluid force on the upper and lower sides of the capillary. This is turn would depend on the pressure difference $\Delta P = P_1 - P_2$. The density of the denser fluid in the capillary due to partial mixing would depend on the pressure difference.

Having identified the two fluxes corresponding to the flux, we can write the linear phenomenological relation as follows:

$$\mathrm{d}(\Delta P)/\mathrm{d}t = -(L_{11}/q)\Delta P + B \tag{11.14}$$

where L_{11} and L_{12} are the phenomenological coefficients and the buoyancy force is given by

$$B = \pi a^2 lg(P_1 - P_2)$$

where a and l are the radius and length of the capillary and g is the acceleration due to gravity.

Further, L_{11} and q are given by

$$L_{11} = \pi a^4/8\eta l, \quad q = \rho lg/\pi b^2$$

where η is the viscosity and b is the radius of the inner vessel. Now, putting $L_{11}/q = kl$, $L_{12}\ \pi a^2 gl = k_2$ and

$$d(P_1 - P_2)/dt = -k_3(\Delta P)$$

since density difference would depend on the pressure difference where k_3 is a constant.

On differentiating Eq. (11.14) with respect to time and after making appropriate substitutions, we get

$$d^2(\Delta P)/dt^2 + k_1 d(\Delta P)/dt + k_2 k_3\ \Delta P = 0 \qquad (11.15)$$

which is a Van der Pol type of equation which predicts oscillations, provided

$$(k_1)^2 < 4k_2 k_3$$

The frequency is given by

$$\omega = \tfrac{1}{2}[-k_1 \pm \sqrt{\{(k_1)_2 - 4k_2 k_3\}} \qquad (11.16)$$

Equation (11.16) is qualitatively supported by experimental data. When the density difference between the two fluids is very small, the magnitude of k_1 would be much smaller and there may be a situation when the inequality may not be satisfied and the oscillations may not occur; this is actually observed experimentally.

The electric potential oscillations wold occur on account of streaming potential. Noting that the surface of glass capillary would be negatively charged, the positive charge in the electric double layer would be carried on in the direction of the flow. The streaming current would be generated during the flow and the corresponding streaming potential would develop. The electric (streaming) potential would be given by [39]

$$\Delta\phi = (\varepsilon\xi/4\pi\eta\chi)\Delta P = K\Delta P \qquad (11.17)$$

where ε is the dielectric constant, ζ the zeta potential, η the viscosity and χ the conductance.

On substitution, one will get the corresponding van der Pol equation for electric potential, assuming that the term K in Eq. (11.17) would be approximately constant, i.e.

$$d^2(\Delta\phi)/dt^2 + k_1 d(\Delta\phi)/dt + k_2 k_3 \Delta\phi = 0 \qquad (11.18)$$

Qualitative support for the above model is provided by testing the dependence of k_1 on a, the radius and length of the capillary, noting that frequency is proportional to k_1, which in turn is proportional to l and inversely proportional to a^4. The experimental results as reported earlier [10, 27] are in conformity with theoretical predication as indicated by Figs. (11.9) and (11.10).

When the time period is plotted against the length of the capillary, positive slope is obtained. Furthermore, when time period is plotted against log d, straight lines with a negative slope are obtained where $d = 2a$.

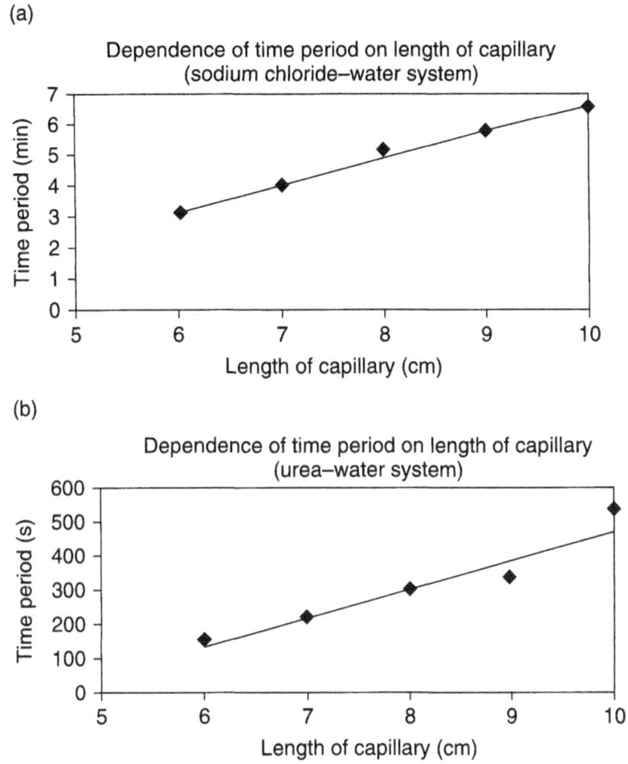

Figure 11.9. Dependence of time period on length of capillary l; (a) NaCl/water system, $a = 1$ mm, [NaCl] = 1 m; (b) urea/water system, $a = 1$ mm [urea] = 3 mm.

Application to biological phenomena – taste sensing mechanism

The concept of density oscillator, based on the development of electric potential due to density difference of two fluids, has been utilized by Srivastava and co-workers to explain the sensing mechanism of sensing of taste [40–43]. It has been shown through laboratory experiments that the potentials are developed when solutions of sweet chemicals/water system are investigated. On plotting the experimental results against taste index, straight lines are obtained in Fig. 11.11. Similarly, when electric potential for salt-water system is plotted against taste index, straightline is also obtained. Thus, there is a clear co-relation between the two, although it is not easy to justify it theoretically. On the basis of the theory developed in the earlier section, it can be argued that zeta potentials developed on account of different chemical species in the taste buds may be related to taste index. Taste buds have receptor cells. Taste buds have diameter ~1/30 mm and length ~1/16 mm.

Insoluble substances have no taste. Sour taste is caused by the acids. Salty tastes are caused by ionizable salts. Cations of the salts are mainly responsible. Most of the

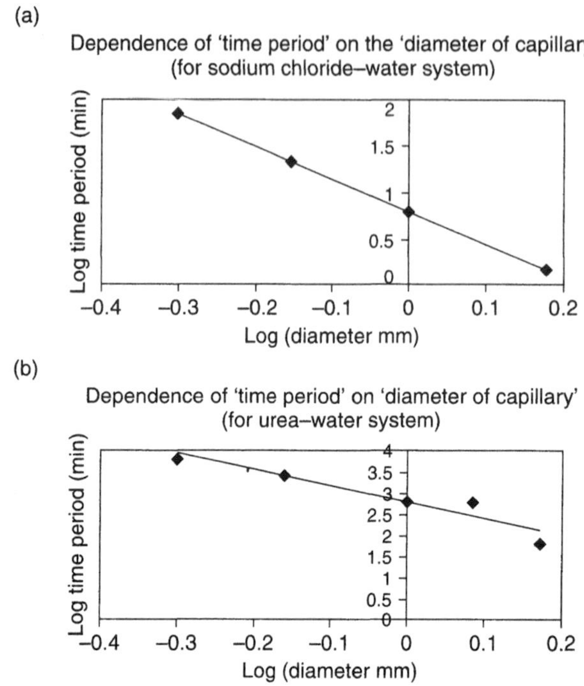

Figure 11.10. Plot of log of time period electric potential oscillations against log a: (a) NaCl/water system, $l = 10\,\mathrm{cm}$, [NaCl] = 1 M; (b) urea/water system, $l = 10\,\mathrm{cm}$, [Urea] = 1 M.

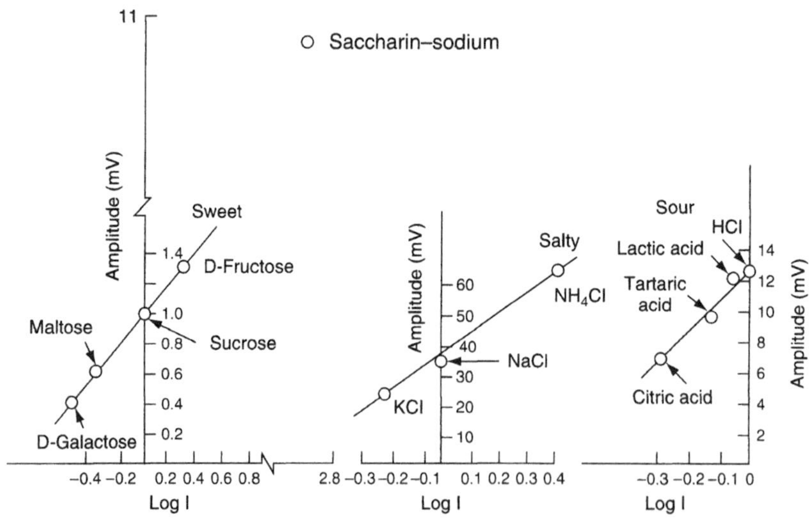

Figure 11.11. Variation of amplitude of electric potential oscillation using aqueous solution of same concentration (1.0 m) of the substances belonging to different taste categories with the logrithm of their relative state indices.

substances that cause a sweet taste are organic chemicals. Bitter taste is also caused by organic chemicals as (i) long-chain organic substances containing nitrogen and (ii) alkaloids.

11.4. Dynamic instability at liquid–vapour interface

11.4.1. Introduction

Interfacial instability is a common phenomenon since in most cases the events occurring at the interface are in the region far from equilibrium [40, 41]. A typical example is Marangoni instability [44–47]. Electric potential oscillations have been observed in biphasic systems [48, 49]. Self-induced oscillations in similar systems have also been observed [10]. Theoretical efforts have also been made to understand the mechanism in some cases [23, 27].

There is considerable interest in the above phenomena from the angle of electrophysiology. The neuronal action involving sense of smell depends on the generation of self-excited electric potential [4, 50, 51]. Hence, a deeper study from fundamental angle is likely to be of considerable use for medicinal or physiological purposes.

11.4.2. Mechanism of Yoshikawa oscillator

Recent studies on electrical potential oscillations across a liquid membrane consisting of an oil layer, 90% oleic acid and 10% l-propanol, containing tetraphenyl phosphonium chloride (TPPC), between aqueous solutions of 0.5 M NaCl and KCl in tow different compartments have been reported by Yoshikawa and Matsubara [52]. TPPC is a cationic lipophilic salt, which can act as a surfactant. On exposure to amine vapour, initially upward deflection occurs, indicating that the KCl solution becomes negative with respect to NaCl solution. Electric potential oscillations did not occur when the concentration of ammonia was below a critical value.

However, their studies are limited to qualitative study of oscillations using different types of amines, only indicating that the phenomenon is general.

A schematic diagram of their experimental setup is shown in Fig. 11.12. The phenomenon is interesting from the viewpoint of olfactory transduction.

A simple mechanism can be postulated as follows. The system contains two interfaces: oil/air (NH_3) and oil/aqueous phases. Along the first interface, ammonia interacts with oleic acid to form ammonium ion which reacts with Cl^- of TPPC and gets transferred to the aqueous phase. NH_4^+ ions combine with Cl^- to produce NH_4Cl which takes place earlier in the compartment having KCl due to higher diffusion coefficient [53] of KCl {1.466} in comparison with NaCl (1.7). Hence the KCl compartment is negative. It has been suggested that concentration difference between sodium and potassium ions is essential for the excitability of biological membranes.

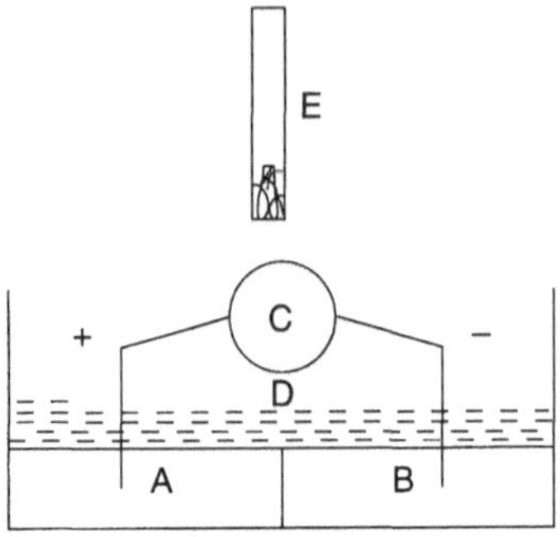

Figure 11.12. Schematic diagram of experimental setup of Yoshikowa oscillator. A = 0.5 M NaCl; B = 0.5 KCl; C = millivoltmeter; D = liquiid membrane (4.5 mm); E = cotton-soaked solution of olfactory agent in a tube.

The electrical potential difference would arise on account of difference in (Cl^-) in the two compartments, which would be controlled by the influx of NH_3 in the form of $NH_4{}^+$ ion and subsequent interaction at the aqueous/liquid membrane interface particularly in compartment B. Hence, there is a larger concentration of Cl^- in compartment A as compared to that in compartment B. Consequently, compartment B (containing KCl) becomes negative while compartment A becomes positive. The positive current one Faraday of electricity will move form left to right which would result in the transfer of t_+g-ion of cations from left to right (A->B) and transfer of Cl^- in the opposite direction. Noting that there are different electrolytes in different compartments, the liquid-junction potential or diffusion potential $\Delta\phi$ would be given by

$$\Delta\phi = (RT/F) \text{ In } \Lambda_2/\Lambda_1 \tag{11.19}$$

where R = gas constant, F = Faraday; Λ_2 and Λ_1 are the conductance of NaCl and KCl solutions, repectively [39]. The ratio of the conductance would be proportional to C_1/C_2 where C_1 and C_2 are the concentrations of Cl^- prevailing in the two compartments A and B, respectively.

From the above considerations, it is clear that $\Delta C = C_1 - C_2$ and $[NH_4^+]$ in compartment B would vary with time in the following manner:

$$d(\Delta C)/dt = -k_1\, \Delta C + k_2[NH_4^+] \tag{11.20}$$

The first term on the left-hand side occurs on account of diffusion of Cl^- from compartment A to compartment B and k_1 is a constant, which would be related to diffusion coefficient and surface area. The second term is related to interaction of Cl^- and NH_4^+ ion, k_2 is a constant which depends on the extent of NH_3 (amine) adsorption. It may be noted that Cl^- would be depleted with increase in (NH_4^+) and consequently ΔC would increase.

Since $[NH_4^+]$ would decrease when ΔC decreases and its increase would be related to influx from organic phase/aqueous solution interface, we can write

$$d[NH_4^+]/dt = -k_3\,\Delta C + k_4[NH_4^+] \tag{11.21}$$

where k_3 and k_4 are constants; k_4 will depend on ammonia/amine adsorption. Using normal mode analysis, Rastogi and co-workers [27] have shown that oscillations would occur when

$$4(k_2k_3 - k_1k_2) > (k_4 - k_1)^2$$

The mechanism and kinetics of adsorption in this case is quite complex in view of the fact that the two interfaces, viz. vapour/organic liquid and organic liquid/aqueous solution interfaces are involved. Kinetics of adsorption in such types of complicated cases have been investigated recently involving kinetics at fluid–fluid interface, non-ionic surfactant solution and surfactant mixtures [54, 55]. Taking a simplistic view, the extent of adsorption would depend on partial pressure of ammonia, which would depend on the concentration of NH_3/amine in the vapour phase and also on the distance and the cotton-soaked solution in the glass tube from the interface.

The amplitude would depend on $\Delta\phi$. The larger the ratio of the concentration difference in the two compartments, the larger would be the magnitude of amplitude. Greater diffusivity of KCl would decide the polarity of the particular compartment as indicated earlier. Consequently, the greater the amine concentration, the greater would be the value of amplitude. In the same way, the nearer the cotton plug, the greater would be the amplitude.

Similar types of experiments using similar experimental setup have been reported Srivastava and co-workers [49], using a membrane which is bipolar in nature. A cationic surfactant was used in compartment A while an anionic surfactant was used in compartment B. In the oil-phase, 2,2′-bipyridine was added to reduce the impedance for diminishing the external noise. Oscillations are observed on exposure to amines and pheromones (*cis*-8-dodecenyl acetate and *trans*-10-dodecenyl acetate). Similar type of mechanism as discussed earlier holds good for Srivastava's oscillator also.

The observed oscillations in both types of oscillators appear to be non-periodic, sometimes chaotic and noisy. This is probably due to random adsorption of amine and pheromone vapours. However, it has been reported that periodicity increases in case of methylamine and pheromones.

Thus instability in the above oscillators occurs on account of adsorption at the vapour–liquid interface, followed by transport of ammonia across in the form of ammonium ion at the oil–water interface. Diffusion of K+/Na+ from the bulk of the aqueous phase to interface influences the relative $[Cl^-]$ in the two chambers which ultimately generated oscillation.

Although the experimental results are in qualitative agreement [27] with the above simplified proposed mechanism, more detailed analysis is needed to account for the adsorption–desorption kinetics since the phenomenon is occurring far from equilibrium.

11.4.3. Applications – sensing mechanism of smell

The vapor–liquid oscillator described above provided an example of self-excited oscillator, which has relevance to biology. Since in living systems self-excited oscillations are quite common, it is tempting to examine its co-relation with sensing mechanism of smell.

Generation of sense of smell by olfactory nerves is a complex phenomenon [56–59]. Although theories of sense excitation have been suggested by physiologists, these are quite complex. The simplest theory suggests that the molecules of odorant proteins open up to become ion channels, allowing mainly large numbers of positively charged sodium ion to flow to the interior of olfactory cell and depolarize it. The second theory postulated the odorant binding protein to become an activated adenylate cyclase, which in turn catalyses the formation of cyclic adenosine monophosphate (cAMP). Subsequently, cAMP acts on other proteins to open up ion channels. The essential steps in sensing smell are the following:

(a) Self-excitation of electrode potential by external sense stimulant.
(b) Depolarization of nerve cell.
(c) Generation of action potential.
(d) Nerve transmission.
(e) Decoding of sensation by
    (i) Intensity of potential
    (ii) Frequency
    (iii) Number of active fibers.
    (iv) Nature of oscillations: periodic, bioperiodic, triperiodic, sequential, intermittency, deterministic chaos, random noise.

A much simpler mechanism based on self-excitation of potentials in relation to generation of smell by olfactory nerves can be postulate as follows.

In case of amines and pheromones, one may speculate that periodic adsorption of amines and pheromones and subsequent interaction of NH_4^+ ions and carbocation with Cl^- could lead to depletion of Cl^-. This would result in migration of Cl^- to the surface of the nerve cell leading to the depolarization followed by development of action potential. Transport of Cl^- through the nano-pores of the cell membrane would be favoured in view of their smaller size as compared to carbo-cations. The stimulus through action potential would travel through the resulting polarized cell. However, further *in vivo* experiments are needed for confirming the proposed mechanism.

11.5. Dynamic instability at solid–gas interface [60–68]

11.5.1. Introduction

Earlier, it was difficult to produce a clean surface and to characterize its surface structure. However, with the development of electronic industry, techniques have been developed to produce clean surface with well-defined properties. It has been possible to investigate catalytic oxidation on metal surface in depth. Example of dynamic instability at gas–liquid interface is provided by such studies. Studies on chemical oscillations during oxidation of CO over surface of platinum group metals have attracted considerable interest [62–68].

11.5.2. Experimental studies

Experimental studies of interaction have been made on well-defined Pt (100) and (1010) surfaces [64, 65]. The experiments were performed in a standard UHV system equipped with facilities for LEED, AES work function measurement () with a Quadruple Mass Spectrometer and self-compensation vibration capacitor method.

11.5.3. Mechanism

Under the influence of the adsorbate, the surface structure may switch periodically between more (=non-reconstructed) and less reactive (=reconstructed) state, whereby the driving forces are the difference in surface free energy of the clean planes on the one hand and the difference in CO adsorption energy on the other. In other words, the reconstructed phase adsorbs CO more rapidly than it is reacted. Thus, the CO coverage increases beyond its critical value for nucleation of the structural transformation into the non-reconstructed state. The latter exhibits an increased oxygen coefficient so that CO is removed more rapidly from the surface. As a result, the CO coverage drops and the surface transforms back to reconstructed states.

The clean Pt (100) and (110) surfaces are reconstructed and then construction is lifted if a critical CO coverage is reacted. Both modifications of the respective planes exhibit different sticking coefficients so that as a net result, the surface switches between the states of high and low reactivities. Thus, the rate of catalytic CO oxidation on defined Pt (100) and Pt (110) surfaces at low pressure under isothermal condition exhibits temporal oscillations which are coupled with periodic transformation of the surface structures between reconstructed and non-reconstructed phases [66].

Reconstructed (1×2) structure is formed when crystal is cut, during the cleaning process at 800°C. During CO oxidation, the surface structure changes from (1×2) to non-structured (1×1) form.

The structure oscillates in between critical CO pressure, which depends on the surface. The oxidation of CO on platinum is well established by the classical Langmuir–Hinshelwood mechanism.

Denoting * as empty site, the adsorption–desorption equilibrium can be represented as

$$CO(g) + {}^* \underset{k_2}{\overset{k_1}{\rightleftharpoons}} CO(ad) \text{ (adsorption–desorption equilibrium for CO)} ----------1$$

$$O_2(g) + 2^* \underset{k_4}{\overset{k_3}{\rightleftharpoons}} 2O(ad) \text{ (adsorption–desorption equilibrium for } O_2)$$

$$CO(ad) + 2O\ (ad) \longrightarrow CO_2(g) \uparrow + 2^* \ ---------2$$

Step 1 describes an adsorption–desorption equilibrium for CO on the surface. Adsorption of oxygen is preceded by the dissociation of molecular oxygen into atoms via step 2; when the two nearest-neighbour sites are CO and oxygen, the reaction takes place with high probabilities via small activation energy barrier. Once a CO_2 molecule is formed, it desorbs immediately into the gas phase due to its low binding energy to the surface.

There are only two essential species on the surface, viz. CO_{ad}, O_{ad}. The sum of their coverage is equal to unity order to satisfy conservation constraint. Surface reaction gives rise to two differential equations for the coverage CO_{ad} and O_{ad}. Analysis of the ordinary differential equations systems (ODE system) indicates that there is hysteresis in one, and cusp in two control parameters as indicated in Fig. 11.13.

The concepts developed in such studies have importance in the development of catalyst technology and nano-technology since it is known that nano-size gold particles are useful catalysts for oxidation of CO.

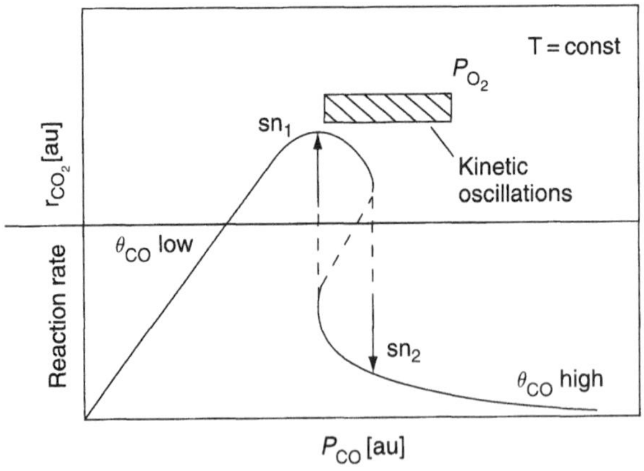

Figure 11.13. Dependence of reaction velocity on partial pressure of CO. P_{CO} = partial pressure of CO, I, II = states, θ_{CO} = number of sites occupied by co in the monolayer.

References

1. R. Gallaghar and T. Appenzeller, Science, 284 (1999) 79.
2. H.C. Pant and B. Rosenberg, Biochem. Biophys. Acta 225 (1971) 379.
3. T. Ishi, Y. Kuroda, T. Omchi and K. Yoshikawa, Langmuir, 2 (1986) 319.
4. T. Ishi, Y. Kuroda, K. Yoshikawa, K. Sebaki, Y. Matsubara and K. Iriyana, Biophys. Biochem. Res. Commun., 123 (1984) 792.
5. K. Yagisawa, M. Naita, K. Gondaira and T. kambana, Biophys. J., 64 (1993) 1369.
6. T. Teorell, J. Gen. Physiology, 42 (1959) 831, 847.
7. R. Larter, Chem. Rev., 90 (1990) 355.
8. P. Meares and K.R. Page, Proc. R. Soc. London, A 339 (1974) 513.
9. P. Meares and K.R. Page, Philos. Trans. R. Soc., London, A 272 (1972) 1.
10. R.P. Rastogi and R.C. Srivastava, Proc. Indian Natl. Sci. Acad., 68A (2002) 73.
11. R.P. Rastogi, R.C. Srivastava and S.N. Singh, Chem. Rev., 93 (1993) 1945.
12. D.C. Mikulecky and S. Roy Caplan, in Ion Exchange, Chapter 1, Marcel Dekker, New York, 1966.
13. R.P. Rastogi, G. Srinivas, R.C. Srivastava, P.C. Pandey, A.P. Misra and A.R. Singh, J. Coll. Interf. Sci., 175 (1995) 262.
14. R.P. Rastogi, G.P. Misra, K. Kumar, K. Bala and P.C. Pandey, J. Coll. Interf. Sci., 217 (1999) 275.
15. K. Bala, K. Kumar, S.K. Saha and R.C. Srivastava, J. Coll. Interf. Sci., 273 (2004) 320.
16. J. Arisawa and K. Misawa, J. Membrane Sci., 32 (1987) 223.
17. J.T. Kim and R. Larter, J. Phys. Chem., J. Phys. Chem., 95 (1991) 7948.
18. J. Arisawa and T. Frurkawa, J. Membrane Sci., 2 (1977) 303.
19. P. Shem, J.T. Kim, R. Larter and K. Lipkowitz, J. Phys. Chem., 97 (1993) 1571.
20. Y. Kobatake and M. Fujita, J. Chem. Phys., 40 (1964) 2212.
21. Y. Kobatake and M. Fujita, J. Chem. Phys., 40 (1964) 2219.
22. R.H. Arnow, Proc. Natl. Acad. Sci., USA, 50 (1963) 1046.
23. R.P. Rastogi, R.C. Srivastava and P. Chand, J. Coll. Interf. Sci., 263 (2003) 223.
24. D.S. Jones, Electrical and Mechanical Oscillations, Free Press, Illinois, 1961, Chapter IV.
25. D. Abbott, Chaos, 11(3) (2001) 526.
26. K. Yoshikawa and Y. Matsubara, J. Am. Chem. Soc., 105 (1983) 5969.
27. R.P. Rastogi, R.C. Srivastava and S. Kumar, J. Coll. Interf. Sci., 283 (2005) 139.
28. K.L. Mittal and B. Lindeman, Surfactants in Solution, Vol. 2, Plenum Press, New York, 1979.
29. H. Haken, Synergetics, Springer, Berlin, 1977, p. 244.
30. S. Martin, Geophys. Fluid Dynamics, 1 (1970) 14.
31. P.H. Alfredson and T. Lagersted, Phys. Fluids, 24 (1981) 10.
32. K. Yoshidawa, S. Nakata, M. Yamanaka and T. Waku, J. Chem. Educ., 66 (1989) 205.
33. K. Yoshikawa, N. Oyama, M. Shoji and S. Nakota, Am., J. Phys., 59 (1991) 37.
34. R.P. Rastogi, G. Srinivas, H.P. Maity and S. Kumar, Indian J. Chem., 40A (2001) 119.
35. S. Upadhyay, A.K. Das, V. Agarwal and R.C. Srivastava, Langmuir, 2 (1986) 319.
36. K. Yoshikawa and Y. Murofishi, Forma 40A (2001) 119.
37. K. Yoshikawa, S. Maeda and H. Kawakami, Ferroelectrics, 86 (1988) 281.
38. O. Steinbock, A. Lange and I. Richberg, Phys. Rev. Lett., 81 (1998) 798.
39. D.A. McInnes, Principles of Electrochemistry, Dover, New York, 1939, Chapter 13.

40. R.P. Rastogi and R.C. Srivastava, Adv. Coll. Interf. Sci. 93 (2001) 1.
41. R.C. Srivastava and R.P. Rastogi, Transport Mediated by Electrified Interface, Elsevier, Amsterdam, 2003.
42. R.C. Srivastava, V. Agarwal, A.K. Das and S. Upadhyay, Indian J. Chem., 33A (1994) 978.
43. U. Roy, S.K. Seha, C.R. Krishnapriya, V. Jayshree and R.C. Srivastava, Instrumentation Sci, Technol., 31 (2003) 425.
44. L.E. Seriven and C.V. Sternling, Nature (London), 187 (1960) 978.
45. E. Nikache, M. Dupeyrat and M. Vigres-Alder, J. Coll. Interf. Sci., 94 (1983) 187.
46. M. Yoneyama, A. Fuje and J. Maeda, J. Am. Chem. Soc., 115 (1993) 11630.
47. N. Magome and K. Yoshikawa, J. Phys. Chem., 100 (1996) 19102.
48. K. Yoshikawa and Y. Matsubava, Langmuir, 1 (1985) 230.
49. R.C. Srivastava, V. Agarwal, A.K. Das and S. Upadhyay, Indian J. Biochem. Biophys., 33 (1996) 195.
50. A.L. Hodgins and A.F. Huxley, J. Physiol. (London), 116 (1952) 493, 117 (1952) 500.
51. R.P. Rastogi, R.C. Srivastav and Y.K. Agarwal, J. Coll. Interf. Sci., 297 (2006) 711.
52. K. Yoshikawa and Y. Matsubara, Langmuir, 1 (1985) 230.
53. L.I. Antropov, Theoretical Electrochemistry, Revised Edition, Mir, Moscow, 1975.
54. H. Diamant, G. Ariel and D. Andelman, Coll. Surfaces, A 183–185 (2001) 259–276.
55. A.F.H. Ward and L. Tordai, J. Chem. Phys., 14 (1946) 453.
56. G. Laurent and H. Davidewitz, Science, 265 (1994) 1872.
57. D.W. Tank, A. Gelperin and D. Kleinfeld, Science, 265 (1994) 1819.
58. D. Lancet, Nature (London), 372 (1994) 321.
59. A.C. Guyton, Text Book of Medical Physiology, Saunders, Philadelphia, 1981, Chapter 53.
60. H.G. Schuster, Deterministic Chaos, Physik Verlag, Weinhein, 1984.
61. P. de Kepper and J. Biossonede, J. Chem. Phys., 75 (1981) 189.
62. R.A. Schmitz, K.R. Graziami and J.L. Hadson, J. Chem. Phys., 67 (1977) 189.
63. L.F. Razon and R.A. Schmiz, Catalysis Rev. Sci. Eng., 28 (1986) 189.
64. R. Imbihl, M.P. Cox and G. Ertl, J. Chem. Phys., 84 (1986) 3519.
65. M. Eisiwirth and G. Ertl, Surface Sci., 177 (1986) 90.
66. M. Eiswrith, P. Moller, K. Wetzl, R. Imbihl and G. Ertl, J. Chem. Phys., 90 (1989) 510.
67. M. Eishwirth, R. Schhwarkor and G. Ertl, J. Phys., Chem., 44 (1985) 59.
68. R.J. Shwarkor, M. Eiswirth, P. Moller, K. Wetzl and G. Ertl., J. Chem. Phys., 87 (1986) 742.

Chapter 12

COMPLEX OSCILLATIONS AND CHAOS

12.1. Introduction

In Chapter 9, it has been indicated that periodic oscillations refer to regular motion. Chaos refers to irregular motion. Irregular motion governed by deterministic equations is characterized by the term "Deterministic Chaos". There are highly irregular motions like Brownian motion which do not belong even to this category and are not governed by deterministic mathematical equations. Such motion belongs to the class of random motion which occurs due to random causes.

Complex systems are now getting increasing attention. These are multi-variable systems involving several processes, which cannot be understood by using the traditional reductionist approach. In fact, study of complex system involves cross-disciplinary science making use of interacting disciplines. Consequently, in a wider sphere, the science of complexity is becoming a bridge between soft sciences and hard sciences.

Biological and physiological systems are typical complex systems, which provide examples of aperiodicity and chaos [1–7]. Aperiodic cardiac oscillations are reflected in ECG for different cases of arrhythmia Fig. (12.1). Similarly, chaotic, aperiodic and noisy oscillations are observed in EEG in specific cases as shown in Fig. (12.2). Closely allied with chemical oscillations are membrane oscillations which have considerable relevance in physiological processes including neurological and cardiac disorders in the context of detection and control.

The term chaotic, as it is now widely used, describes the non-periodic behaviour that arises from the non-linear nature of deterministic systems, not the noisy behaviour arising from random driving behaviour [8].

Before proceeding further, it would be desirable to discuss the types of complex and chaotic oscillations and the corresponding phase–plane plots.

12.2. Complex oscillations

Complex oscillations can be of the following types:

(a) multi-periodicity (Fig. [12.3(c) and (d)]);
(b) bursting/Intermittency [9(a)] Fig. (12.4);

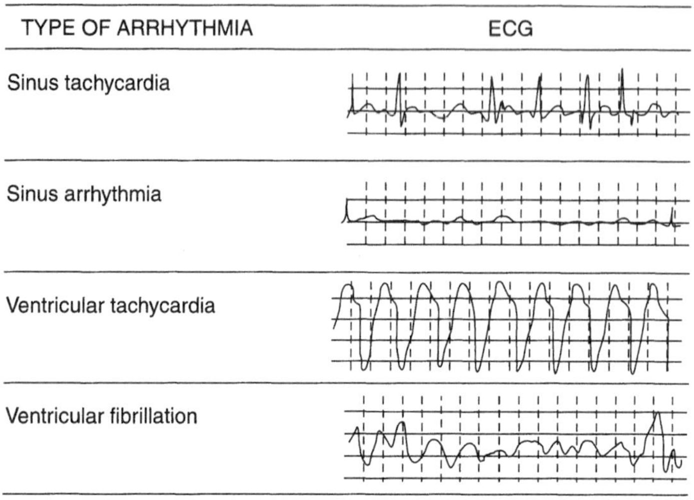

Figure 12.1. Some typical arrhythmia.

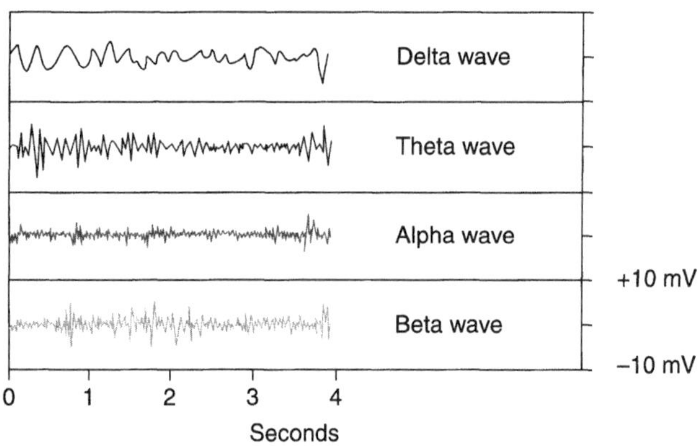

Figure 12.2. Classification of EEG waves. See also Colour Plate Section.

(c) relaxation oscillations;
(d) sequential oscillations of different time period and amplitude [9(b)] Fig. (12.5);
(e) coupled oscillations; and
(f) Chaos.

Complexity of the above types can also be observed in physical, chemical and biological systems.

A dynamical system can be described by a set of N non-linear differential equations if n state variables are involved. If $X(t)$ represents a system of state variables

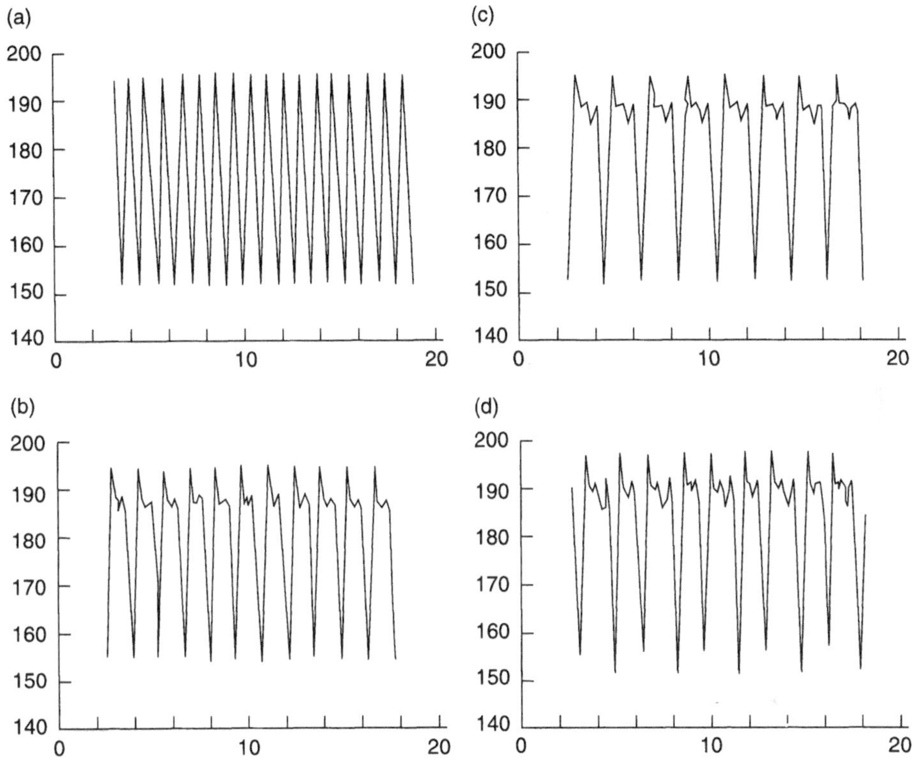

Figure 12.3. Complex oscillations (a) periodic; (b) biperiodic; (c) with period –3; and (d) with period –4 [1-large, 3 small].

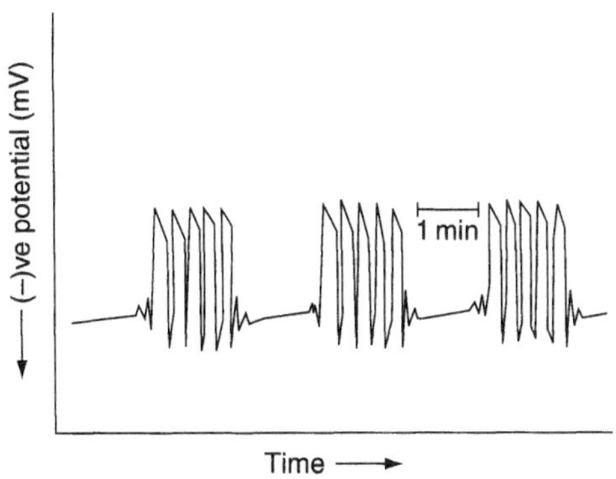

Figure 12.4. Intermittency : Br^- potential oscillations for the system [Fructose] = 0.4 M; [BrO_3] = 0.06 M; [Ce^{4+}] = 1.45×10^{-3} M : [H_2SO_4] = 1.5 M; in CSTR at flow rate = 4.6 ml/min, at Temp. 31 ± 1°C.

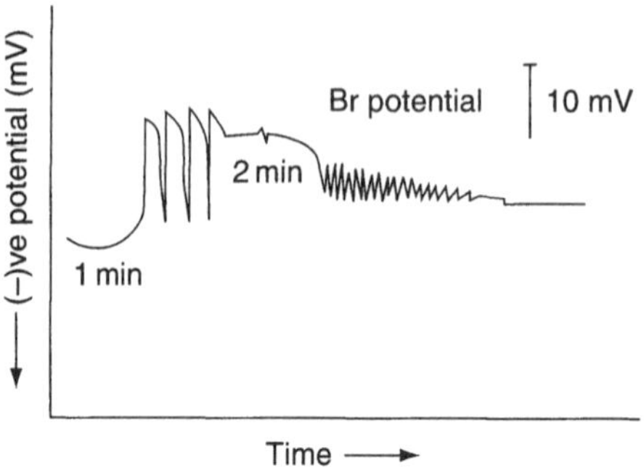

Figure 12.5. Sequential oscillations in Fructose + acetone oscillator $[BrO_3^-] = 4.99 \times 10^{-2}$ M; $[Ce^{4+}] = 6.3 \times 10^{-4}$ M; [Acetone] = 0.277 M; $[H_2SO_4] = 1.5$ M; [F] = 0.095 M, at Temp. 29 ± 1°C.

$X_1(t), \ldots\ldots, X_n(t)$, then the differential $X(t)$ is another linear function $X(t)$, so that,

$$X(t) = F[X(t)].$$

If the initial state is known, the new state can be calculated a moment later and so on. One can describe the temporal oscillations in the form of a trajectory in the n-dimensional state-space.

Before we proceed to discuss complex oscillation further, let us consider the simple example of periodic oscillations governed by the two variable differential equations such as

$$\mathrm{d}x/\mathrm{d}t = -K_1 y \tag{12.1}$$

$$\mathrm{d}y/\mathrm{d}t = K_2 x \tag{12.2}$$

Therefore,

$$\mathrm{d}y/\mathrm{d}x = -[K_2/K_1]y/x \tag{12.3}$$

here K_1 and K_2 are constants. On integrating Eq. (12.3), we get

$$K_2/2c\ x^2 + K_1/2c\ y^2 = 1 \quad \text{(where } c \text{ is the constant of integration)}$$

which is an equation of a circle or ellipse depending on the values of K_1 and K_2.

For a typical case, the computed oscillations have been recorded in Fig. 12.6(a). The phase–plane plot (x–y) is recorded in Fig. 12.6(b). When the initial conditions are changed a different closed curve is obtained.

For a two variable system, the phase–plane plot (x–y) yields a point, which is an attractor, when the system approaches equilibrium state. Similarly for periodic oscillations, closed cycle is the attractor. Limit cycle is a specific class of such cycles, when

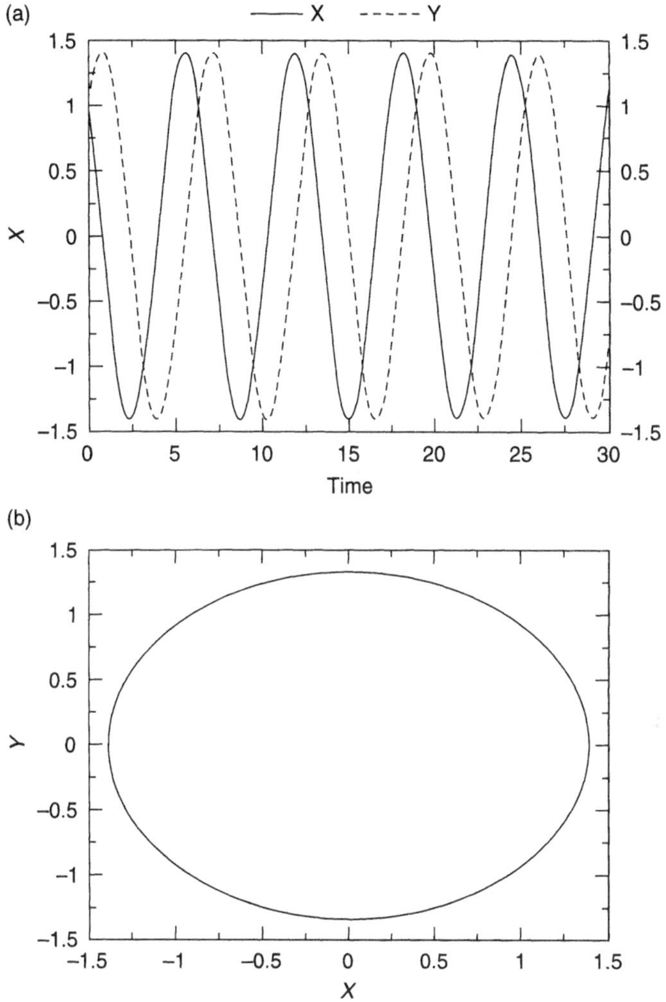

Figure 12.6. (a) Periodic oscillations; (b) Corresponding attractor of the above time series.

the system approaches it irrespective of the initial values of the variables. A typical example is discussed below.

In order to understand the concept of limit cycle further, let us consider the following model reaction scheme in a CSTR [10].

(a) $A + B \xrightarrow{k_1} 2B$

(b) $B \xrightarrow{k_2} C$

here k_1 and k_2 denote the rate constants of the corresponding steps and where the rate of the second step is given by $k_2B^{1/2}$.

In other words the reaction (b) obeys half-order kinetics. The kinetic equations are given by

$$dA/dt = -k_1 AB + k_f(A_0 - A) \tag{12.4}$$

$$dB/dt = k_1 AB - k_2 B^{1/2} + k_f(B_0 - B) \tag{12.5}$$

where the flow rate k_f is the inverse of residence time t_{res} in the reactor, and A and B denote the initial concentration of the reactants whereas A_0 and B_0 are concentrations of the reactants in the reservoir.

On reducing the above equations to a dimensionless form

$$dx/dt = -yx + (1 - x)/t_{res} \tag{12.6}$$

$$dy/dt = xy - y_n/t_2 + (y_0 - y)/t_{res} \tag{12.7}$$

Using the reduced variables as given by,

$$x = A/A_0,$$

$$y = B/A_0,$$

$$y_0 = B/A_0,$$

$$t_o = k_1 A_0.t$$

$$t_{res} = k_1 A_0/k_f \quad \text{and} \quad t_2 = k_1 A_0^{3/2}/k_2$$

and solving between x and y on the computer, the phase–plane plot, as shown in Fig. (12.7), is obtained. T represents time while t_o denotes reduced residence time.

The Fig. (12.7) shows that the initial values of x_0, y_0 may be different at the points A, B, C, D, but ultimately the trajectories converge to the same closed cycle. Such a cycle is called a limit cycle.

In contrast with the above, chaotic dynamics evolves an asymptotic trajectory in phase space, tracing out a strange attractor. "For continuous dynamical system, there are an infinite number of unstable limit cycles embedded in such an attractor, each characterized by a distinct number: of oscillations per period." We may note that there are two types of plots which provide information about the attractor.

(1) $x-y$, $y-z$, $z-x$ plots.
(2) time – delay plots involving t, $t+\tau$, $t+2\tau$, where τ is the delay.

Whereas attractor in the steady state has zero dimensions and Euclidian dimension of the limit cycle is two, it is not possible to define the Euclidian dimension of the strange attractor. However, using the concept of fractal geometry, it is possible to define the dimension of such an attractor, which is not an integer.

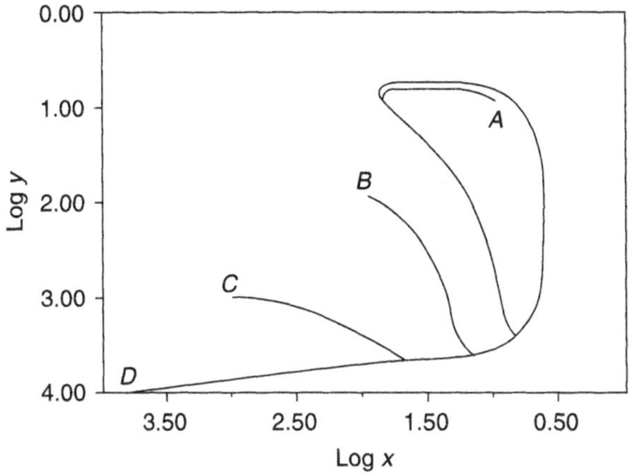

Figure 12.7. Limit cycle oscillations for the model of half order kinetics ($n = 0.5$; $t_{res} = 700$; $t_2 = 100$; (i) A; $x_0 = 0.001$, $y_0 = 0.001$; (ii) B; $x_0 = 0.01$, $y_0 = 0.01$; (iii) C; $x_0 = 0.001$, $y_0 = 0.001$; and (iv) D; $x_0 = 0.0001$, $y_0 = 0.0001$).

Deterministic chaos is a special type of chaos, which can be predicted on the basis of a set of differential equations. In the next section, deterministic chaos will be discussed in detail.

12.3. Deterministic chaos

The phenomenon of deterministic chaos was established by the work of Lorenz who showed for the first time that chaos is not unnecessarily an unpredictable phenomenon. A certain class can be governed by mathematical equations suggesting it's predictability. This type of chaos is called deterministic chaos.

We give below some typical examples of deterministic equations which yield chaotic oscillations and specific class of trajectory called "Strange attractor" on computer solution:

(a) *Lorentz attractor*: This involves the following set of equations:

$$dx/dt = -\sigma x + \sigma y$$

$$dy/dt = -xz + rx - y$$

$$dz/dt = xy - bz$$

For $\sigma = 10$, $b = 8/3$, Lorenz found that the set of equations yield chaotic oscillations time series whenever r exceeds a critical value $r \sim 24.74$. The Lorenz attractor can be reconstructed from a time series with a delay time ($\tau = 0.1$).

(b) *Rössler Attractor*: A good example of deterministic chaos is provided by a set of the following equations proposed by Rössler [11]:

$$dx/dt = -(y+z)$$

$$dy/dt = x+ay$$

$$dz/dt = b+xz-cz \tag{12.8}$$

with initial conditions (−1, 0, 0) and $a = 0.2$, $b = 0.2$ and $c = 5.7$, a chaotic time series in x, y, z, is generated on solution.

The phase–plane plot in three dimensions is further shown in Fig. (12.8). The phase–plane plot in three dimensions has fractal geometry and the attractor is called the strange attractor.

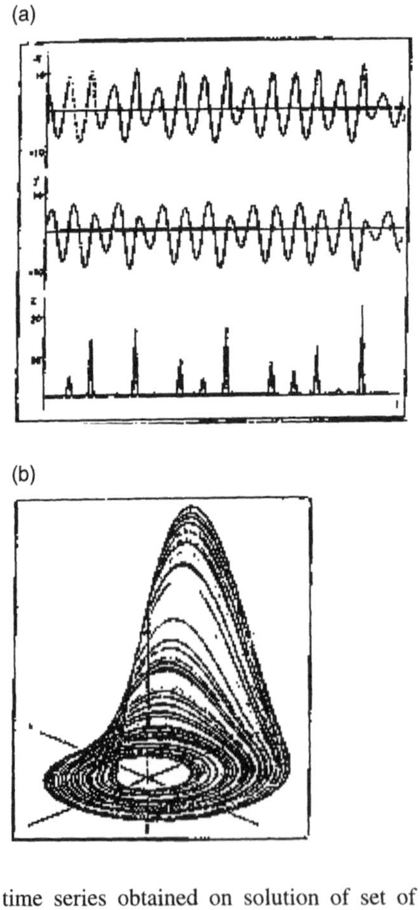

Figure 12.8. (a) Chaotic time series obtained on solution of set of equations [8] with initial conditions (−1, 0, 0) and $a = 0.2$, $b = 0.2$, $c = 5.7$; and (b) Trajectory in three dimensions.

Chaotic or strange attractors have unusual geometry. These are described by

(a) fractal dimensions (see Chapter 13); and
(b) liapunov exponents.

Strange attractor has different properties and cannot be described by the following features:

(a) a set of finite points;
(b) a closed curve;
(c) a smooth or piecewise smooth surface; and
(d) a volume bounded by a smooth or piece-wise smooth surface.

In the case of non-deterministic chaos (noisy) oscillations, the attractor is found to be as shown in Figure 12.9(b) corresponding to the time series generated from thiophenol + Bromate oscillator [12], which is quite different as compared to the strange

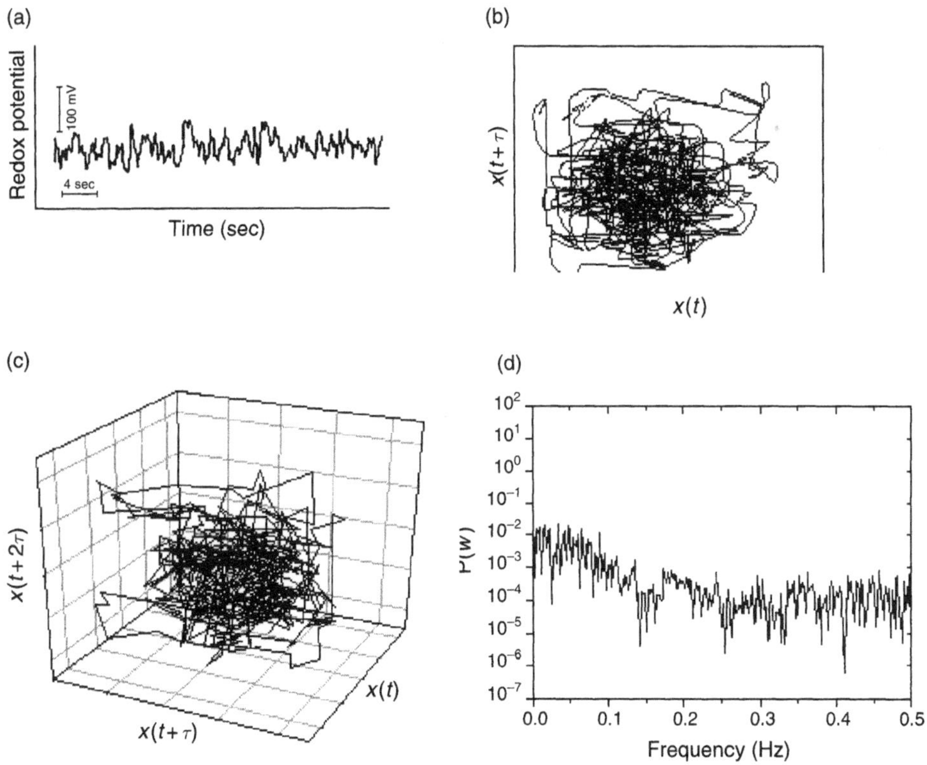

Figure 12.9. (a) Time series generated from thiophenol + bromate oscillator; (b) 2-D phase plane plot; (c) 3-D phase plane plot; (d) Power spectra. See also Colour Plate Section.

attractor. The phase–plane plots of thio-phenol oscillator based on experimental results Figure 12.9(a) display an unusual attractor since noise in oscillation is produced by heterogeneous reactions.

12.4. Routes to chaos

How do periodic oscillations get transformed to chaotic oscillations? This happens when the magnitude of bifurcation parameter is changed. In the case of chemical reactions in CSTR, bifurcation is the flow rate. Another question is: In what steps periodic oscillations are converted to chaos?

There are three main routes to chaos in dissipative dynamical systems as given below:

(1) Feigenbaum route;
(2) Manneville–Pomeau (intermittency route); and
(3) Rulle–Takens–Newhouse route.

These are associated with periodic doubling or regular oscillation intercepted by occasional bursts of noise. The features of three routes are summarized below:

Route	Features	Example of typical cases
Feigenbaum route	Infinite cascade of periodic doubling	Chemical reaction Benard experiment
Manneville–Pomeau	Intermittent transition	Chemical reaction Benard experiment
Rulle–Takens–Newhouse	Strange attractor after three transitions	Benard experiment Taylor experiment

The last route is illustrated in Fig. (12.10).

12.5. Characterization of chaos

A chaotic system has following properties:

(a) The system can at least under some circumstances display highly disordered behaviour.
(b) A very small change in parameters and values can lead to dramatic changes in the behaviour of the system.

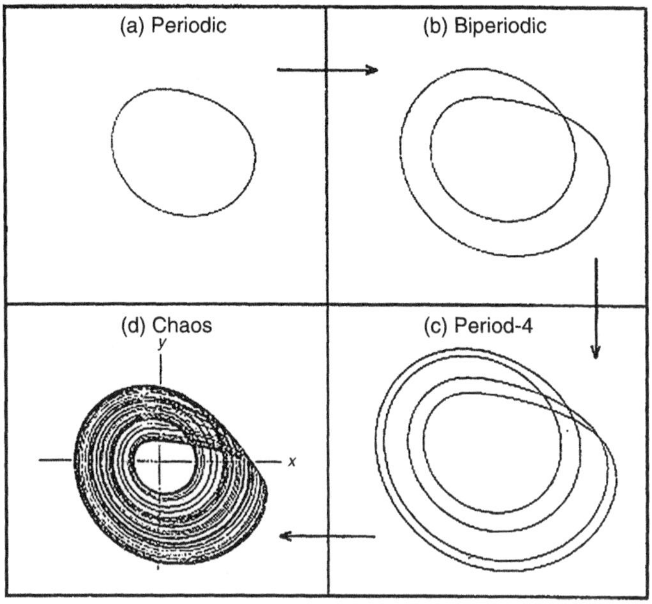

Figure 12.10. Ruelle–Takens newhouse route to chaos (a) periodic; (b) biperiodic; (c) period-4; and (d) chaos.

(c) The chaos is deterministic if it obeys some laws that completely specify its motion. Experimental measurements for unequivocal assessment of chaos requires attention to the following aspects:

Accuracy of voltage measurements: It should be noted that the resolution of the digitizer that transforms an analog voltage signal into a numerical value is a crucial factor so far as accuracy is concerned.

Sampling of data: In this context, it is often recommended that oversampling can seriously bias analysis for chaos.

No. of data points: Analytical studies for detecting chaos require extremely large data sets.

Stability of the system: Electronic systems are stable systems while biological systems are constantly changing.

Choice of signal filters: For distinguishing chaos from noise.

For confirmation of chaos, several analytical techniques are employed which are as follows:

(a) Next amplitude or next maximum maps from the time series, which are special cases of a more general concept, the return map;
(b) Phase–plane plot;

(c) Poincare's Section;
(d) Power spectra;
(e) Auto-correlation function; and
(f) Lyapunov exponent.

In the next amplitude (maximum) plots, the amplitude (maximum) b_{n+1} of the $(n+1)$th peak is plotted against the amplitude (maximum) b_n of the nth peak. For regular periodic oscillations, the plot will reveal a finite number of discrete points, whereas for chaotic signals such a plot will show a sharp peak.

Phase–plane plot provides a powerful method for detecting chaos since as explained earlier, one would get a strange attractor in contrast with a limit cycle.

Further confirmation of chaos is obtained by the Poincare's map, wherein the points on the Poincare's section represent repeated intersections at the ends of each circuit of the attractor.

Power spectra P(w) provides an equally strong proof of chaos which is defined as follows [1]

$$\mathrm{P}(w) = |x(w)|^2 \tag{12.9}$$

$$x(w) = \lim_{T\to\infty} \int \mathrm{d}t\, e^{iwt}\, x(t) \tag{12.10}$$

Here $x(t)$ is the signal, w is the frequency, t is the time and T is the time period. For multiple periodic motions, the power spectrum consists only of discrete lines of the corresponding frequencies. On the other hand, for chaotic motion which is completely aperiodic, broad noise is indicated in the power spectra, which is mostly located at low frequencies.

The correlation function measures the correlation between subsequent signals. It remains constant or oscillates for regular exponential tail when the signal becomes uncorrelated in the chaotic region.

Chaos is characterized by sensitivity to initial conditions and consequent exponential diversions of initially adjacent phase space trajectories. $\lambda(x)_0$, which is called the Lyapunov exponent (LE), is a measure of such divergence which is defined by

$$\begin{aligned}\lambda(x_0) &= \lim_{N\to\infty} \lim_{\varepsilon\to 0} \frac{1}{N} \log \frac{|f^N(x_0+\varepsilon) - f^N(x_0)|}{\varepsilon} \\ &= \lim_{N\to\infty} \frac{1}{N} \log \frac{|\mathrm{d}f^N(x_0)|}{d(x_0)}\end{aligned} \tag{12.11}$$

where $x_{n+1} = f(x_n)$ and N is the no. of iterations. The Lyapunov exponent (λ), which measures the exponential separation, can also be expressed as follows

$$\varepsilon\ e^{N\lambda(x_0)} = |f^N(x_0+\varepsilon) - f^N(x_0)| \tag{12.12}$$

Thus $e^{N\lambda(x_0)}$ is the average factor by which the distance between closely adjacent points becomes stretched after iteration. The Lyapunov exponent can be derived from the dynamics of the differential equations. The Lyapunov exponent corresponding to different types of attractors has the following characteristics:

(a) Point attractors have only negative exponents, while cycles and tori have negative Lyapunov exponents and one or more exponent of zero value.
(b) Chaotic attractors have, in addition, exponents with a positive value.
It may be noted that the Lorentz attractor has one zero and one positive value. Further, a negative exponent indicates an exponential approach of the initial condition on the attractor.
(c) A positive exponent λ_l expresses the exponential divergence on an otherwise stable attractor. Thus a positive Lyapunov exponent indicates the presence of chaotic dynamics.
(d) If in the case of systems of the above type, the largest λ is negative then there is either a stable steady state or a stable limit cycle. On the other hand, a positive λ is often considered as indicator of chaos. However, it should be noted that it can give erroneous results in the presence of large time derivatives.

In case of random oscillations generated by computer simulation (software ref ORIGIN Microcal), the phase portrait, Poincare's map and power spectra are found to be as shown in Fig. (12.11).

12.6. Modelling and test of reliability

Mathematical modelling of oscillation involves:

(a) identification of variables and reaction steps/processes;
(b) setting up of appropriate differential equations in the variables;
(c) assignment of values to constants/parameters of the differential equations;
(d) assignment of boundary conditions; and
(e) choice of appropriate algorithm for computer solution.

Experimental test of model involves comparison of theory and experiment with respect to numerical simulation. A falsifier, as postulated by Karl Popper, would be needed for a still more rigorous test. Even if it is not possible, the exercise in modelling does give an insight in to the physics of the phenomena.

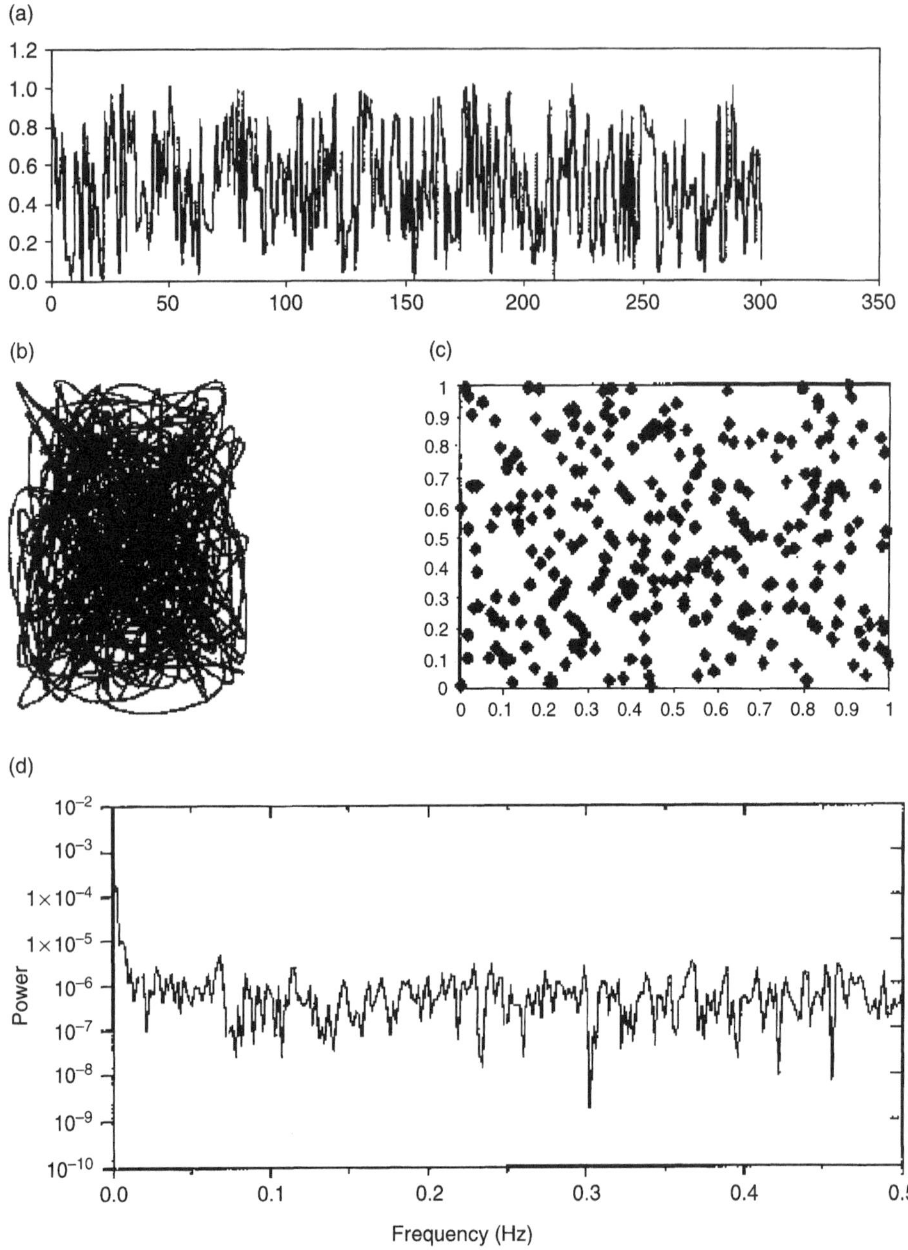

Figure 12.11. (a) Time series from computer generated random signal; (b) 2d phase plane plot; (c) Poincare's map; and (d) power spectrum.

12.7. Control of chaos

In recent years, some studies on maintenance of chaos and control of chaos had been undertaken. The former has recently been experimentally demonstrated in a magneto-mechanical system demonstrating intermittency. There is interest in such studies in view of the likelihood that "Pathological destruction of chaotic behaviour possibly due to some underlying disease may be implicated in heart failure and brain seizers".

Control of chaos has interest since "one may want a system to be used for different purposes or under different conditions at different times. Such multipurpose flexibility is essential to higher life forms" Ott, Grebegogi and Yorke (OGY) suggested a method [13] according to which one can convert a chaotic attractor to any one of a large number of possible attracting time-periodic perturbation of an available system parameter.

Control of chaos can be exercised by

(a) OGY method (linear control) using a small time independent perturbation to the system;
(b) non-linear extension of reduced OGY method; and
(c) continuous control method linear delayed feedback of Pyrag.

It should be noted that chaos control can only be obtained if deterministic chaos is involved. In case of (i) chaotic laser; (ii) diode; (iii) hydrodynamic and magneto-elastic systems; and (iv) more recently myocardial tissue, feedback algorithm has been successfully applied to stabilize periodic oscillations. Quite recently, in order to stabilize periodic behaviour in the chaotic regime of oscillatory B–Z reaction, Showalter [14] and co-workers (1998) applied proportional feedback mechanism. Feedback was applied to the system by perturbing the flow rate of cesium-bromate solutions in the reactor keeping the flow rate of malonic acid fixed in these experiments. This experimental arrangement helped the stabilization of periodic behaviour within the chaotic regime.

Chaos in discrete neural networks has also received attention recently. Low dimensional strange attractors have been often observed in brain dynamics. Practical application to the treatment of epileptic foci has also been conjectured.

In the context of cardiac arrhythmia, the strategy adopted by Alan Garfinkel [15] *et al.* (1992) was based on the following strategy: "Sensitivity of the chaotic systems makes them highly susceptible to control, provided that the developing chaos can be analyzed in real time and that analysis is then used to make small control interventions, to stabilize cardiac arrhythmia induced by the drug, ovalin in rabbit ventricles, adopted this strategy". An electrical stimuli to the heart at irregular times as determined by the chaos theory was administered. In this manner arrhythmia was converted to periodic beating. However, it is difficult to say whether the chaos control strategy in in vitro cardiac ventricle can be successfully applied to the in vivo heart.

In the case of non-deterministic chaos (noisy) oscillations, the attractor is found to be as shown in Fig. 12.9(b) corresponding to the time series generated from thiophenol + bromate oscillator [12], which is quite different as compared to the strange attractor. The phase–plane plots of thio-phenol oscillator based on experimental results Fig. 12.9(a) displays an unusual attractor since noise in oscillations is produced by the heterogeneous reaction.

12.8. Noise

Aperiodic oscillation chaos and noise are receiving a good deal of attention and have contributed to the development of new paradigms [16–19]. Phase–plane plots and the nature of attractors can be used for differentiating deterministic chaos and noise. An important feature of some of the oscillatory reactions is noise [16] which can result from experimental noise or local inhomogeneity resulting from gas bubbles, colloidal particles or mixing effects. Noise is expected to occur when heterogeneous reactions are components of the reaction network due to gas evolution at different centres, which are time and space dependent.

Noise [17] can be of two types: (1) interactive, which may be due to nature of coupled reactions and fluctuations in flow rate, temperature and concentration variables; and (2) non-interactive. Noise also plays an important role in physiological processes. It has been shown that noise makes sense in neuronal computing [18]. The contribution of noise in contrast tuning in cat visual cortex is also reported [19]. Noise is relevant in cardiology in the context of mapping and control of complex cardiac arrhythmia [20]. Noise-resistant circadian oscillators have also been recently investigated [21]. Gastric disturbance creates abnormal problems such as distortion, oscillation, agitation, tremors and convulsions [22].

In non-deterministic regime, in electrical and electronic systems (communications), we can have "internal noise" which is inherent in the system and "external noise" which is the disturbance caused by external factors. It may be noted that the study of effect of noise on oscillatory behaviour is of relevance in "neuro-sciences" and information processing [20–22]. It has been pointed out that in linear systems, noise plays a destructive role, whereas in non-linear systems, the noise can play a constructive role in some cases.

Detailed studies on the following oscillatory reactions: (i) thiophenol + BrO_3^- + H_2SO_4; (ii) fructose + $BrO_3^- + H_2SO_4 + Ce^{4+}$; and (iii) iodate + arsenite have been undertaken recently [12(b)]. Analysis of phase–plane plots yields complex attractors, which are quite different from the strange attractor indicating the occurrence of random noise. Power spectra also indicates chaos with a broad band, not necessarily the deterministic chaos. For the reactions (i) and (iii), a mechanism has been proposed involving precipitation of C_6-H_5-S-C_6H_5 and iodine, respectively; while in case of reaction (ii), gaseous evolution of Br_2 has been invoked by adding a noise term $\xi[Br_2]$ to the rate for $d[Br_2]/dt$. $\xi[Br_2]$ is a function of prevailing $[Br_2]$, which will fluctuate rigorously under the prevailing circumstances. Thus in such case, heterogeneous reactions play an important role in generating chaos and noise in oscillatory chemical reactions.

It may be noted that the terms $\xi[C_6H_5\text{-S-S-}C_6H_5]$ in the case of thiophenol–bromate oscillator and $\xi[Br_2]$ in fructose oscillator are not of exact stochastic character of differential equation of type [23].

$$X(t) = T(t) + S(t) + R(t)$$

where $T(t)$, $S(t)$ and $R(t)$ are the trend, seasonal and irregular components of the variable $X(t)$, respectively. Recurrence quantitative analysis in the context of detection of short complex signals has been reported recently [24].

12.9. Turbulence

Laminar flow corresponds to the state of the system near thermodynamic equilibrium while turbulent flow occurs only for conditions sufficiently far from equilibrium when non-linearity due to inertial effect becomes dominant [25]. Readable [26] accounts of developments in the area are available.

When a fluid flows rapidly, its flow pattern typically exhibits a subtle mixture of order and chaos, and it is this structured chaotic motion that is referred to as turbulence [27]. Turbulence can be defined as a property of an incompressible fluid flow at a very high Reynolds number given by

$$\text{Reynolds number} = UL/\nu$$

where

L = diameter of cylinder;
U = uniform upstream speed; and
ν = kinematic viscosity.

12.10. Future perspectives

Study of complex processes, as described above, would be an area of continued interest which would be stimulated by the need for understanding cardiac and biological control mechanism required for clinical detection and control and their applications as indicated earlier.

Deeper understanding of similar phenomena in socio-political systems and fluctuations in Finance is also needed.

References

1. H.G. Schuster, Deterministic Chaos, Second revised ed., VCN Verlagsgesellshaft GmbH, Weinsheim, FRG 1988, (a) p 9 (b) p 181.
2. S.K. Scott, Chemical Chaos, Oxford Science Publications, Clarendon Press, Oxford, 1991.
3. David Ruelle, Chance and Chaos, Princeton University Press, Penguin Books, 1993.
4. S.K. Scott and P. Gray, Chemical Oscillations and Instabilities, Non-Linear Chemical Kinetics, Clarendon Press, Oxford, 1990.
5. O. Hein, O. Peilgen, H. Jirgens and D. Saupe, Chaos and Fractals, New Frontiers of Science, Springer-Verlog, New York, 1993.
6. D.L. Turcolte, Fractals and Chaos in Geology and Geophysics, Cambridge University Press, Cambridge, 1992.
7. R.J. Field and L. Gyorgy (eds), Chaos in Chemistry and Biochemistry, World Scientific, Singapore, 1993.
8. F. Argoul, A. Arnedo, P. Richeti and J.C. Roux, Acc. Chem. Res., 29 (1987) 436.
9. (a) R.P. Rastogi, R. Khare, G.P. Misra and S. Srivastava, Indian J. Chem., 36A 19 (1997). (b) R.P. Rastogi, M.M. Husain, P. Chand, G.P. Misra and M. Das, Chem. Phys. Letts., 353 (2002) 40.
10. R.P. Rastogi and R. Khare, Indian J. Chem., 33A (1994) 1–3.
11. (a) O.E. Rössler, Phys. Lett., 57A (1976) 397–398, (b) O.E. Rössler, Z. Naturforsch., 31a (1976) 259–264.
12. (a) R.P. Rastogi and P.K. Srivastava, in Spatial in Homogeneities and Transient Behavior in Chemical Kinetics, P. Gray, G. Nicolis, F. Baras, P. Borckman and S.K. Scott (eds), Manchester University Press, Manchester and New York, 1990, p. 653. (b) R.P. Rastogi, I. Das, P.K. Srivastava, P. Chand and S. Kumar, Indian J. Chem., 45A (2006) 358–364. (c) R.P. Rastogi, I. Das and S. Kumar, Nat. Acad. Sci. Lett., 28 (2005) 303.
13. E. Ott, G. Grebog and J.A. Yorke, Phys. Rev. Letts., 64 (1990) 1196.
14. V. Petrov, V. Goldspar, J. Mareve and K. Showalter, Nature, 361 (1993) 240.
15. A. Garfinkel, M.L. Spano, W.L. Ditto and J.W. Weiss, Science, 257 (1992) 1230.
16. I.R. Epstein and J.A. Pojman, An Introduction to Non-Linear Chemical Dynamics: Oscillations, Waves, Patterns & Chaos, First edition, Oxford University Press, New York, 1998.
17. (a) J.M. Lebender and W. Schneider, J. Phys. Chem., 99 (1995) 492. (b) Z. Nosticzius, W.D. McCormic and H.L. Swinney, J. Phys. Chem., 91 (1989) 2159.
18. M. Vaigushev and V.T. Eisel, Science, 218 (2000) 1908.
19. J.S. Anderson, I. Lample, D.C. Gillespie and D. Ferster, Science, 290 (2000) 1968.
20. J. David, D.J. Christine and G. Leon, Chaos, 12 (2002) 732.
21. R.U. Hiroke, H. Kenzo and I. Masamitsu, J. Theor. Biol., 216 (2002) 501.
22. (a) N. Frank, Nature, 391 (1998) 743. (b) A. Derek, Chaos, 11 (2001) 526.
23. J. Gareth and S. Louise, Time Series, Forecasting, Simulations & Applications, Ellis Harwood Limited, UK, 1993.
24. P.Z. Joseph, G. Allesando and L.W. Charles Jr, Phys. Lett. A, 267 (2000) 174.
25. P. Glansdorff and I. Prigogine, Thermodynamics Theory of Structure, Stability and Fluctuations, Willy Interscience, New York, 1971.
26. U. Frishch, Turbulence, Cambridge University Press, Cambridge, 1977.
27. M. Nelkin, Science, 255 (1992) 566.

Chapter 13

COMPLEX PATTERN FORMATION

13.1. Introduction

In the earlier chapter we have discussed the formation of colour bands in moving wave fronts, stationary structures which are governed by coupling of reaction and diffusion. In this chapter we will be concerned with pattern formation governed by the process of mass flow and diffusion related to precipitation as crystals, electro-deposits, bacterial colonies and diffusion. Just as in the former case, in the present case also we come across very complex patterns, depending on the experimental conditions. However in the present case it is possible to rationalize the complex structure with use of new mathematical concepts of fractal geometry.

A fractal is a geometrical structure that at first seems to be complicated, irregular and random. A fractal pattern is one that repeats itself at smaller and smaller scales. When viewed carefully, one begins to realize the presence of tractable properties that are inherent in it and helps us to systematically study them. Following are the principal objectives of fractal growth studies: (i) characterization and quantification of hidden order in complex pattern and (ii) analysis of correlation in the development of order in seemingly disordered state. Ferns are one example. They are made up of branches that also look like individual ferns and in turn each of these is made up of even smaller branches that also look the same and so the patterns goes on.

The process of aggregation of small particles to form large clusters and the structures that result are technologically and scientifically important [1–6]. Random aggregates occur in many fields of science. Archaeologists find fractal patterns useful because they can quantify patterns. A useful way of measuring social dynamics of the past and present, fractal analysis could help in the design and construction of the building and towns of the future. Interest in the complex structures under non-equilibrium conditions has been stimulated by several recent developments including dissipative structures. The study of growing patterns has developed mainly due to the introduction of the concept of fractal geometry, faceted tip splitting, dendritic growth, dense radial morphology, spherulitic growth and fluid–fluid displacement in Hele-Shaw cells. Bacteria such as *Bacillus subtilis, E. coli, Pr. vulgaris* and *K. ozaenae* have also been shown to grow with various morphologies under different conditions on agar plates.

A typical property of fractals is that they are locally (asymptotically) self-similar of small-length scales. Fractals are shapes that look more or less the same on all or many scales of magnification. Self-similarity of pattern can be characterized by any one of the following:

(a) Hausdorff dimension D
(b) generalized dimension D_q
(c) correlation dimension D_2

Fractal is a subject associated with the discipline of non-linear dynamics. Fractals enjoy widespread attention not only in science but increasingly in popular culture. The reason why the fractal geometries are of physical interest is that in nature, a number of random processes build such geometries naturally. A fractal is composed of similar structure of finer details. Its length is not well defined. A fractal consists of geometric fragments of varying size and orientation but similar shape. This property of self-similarity or scale invariance means that if we take part of a fractal object and magnify it by the same magnification factor in all directions, the magnified picture is indistinguishable from the original. The dissemination of the concept of fractal geometry and related ideas have provided us the ways of describing a very broad range of irregular structures generated by living and non-living systems as shown in Fig. 13.1.

13.1.1. Fractal theory and fractal dimension

One quantitative measure of the structure of such objects is their fractal dimension D. Mathematicians calculate the dimension of fractal to quantify how it fills space. The familiar concept of dimensions applies to the object of classical or Euclidian geometry. Fractals have non-integer (fractional) dimensions whereas a smooth Euclidean line precisely fills a one-dimensional space. A fractal line spills over a two-dimensional space. Figure 13.2 shows subjects with increasing fractal dimension.

For linear like structures $1 < D < 2$ (Fig. 13.2a). For fractally rough structures $2 < D < 3$ (Fig. 13.2c); however, for uniform non-fractal objects $D < 3$ (Fig. 13.2d).

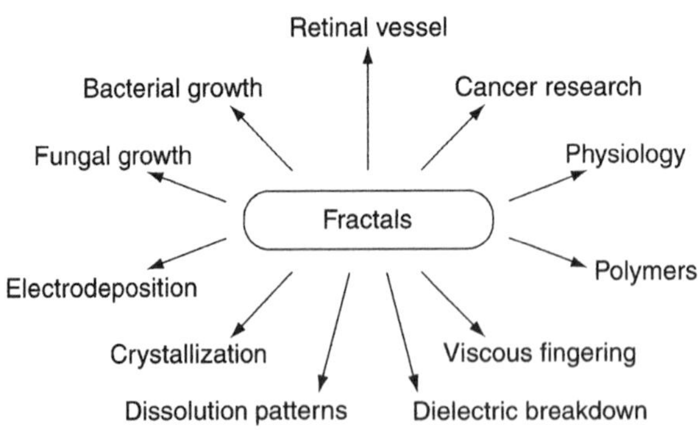

Figure 13.1. Fractals observed in different areas.

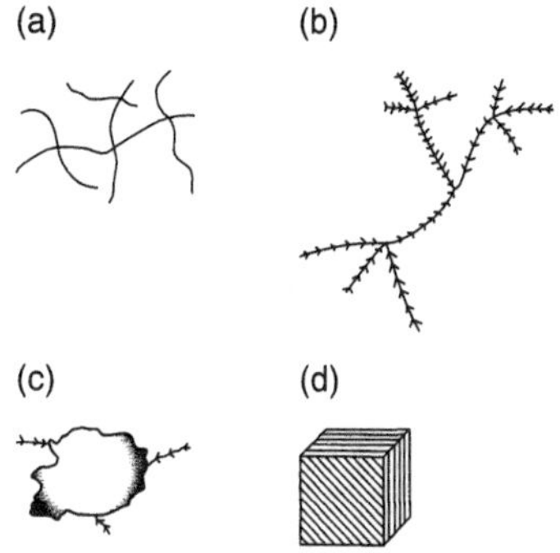

Figure 13.2. Subjects with increasing fractal dimensions.

Fractal geometry is a new language used to describe the model and analyse the complex forms found in nature. The methods of fractal geometry allow the classification of non-equilibrium growth processes. The classification and computer simulation give insight into a great variety of complex structures [7]. The basic concepts of fractal geometry are reviewed by Sagues *et al.* [8]. Readable account of concepts of fractal geometry is now available (Fig. 13.3).

It has been shown that every simple non-linear system and model can lead to complex, often chaotic behaviour that can frequently be described in terms of fractal geometry. In addition, it has been shown that simple models for growth and aggregation processes frequently lead to disorderly structures that exhibit spatially chaotic fractal geometry. Fractal-like structures [9] play a vital role in the healthy mechanical and electrical dynamics of the heart. Multi-fractal measures are one of the hottest subjects in the current scientific discussion of fractals. The concept of topology is relevant for the analysis and synthesis of models, which attempt to explain the continuous distortion of forms, represented by fractal geometry [10]. Physiologists quantified observations of the branching pattern and discovered that lung tree networks of blood vessels, nerves and ducts have fractal-like structures. Many other organ systems also appear to be fractal,

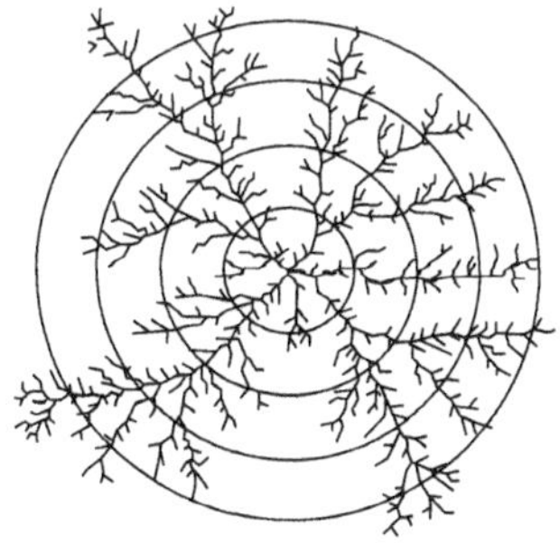

Figure 13.3. Branched structure within the circles of different radii.

although their dimensions have not been quantified. Devastation caused by forest fires and wars can be characterized by similar type of fractal dimensions. Forest fires are known to follow behaviour patterns known as self-organized criticality. This has also been recognized as the fundamental driving force behind earthquakes, solar fares and even the muscle contraction of a woman giving birth [11].

Self-similar objects can be divided into N parts. Consider a two-dimensional ℓ

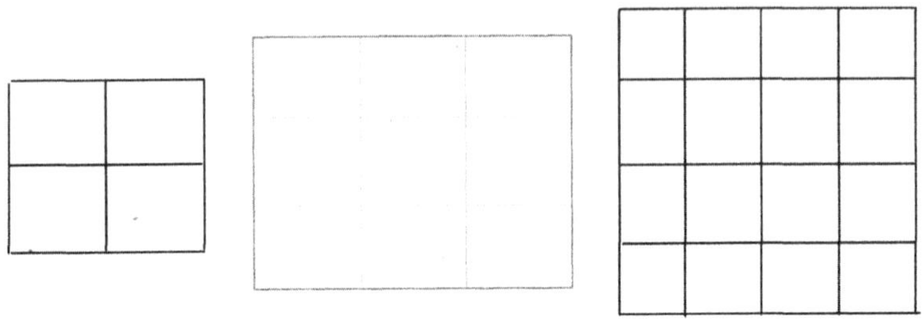

$N=4$ $\ell=2$ $N=9$ $\ell=3$ $N=16$ $\ell=4$

square of unit area divided into N equal subsquares of length ℓ. In fractal geometry N is supposed to be related to ℓ by the relation

$$N = 1/\ell^D$$

or

$$D = \frac{\log N}{\log 1/\ell}$$

where D is a fractal dimension. It does not conflict with Euclidean dimension as illustrated below. Let us divide a two-dimensional square of unit area into N equal subsquares of length ℓ where $N = 16$ and $\ell = 1/4$. It is obvious that

$$D = \frac{\log\ 16}{\log\ 4} = 2$$

which is just the Euclidean dimension.

Let us consider the following simple structures, where D is non-integer, called fractals. Let us first consider contour set. We can have a set of lines as shown in Fig. 13.4. A line is divided into three parts. One middle part is removed. Then again, we divide the remaining two parts into three subparts and the middle part in the two cases is also removed. The process is repeated again and again and finally the set of lines called contour set is obtained.

We can also apply these ideas to the generation of Koch curves. Let us start with a line one unit long (Fig. 13.5a). The middle third of the line is removed and replaced

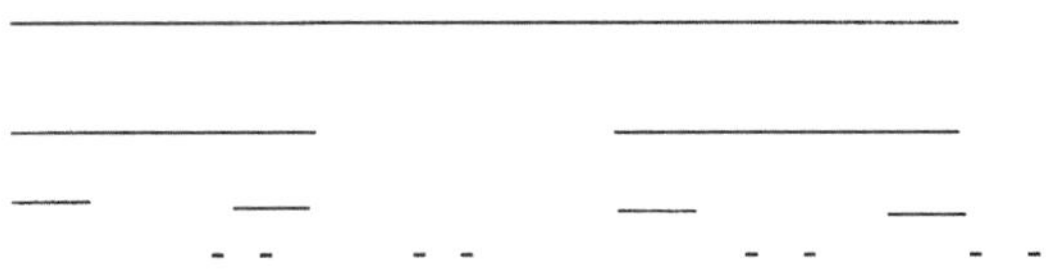

Figure 13.4. Simple example of contour set.

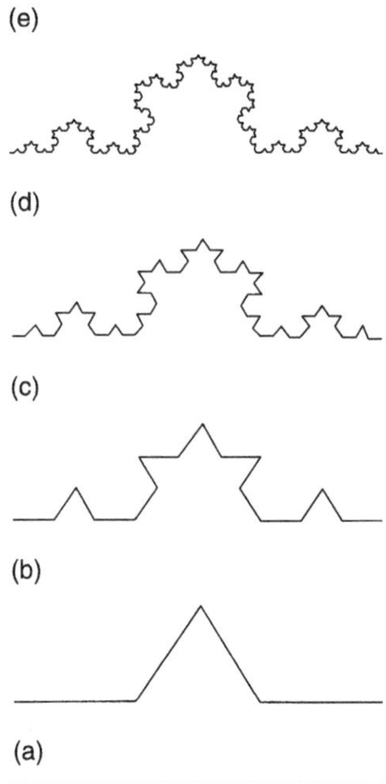

Figure 13.5. Different stages in the growth of an exact fractal, the triadic koch curves.

by two lines of length 1/3 (Fig. 13.5b) and the total length of the curve is 4/3. In the next stage each of the segments of length 1/3 is divided into lines of length 1/9 and the process is repeated (Fig. 13.5c and d). These stages can be extended an infinite number of times. Such a curve is known as triadic Koch curve, "Koch" after the mathematician Von Koch who first described it.

Likewise, Sierpinski gasket and Sierpinski carpet are illustrated in Fig. 13.6a and b, respectively.

Thus, the Koch curve has a dimension between a line ($D = 1$) and the plane ($D = 2$). Since D is non-integer, such structures are called fractals. An approximate value of D can be obtained for such structures by drawing concentric circles of different radii (R) and then counting the number of branches in each circle. A plot of $N(R)$ against log (R) yields a straight line, the slope of which yields the fractal dimension D. In the same way if we look at the branched structure within a circle of different radii, they all look alike.

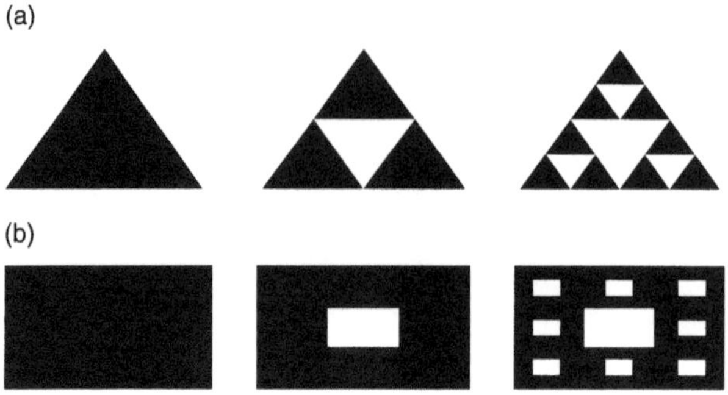

Figure 13.6. (a) Sierpinski gasket and (b) Sierpinski carpet.

The Simplest example of fractal is the contour set. We have taken the simplest example to illustrate the meaning of fractal dimension and its estimation. It should be noted that the fractal dimension is a global property of the cluster and it does not provide a deep insight into the structural details of the aggregate. Euclidean dimensions for different geometries are recorded in Table 13.1 while self-similarity dimensions of some deterministic fractals are recorded in Table 13.2.

13.1.2. Methods for fractal dimension calculation

The patterns produced by the diffusion-limited aggregation (DLA) processes are characterized by the open random and tree-type structures and can be well described as fractals. Computer simulations of fractal growth have been shown to produce structures

Table 13.1. Self-similarity dimension of some deterministic fractals.

Object	Scale	Pieces	Dimension D_s
Contour set	1/3	2	log 2/log 3 = 0.6309
Koch curve	1/3	4	log 4/log 3 = 1.2619
Sierpinski gasket	1/2	3	log 3/log 2 = 1.5810
Sierpinski carpet	1/3	8	log 8/log 3 = 1.8910

Table 13.2. Euclidean dimension for different geometries.

Geometry	Dimension
Point (.)	Zero (Euclidean)
Line (–)	1 (Euclidean)
Circle or square (o, □)	2 (Euclidean)
Cube	3 (Euclidean)

that closely approximate the structure obtained by these natural processes. Fractal dimension calculator estimates the fractal dimension of an object represented as black and white images . An object to be analysed is assumed to be made up of the black pixels. This is accomplished by an algorithm called "box counting". Mathematicians have come up with some 10 different notions of dimension which are all related. Some of these are: (i) topological dimension, (ii) Hausdorff dimension, (iii) fractal dimensions, (iv) self-similarity dimension, (v) box counting dimension, (vi) capacity dimension (vii) information dimension, (viii) Euclidean dimension. The relation between the total number of points N inside a circle of radius R and radius R itself should be a power law with non-integer exponent D as $N \sim R^D$ where thickness of branches is considered as zero dimensional. Therefore a careful counting of number of branches for different values of R as indicated in Fig. 13.6 provides approximate information about the exponent D, the fractal dimension. Using this concept, fractal dimensions for various aggregates described here may be calculated. The method has been checked on a number of systems reported earlier. Their values were calculated and compared with the computer simulation results and found to be in good agreement. Besides this there are other methods for fractal dimension calculation, namely

(a) Mass radius method in which mass of the object particle within a circle of radius R is plotted as a function of radius R. The slope of the linear plot gives the value of fractal dimension D using equation $M = R^D$.
(b) Weight angle method is applicable for circular geometry such as patterns obtained by dropping coffee onto a paper. In this method log–log plot of $W(\theta)$ and θ is made and the slope of best-fitted curve provides its fractal dimension [12]. $W(\theta)$ represents weight within two surface points separated by an angle such that $W(\theta) \sim \theta$.

Fractal dimensions of some non-living and living systems including crystallization patterns, electro-deposited aggregates, polymers, chemical dissolution patterns, dielectric breakdown, sputter deposited film of $NbGe_2$, retinal vessel and bacterial growth are given in Table 13.3.

13.1.3. Growth models for different types of fractals

Mandelbrot [2, 3] systematized and organized mathematical ideas concerning complex structures such as trees, coastlines and non-equilibrium growth processes. He pointed out that such patterns share a central property and symmetry which may be called scale invariance. These objects are invariant under a transformation, which replaces a small part with bigger part that is under a change in a scale of the picture. Scale-invariant structures are called fractals [7]. More recently the relevance of natural and mathematical structure has become clearer with the help of computer simulation. Self-similarity turns out to be a general invariance principle of these structures.

Table 13.3. Fractal dimensions of typical systems.

System	Fractal dimension D
(i) Crystallization patterns	
Oxalic acid [77]	1.65
o-Toluic acid [77]	1.61
Mandelic acid [77]	1.84
Glucose [25]	1.89
Cellobiose (0.1 M) containing 0.1% agar-agar [25]	1.74
Lactose (0.1 M) containing 0.1% agar-agar [25]	1.58
Dextrose (0.1 M) containing 0.1% agar-agar [24]	1.89
(0.1 M) containing 0.2% agar-agar [24]	1.48
Saccharin (0.007 M) containing 0.075% PEG [25]	1.26
(0.014 M) containing 0.075% PEG [25]	1.28
KCl admixed with 0.2% agar-agar [22]	
[KCl] = (0.001 M)	1.42
(0.01 M)	1.62
(0.05 M)	1.76
$SrCl_2$	1.52
(ii) Electro Crystallization patterns	
Silver electrodeposits from silver nitrate solution [48]	1.95
2-D electrochemical polymerization of pyrrole [27]	1.74 ± 0.03
Lead-Zinc binary system [48]	1.87
(iii) Living systems	
Retina [62]	1.7
Bacillus subtilis [66]	1.72
E. coli [70]	1.55
K. ozaenae [72]	1.61
(iv) Some other systems	
Koch curve [7]	1.26
Fractal patterns from chemical dissolution [36]	1.6 ± 0.01
Platinum and gold electrodeposits [78]	1.50 ± 0.1

The methods of fractal geometry allow the classification of non-equilibrium growth processes according to their grading properties. This classification and computer simulations give insight into a great variety of complex structures. An interesting and instructive microcomputer simulation of fractal electro-deposition has been described by Sagues and Costa [13]. Importance of fractal geometry has been emphasized in the context of strange attractors. Fractal concept has also been used in investigating turbulence [14]. For simulation, stochastic growth models are postulated and compared with experimental growth patterns wherever possible.

Diffusion-limited aggregation

A large number of growth models have been studied from the above viewpoint. Some of the typical examples are:

Random walk: Random walk is a stochastic process. Random walk model leads to master equation which can be used for different purposes [15]. In this model we consider a two-dimensional lattice. We choose a walker, which randomly chooses one of the nearest neighbour sites. After many steps, the set of points visited by the walker becomes space filling so that the fractal dimension attains the value of 2. Different types of random walk models have been used, such as (i) self-avoiding random walk (SAW), (ii) plane-filling walker and (iii) epidemic growth models. In (i) the restriction is that each lattice site may be visited once during the walk. In case (ii) we consider a two-dimensional lattice and a walker which at each step chooses randomly one of the nearest neighbour sites. After a series of steps, the set of points visited by the walker becomes plane-filling and hence has the fractal dimension $D = 2$. In case of epidemic growth, one regards the lattice sites as cells, which may get infected or become immune. The infection starts with one cell, which acts as a seed. The growth proceeds by choosing at random, at each step, fresh cell adjacent to the boundary of the cluster. This cell either gets infected with a certain probability and hence joins the cluster or becomes immune. The model has been used in the case of percolation and forest fire. It has to be emphasized that all irregular structures are not ultimately fractals.

Self-avoiding random walk model (SAW): Here one imposes the condition that each lattice site may be visited only once during the walk.

In both the cases, mean squared end-to-end distance $<R^2>$ scales as a function of the number of steps N,

$$<R^2> \sim N^{\alpha}$$

where $\alpha = 1$ for random walk. For SAW it is not equal to 1 but depends on the dimension D of the embedding space for the walk [16].

Several modified models like (i) self-attracting self-avoiding walk (SASAWS) and (ii) true self-avoiding walks (TSAWS) have been reported in the literature [17].

The rapid growth of computing and applications has helped cross-fertilization of mathematical sciences yielding an unprecedented abundance of new methods, theories and models. Mathematical science has become the science of patterns, with theory built on relations among patterns and on applications derived from the fit and observation.

We discuss below the DLA model in detail which was proposed by Witten and Sander [16, 18]. The DLA is a particular model of a random irreversible growth. The growth process starts from a seed particle. A second particle is launched far enough from the seed and makes a random walk. If it visits a position next to the seed, it is stuck to it and both form a two-particle cluster extending the initial seed. Then a third particle is launched and moves randomly around this cluster. It may join the two-particle cluster if it passes on one of the neighbouring positions. By repeating the process, one observes the growth of random object. The salient features of this model are as follows:

(a) There is a seed particle at the origin.

(b) Another particle is allowed to walk at random (diffuse) from far away until it arrives at one of the lattice sites adjacent to the occupied site.

(c) Then it is aggregated.

(d) Finally another particle is launched and the process is repeated.

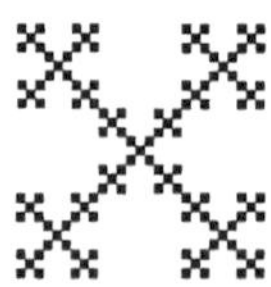

The role of diffusion (Brownian motion) in colloidal aggregation had been emphasized much earlier [19]. Brownian movement can be understood as follows. When the movement of a small particle in a liquid is observed under a microscope, zig-zag motion of the particles is observed under a microscope. In the model purposed by Haken, it is assumed that a particle moves along one-dimensional chain where it may hop from one point to one of its neighbouring points with equal probability. The model is used for estimating its probability that after n steps the particle reaches a certain distance x, from the point where it had started. Large-scale computer simulations have recently been used to provide in-depth studies of growth processes. Witten and Sander [17] showed that a simple diffusion-limited growth model leads to complex patterns for aggregates based on the concept that growth results from particles diffusing to a cluster and getting attached to it. In their simulation a particle seed is placed at the origin. Then particles are added one by one, which are released far from the seed. They join the aggregates by performing a random walk as shown in Fig. 13.7.

The mean square displacement of $<R_t^2>$ is given by

$$<R_t^2> = [2kT/\zeta]t$$

where ζ is the friction coefficient and t is the time. The walk terminates when the particle reaches a site nearest neighbour to the DLA formed earlier.

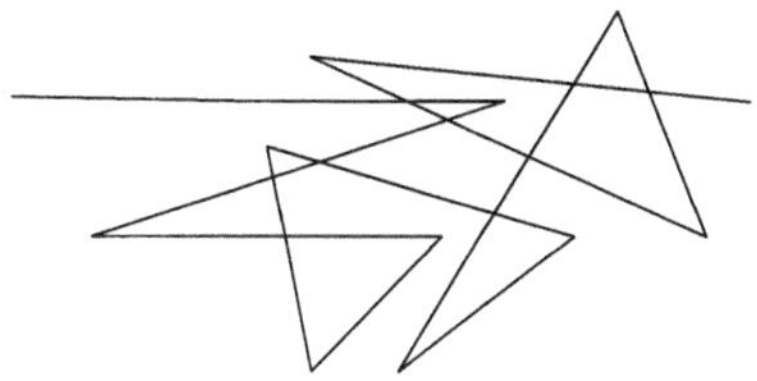

Figure 13.7. Brownian motion.

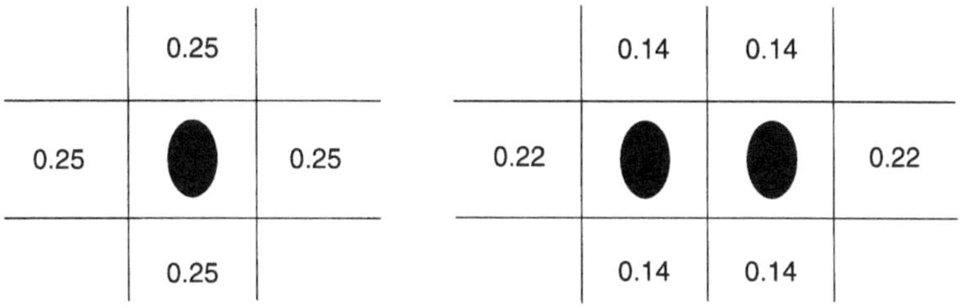

(a) Square lattice DLA at time $t = 1$ (b) DLA at time $t = 2$ with six growth sites

Figure 13.8. Diffusion-limited aggregation at a (time $t = 1$) and b (time $t = 2$).

Like many models in statistical mechanics, the rule defining DLA is simple. At time l, we place in the centre of the computer screen a white pixel (mass M) and release a random walk surrounding the white pixel as shown in Fig. 13.8.

The four-parameter sites have an equal a priory probability p_l to be stepped up by random walk. Thus in case of Fig. 13.8a

$$p_l = {}^1\!/\!_4 \; (I = 1, \ldots, 4)$$

In Fig. 13.8b the number of sites $N_p = 6$ are called growth sites, but now the probabilities are not identical, because the third particle is more likely to stick. It does not mean that the next particle will stick on the top. It is comparatively more convenient to let the particle follow a random walk [20], so when the particle is released, it moves with a probability ¼, left, right, up or down to a neighbouring square, continuing until it leaves the circle or occupies a square next to the shaded one. For illustration, if the model is run for 10 000 shaded squares one obtains a highly branched structure (Fig. 13.9).

Sometimes a situation may arise when late-arriving particles are less prone to reach the central region, thereby attaching themselves at the periphery of the cluster. For analysis one uses mathematical properties, such as the number of distinct sites visited at least once in a t-step walk S_t, the mean square displacement R_t^2 to monitor such processes [21]. One of the reasons for the success of DLA model is the large number of experimental systems displaying a structure similar to the DLA aggregate. A typical DLA cluster is shown in Fig. 13.9.

Figure 13.9. A typical DLA cluster.

Different types of DLA models have also been proposed. These include two- and three-dimensional DLA models. Sometimes, tree-like shape fractals can occur due to the fact that later arriving particles may not reach the central region and thus attach rather at the periphery of the cluster. In DLA all the particles that stick to a cluster are of the same size and the cluster that is formed is motionless.

Cluster–cluster aggregation

The limit in which clusters of all sizes can bind themselves and the growth is called cluster–cluster (CC) aggregation. Instead of growth of one cluster, many aggregation processes involve cluster-cluster aggregation. Another model depends on kinetic gelation, according to which growth process is initiated by activating several monomers. Activated monomers can link themselves to their nearest neighbours provided both monomers can functionally undergo a bond. The process continues by having the newly bonded monomers activated.

Kinetic growth models

Kinetic growth models describe polymerization and aggregation processes that occur far from equilibrium. In chemical systems these models are realized when union between growing species are strong and irreversible, i.e. bonds once formed do not dissociate. Other thermodynamic models relevant for nucleation and growth close to equilibrium discussed earlier are not expected to produce polymeric structure. Kinetic models are distinguished by assumption concerning mass transport and accretion. Mass transport refers to the process by which reacting molecules approach each other. If the process is Brownian, then the aggregation is diffusion-limited. We may call it DLCA (diffusion-limited cluster aggregation). In case the transport plays no role at all, the role of aggregation mechanism is most important so that all available sites are equally convenient for aggregation.

In the reaction-limited process (RLCA), transport plays no role at all; so all available sites are suitable for accretion, leading to more ballistic aggregation, which describes growth of clusters that approach each other along linear trajectories. There are two types of RLCA processes, depending on the mode of accretion. When the dominant growth process occurs by adding of small species to large clusters, one has monomer-cluster growth. On the other hand, when clusters of all sizes can bond we have cluster–cluster bond.

Deterministic fractals: Deterministic fractals are based on computer simulation. Distribution of sites is determined by an unambiguous, non-random prescription such as that involved in Sierpinski-type fractals.

13.2. Experimental studies of complex patterns

Simple experimental description for the development of random fractals.

The development of fractal patterns during crystallization: A simple experimental setup shown in Fig. 13.10 was employed to study the crystal growth in a two-dimensional configuration.

A known volume of solution is poured and spread uniformly over the surface of micro-slide (7.5 cm × 2.5 cm) and placed in an air thermostat, the temperature of which could be controlled to ±0.1°C. Evaporation starts and after a certain time of initiation crystallization takes place. To examine the morphology, photographs are taken with the help of a camera attached to a microscope.

Recently, Rastogi *et al.* [23] have investigated a variety of systems that exhibit different types of morphology when allowed to grow in thin films of solutions containing a denser matrix such as agar-agar or PVA polymer. Irregular fractals like growth of potassium chromate, strontium chloride, potassium chloride in agar-agar, tree-like geometry in the growth of potassium dichromate were reported. A typical result is shown in Fig. 13.11.

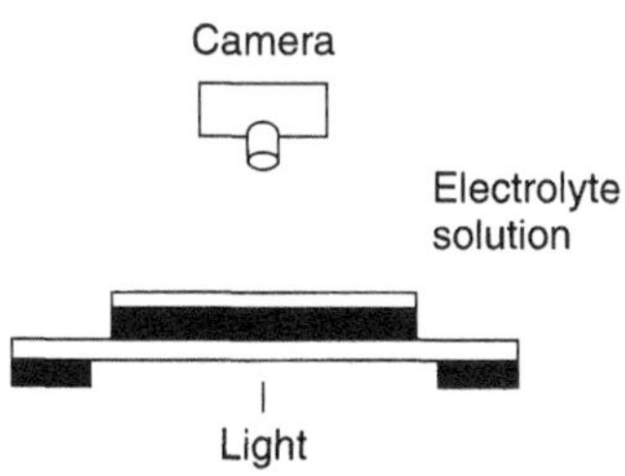

Figure 13.10. Experimental setup for observing the development of fractal geometry during crystallization [22].

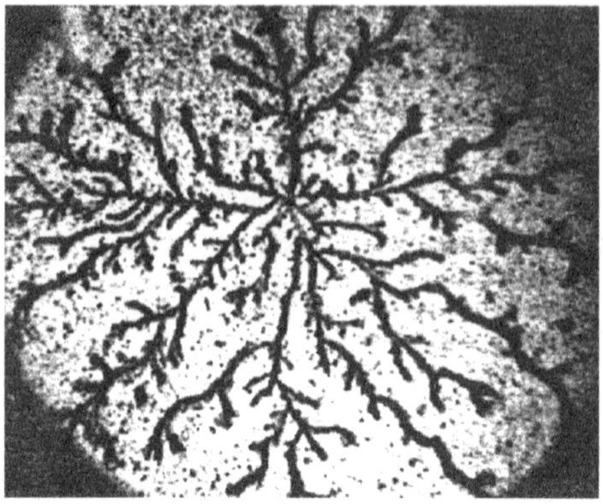

Figure 13.11. Fractal pattern of potassium chloride containing 0.1% agar-agar medium at 25°C, 0± 0.1°C [23].

A variety of patterns of dextrose were also observed. Gel and dextrose concentrations influence the morphology to a great extent. Later on, Rastogi *et al.* [24] have described supplementary studies for the development of different morphologies of dextrose/agar-agar, ascorbic acid from aqueous solution containing agar-agar as well as methanolic and ethanolic solutions. Das *et al.* [25] have described the non-equilibrium growth patterns of mono and disaccharides, and sweetener saccharin has been developed on micro-slides in the presence of a dense matrix. DLA-like growth patterns were observed in both mono and disaccharides. Scanned pictures were analysed and fractal dimensions calculated by a box counting method admixed with agar-agar. Glucose, a mono-saccharide with molecular formula $C_6H_{12}O_6$ in water shows an equilibrium between cyclic structures.

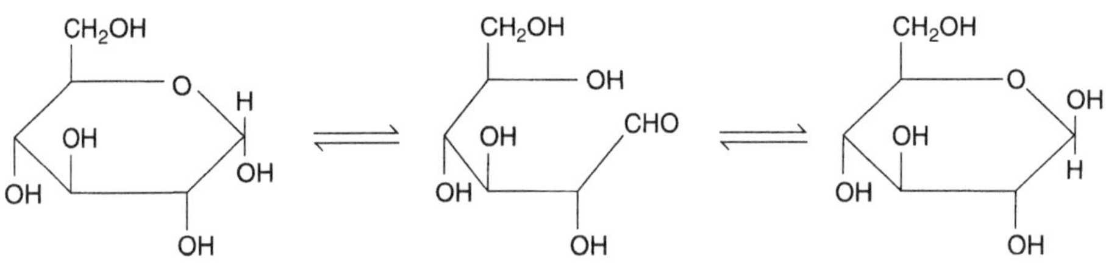

It shows a fractal structure (ringed spherulite) with a fractal dimension $D = 1.89$. A decrease in the fractal dimension ($D = 1.48$) was observed with an increase in the agar-agar concentration. Transition from the fractal ($D = 1.48$) to dense radial morphology occurred when the glucose concentration or the concentration of dense matrix was increased. Maltose having a 4-*O*-β-D linkage does not show any pattern. It is observed that the morphology depends on monomer units of the disaccharide.

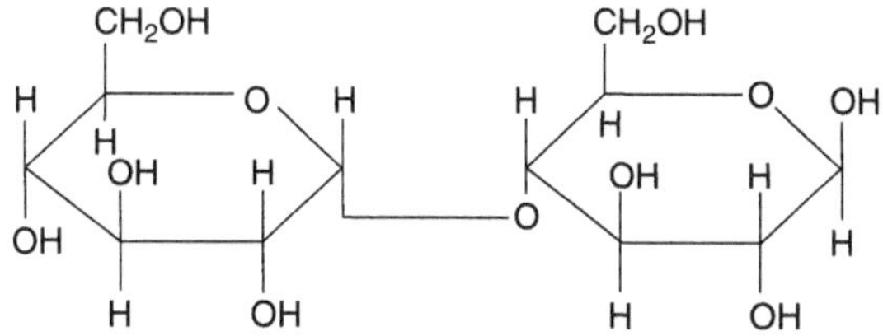

4-*O*-α-D-Glucopyranosyl-D-glucopyranose

Lactose (4-*O*-β-D-galactopyranosyl-D-glucopyranose) in which glucose and galactose units are linked together by 4-*O*-β-D linkage showed a pattern with fractal dimension $D = 158$, in agreement with two-dimensional DLA model. In case of disaccharide, the morphology depends on a linkage between the monomer units. The crystallization pattern of mandelic acid was found to depend on the solvent/medium and thereby the rate of evaporation. Mandelic acid shows fractal-like morphology with fractal dimension $D = 1.84$ when the rate of evaporation was fast. However, it crystallizes in the form of concentric rings in an aqueous medium containing agar-agar, which reduces the rate of evaporation. Isonicotinic acid hydrazide (INH) is one of the most effective anti-tubercular drugs besides its bactericidal properties while nicotinic acid hydrazide (NH) shows only bactericidal activities. These hydrazides crystallize differently. INH (0.05M) crystallize in the form of needle from its aqueous solution while NH (0.05) crystallize in the form of fractal with right-handed twisting in the presence of 0.2% agar-agar, as shown in Fig. 13.12b with fractal dimension $D = 1.77$.

Rastogi *et al.* [22] have reported the observation of fractal-like, dendritic and periodic morphologies experimentally in thin films of KCl solution containing a denser matrix such as agar-agar or polyvinyl alcohol. In the case of fractal-like structures, calculated values of fractal dimensions were in good agreement with two-dimensional DLA model. A transition-like behaviour from fractal type to dendritic growth with increase in

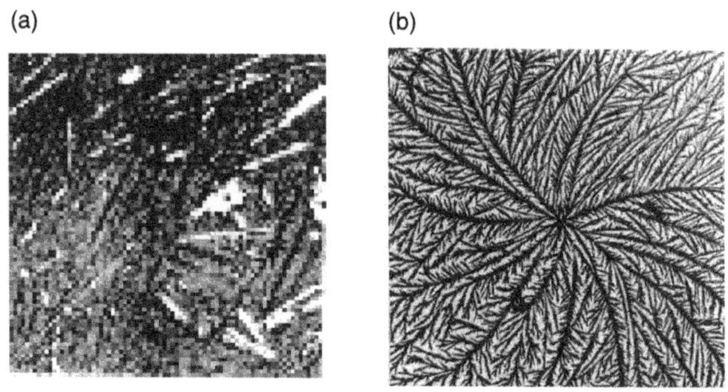

Figure 13.12. Polarized light microphotographs of [INH] = 0.01 M, crystallized from aqueous solutions containing 0.2% agar-agar.

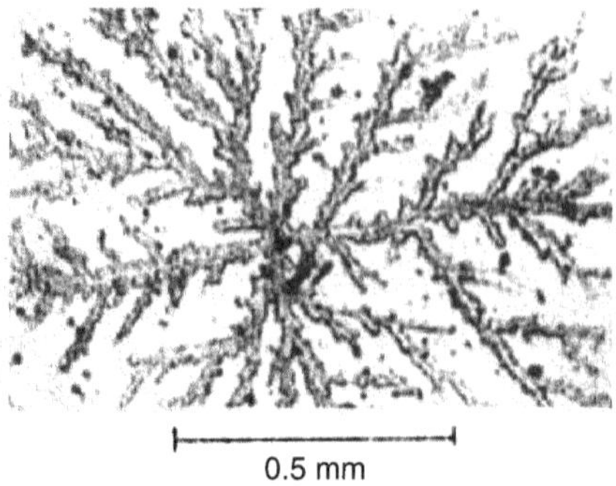

Figure 13.13. Microphotograph of KCl crystallized from its aqueous solution (0.01 M) containing 0.2% agar-agar at 25.0 ± 0.1°C [22].

KCl concentration has been observed. Typical fractal pattern of KCl crystallized from aqueous solution containing 0.2% agar-agar is shown in Fig. 13.13.

13.2.1. Complex patterns in polymers

Polymers belong to an important class of materials with a wide range of applications including paints, rubbers, oils, plastics and composite materials [26]. Fractal aggregates can be obtained through polymerization of certain monomers under suitable conditions. Two-dimensional polymerization of pyrrole was carried out in a thin layer electrical cell consisting of two glass plates [27]. The solution in the cell was 0.1 M pyrrole/0.1 M AgO toluene sulphonate/acetonitrile. The growth of polypyrrole involved electrochemical oxidation of neutral pyrrole monomers. Two-dimensional DLA-like clusters with a fractal dimension of 1.74 ± 0.03 were obtained.

Fractal pattern of polyethylene oxide in a radial Hele-Shaw cell was reported by Maher and Zhao [28]. A very rich variety of patterns were observed as water advances against an aqueous solution of polyethylene oxide in radial Hele-Shaw cell. The mass fractal dimensions of patterns varied from 1.2 to 1.8. A wide variety of substance ranging from naturally occurring minerals to stereochemically regular synthetic polymers crystallize in the form of spherulites. Spherulites are polycrystalline aggregates. They exhibit the super molecular structure. Spherulites as shown in Fig. 13.14 are regular birefringent structures with spherical symmetry in three dimensions. They are composed of crystallite fibrils that grow radially outwards from the nucleus.

Spherulitic growth may appear whenever crystallization develops with repeated branching. An oak tree is a good example of spherulitic form. It should be noted that spherulites might be fibrilar as well as lamellar in structure. Spherulites are most easily recognized by their characteristic birefringence. It follows from the spherically symmetrical disposition of their components units that this birefringence can be defined in terms of two refractive indices, corresponding to directions parallel and perpendicular to

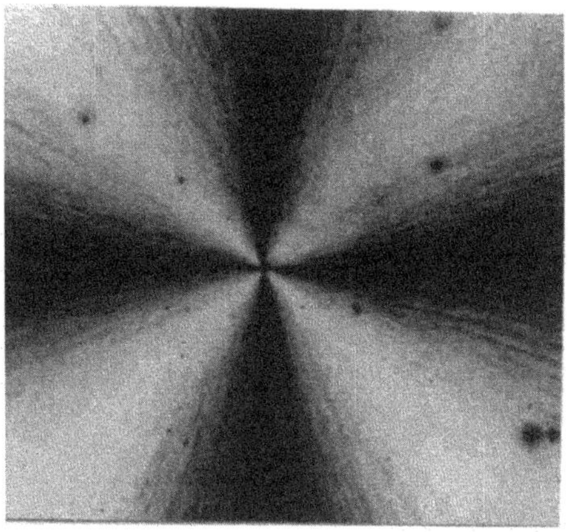

Figure 13.14. Maltese cross in a polymer spherulite.

radius. The most conspicuous property of the spherulites is their maltese cross extinction pattern. Spherulitic birefringent structures range in diameter from approximately 1 μm to 1 mm, depending on nucleation density. They are not visible in ordinary light but can examined in thin section between crossed polars using a polarized light microscope. It has also been observed that the birefringence and morphology of spherulites depend on chemical identity, unit cell and crystallization temperature. Certain polymers such as linear polyethylene and poly(3-hydroxy butyrate) (PHB) exhibit regularly spaced concentric extinction rings in addition to radial extinction. Such a banding in spherulites is a common observation that has been studied for more than a century. In early work on minerals such as chalcedony and many organic and some inorganic compounds, correlation was found in almost all cases between such banding and chirality of molecular structure. Banding appears in polyethylene samples, which have been crystallized at a relatively high degree of super cooling, i.e. at low temperature.

Morphology of polyethylene glycol (PEG) before and after UV irradiation was investigated by Das and Gupta [29]. A drastic change in the morphology from spherulitic to rhythmic pattern has been observed under polarized light microscope when exposed to UV light (Fig. 13.15).

Dendrimer and hyper-branched polymers are important in scientific, engineering and medical applications [30, 31]. An analytic solution for growth of branched aggregates or polymers with distributed cluster (dendrimer) size was proposed by Benjamin [30]. Exact solutions showed how the properties of fractal clusters grew with time. There is considerable scientific and technological interest in conducting polymers. Polyaniline is a conducting polymer used in battery industry. Polyamine has attracted the interest of researchers due to its satisfactory environmental stability, high degree of electrical conductivity with reasonable mechanical properties. It is the only known conducting polymer with nitrogen atom occupying the bridging position in its backbone structure. Electro-polymerization of aniline was carried out by Das *et al.* [32] in the presence of

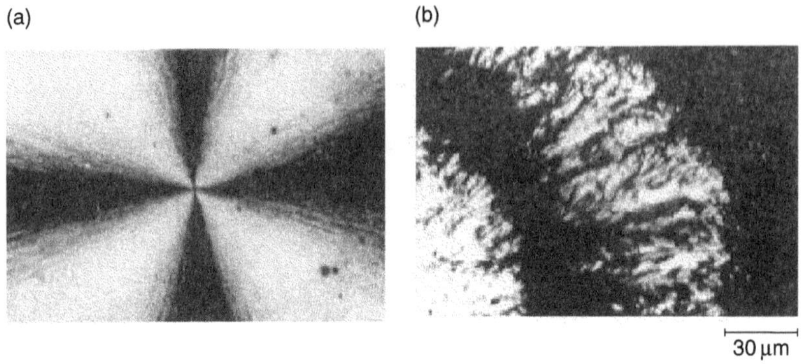

Figure 13.15. Microphotographs of PEG taken at different times of UV exposure (a) 0 h and (b) 7 h.

light. The kinetics of polymer growth was studied in terms of weight of the aggregate as a function of time in the absence and presence of $ZnSO_4$, which inhibits the polymer growth by incorporating itself in the polymer network. An empirical equation $w = ce^{mt}$ was obeyed in acid–aniline system while a different equation $w_1^2 = m_1 t + c$ was obeyed when $ZnSO_4$ was admixed. Further $ZnSO_4$ increased the stability of the polymer. Polymerization was found to be reduced by 70% in complete darkness.

13.2.2. Viscous fingering

Viscous fingering phenomenon is of practical importance in the recovery of oil. It is observed that when a fluid displaces another fluid with a higher viscosity, the interface is unstable and the driving fluid intrudes into the viscous fluid in the form of "fingers". Horvath and co-workers [33] studied viscous fingering in the radial Hele-Shaw cell with uniaxial anisotropy versus the driving force in the growth of diffusion-limited patterns [34, 35]. A typical viscous fingering pattern is shown in Fig. 13.16.

Figure 13.16. A typical viscous fingering pattern.

13.2.3. Fractal patterns from chemical dissolution

Daccord and Lenormand [36] have shown experimentally that the dissolution patterns obtained by injecting water through pure master were fractal. These results are of interest in different areas where chemical dissolution of porous media by a flowing fluid occurs. In nature, the formation of caves and the oil industry are examples of this process. In two dimensions, these dissolution patterns are remarkably similar to patterns associated with DLA, which includes dielectric breakdown. The two-dimensional dissolution patterns are remarkably similar to patterns associated with DLA.

13.2.4. Surface fractals

In a similar fashion we can have surface fractals, which are uniformly dense but have rough structures such as in the case of colloidal particles. The surface similarity in this case is represented by the following analogous equation:

$$S \propto R^D$$

Aggregates are further classified as surface fractals, volume fractals or non-fractal objects. Kolb and Herrmann [37] have introduced a mechanism to model aggregation of clusters at high concentration and investigated it by Monte-Carlo simulations. Surfaces of most materials are fractal on a molecular scale to chemically reactive surfaces. Farin and Avnir define the reaction dimension D_R as characteristic parameter of the reaction which is related to the reaction rate v and particle radius R as

$$v \propto R^D$$

The aggregation processes frequently results in the formation of complex materials which can be described in terms of the concept of fractal geometry [27]. In general, the reaction dimension D_R would be smaller than the surface fractal dimension D because of selective participation of surface site in the reaction. However, Farin and Avnir [38–40] found that heterogeneous reactions were phenomenologically much richer. If the surface of particles is fractal, then the surface area A would be related to the particle radius R as $A \propto R^{D-3}$ which is obeyed by many materials. Another situation may arise when the aggregation occurs during the growth of cluster that approaches each other along linear trajectories. Monomer crystal growth occurs when the dominant growth process involves the addition of small particles to large clusters.

Solid–solid reactions are the example of reaction in which the motion of reacting species is constrained in space that has lower dimensionality in contrast with usual chemical reactions where the reactants are free to move in a three-dimensional space. Such a situation leads to development of surface fractals [41]. In view of the nature of diffusion, one can expect formation of surface fractals in heterogeneous reactions.

Typical inorganic solid-state reactions between (i) $HgCl_2 + I_2$ and (ii) cobalt nitrate + NH_3 have been investigated from this angle [42, 43].

Heterogeneous reaction kinetics in areas such as chemistry, biology, geology, solid-state physics, astrophysics and atmospheric science can be understood in terms of fractal kinetics. Recently, Rastogi *et al.* [42] performed an experiment to visualize the phenomenon in a solid–gas reaction. Mercurous chloride thin film was prepared on micro-slide and put in the iodine chamber. A white continuous interface was covered with yellow HgClI after 15 min (Fig. 13.17). The interface becomes percolation cluster type after 2 h. At the end of the reaction (after 72 h) red crystals are found embedded on the reaction interface. For the reaction to obey fractal-like kinetics, the reaction between ξ thickness of the boundary layer at time t is complex. These authors have further observed an identical relation $\xi^2/t = k^0 t^{-5/7}$ for the above reaction, where k^0 is some constant independent of time.

Das *et al.* [43] have studied the reaction between cobalt nitrate and ammonium hydroxide (solid–vapour reaction) on micro-slides. To study the reaction on a two-dimensional micro-slide, a small amount of finely powdered cobalt(II) nitrate is spread over a micro-slide (~ 0.2 mm thick layer). The surface of the material is made uniform

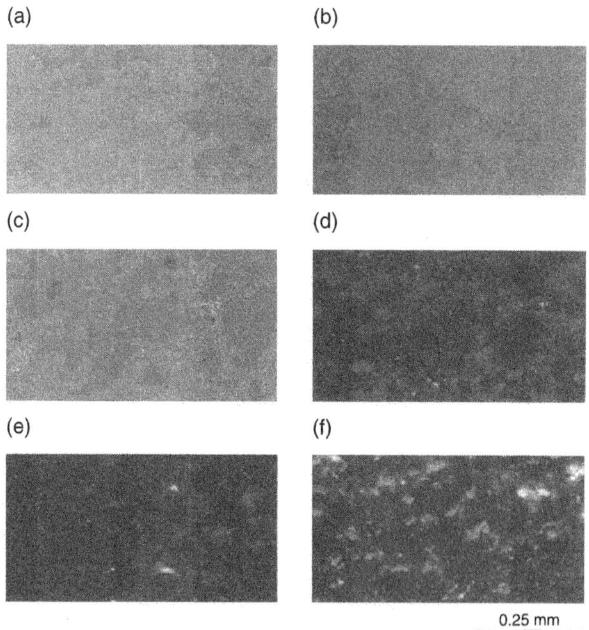

Figure 13.17. Various stages of two-dimensional growth behavior for reaction between mercurous chloride and iodine. Plate (a) indicates the initial stage of mercurous chloride (film thickness 0.4 mm), plates (b–f) represent patterns obtained after 15 min, 2.5, 24, 48 and 72 h, respectively, at 30.0 ± 0.1°C [42]. See also Colour Plate Section.

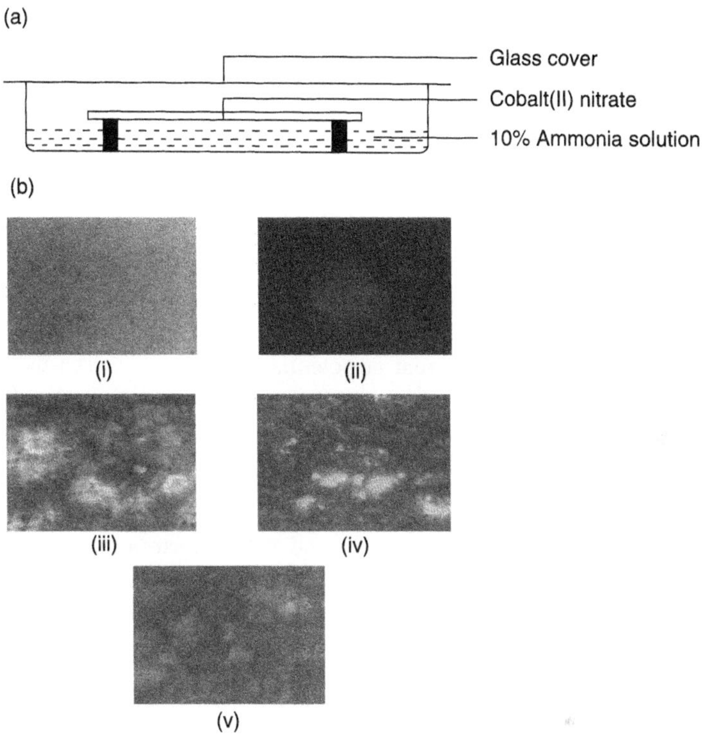

Figure 13.18. (a) Experimental setup. (b) Microphotographs of Co(II) nitrate and the products of the reaction between Co(II) nitrate and ammonia vapour at 20 ± 0.1°C on micro-slides. Plate (i): initial view of Co(II) nitrate; plates (ii) and (iii): patterns obtained after 5 min when viewed from the camera directly and with the microscope, respectively. Plates (iv) and (v) indicate the microscopic views of the products after 10 min and 24 h, respectively [43]. See also Colour Plate Section.

with the help of another glass plate. The micro-slide is then put into a chamber containing 10% ammonia solution (10 ml) and covered with a glass plate as shown in Fig. 13.18a. As a result of diffusion and chemical reaction, different coloured reaction products are formed at different time intervals. During the reaction, the colour of the reaction product undergoes change in the sequence *pink* → *blue* → *brown*.

13.2.5. Electro-deposition

Development of fractal patterns during electro-deposition: Electro-deposition has for a long time been one of the most familiar aggregation phenomena in chemistry. Electro-deposition is a typical example of far from equilibrium growth phenomena governed by diffusion [44, 45]. A number of different morphologies developed by the process of electro-deposition have been reported. In this case, the metal deposition is controlled mainly by a single process, e.g. diffusion. The deposits usually exhibit statistically simple, self-similar, i.e. fractal, structures. Typical examples are the recent studies on electro-deposition of binary systems containing Pb and Zn as well as Cu and Zn [46, 47]. It has recently been recognized as a genuine example of pattern formation process similar to other crystallization or interface displacement phenomena. Compared to other pattern-growing phenomena electro-deposition offers a unique opportunity to show different morphologies. Electro-deposition is an important industrial process and it can lead to rough surfaces and dendritic growth. Electrochemical deposition consists of growing a metallic cluster in an electrochemical cell filled with a solution of a salt of the metal by imposing constant current or potential difference between the electrodes. At the cathode, the reaction is

$$\mathrm{M}^{z+} + \mathrm{Ze} \longrightarrow \mathrm{M}_{\mathrm{deposit}}$$

where M is the metallic ion, Z its charge and e the electron. The morphology of the deposit depends on (i) the current density, (ii) concentration of the electrolyte, (iii) temperature, (iv) colloidal matter and (v) nature of the electrolyte. As the current density is raised the rate of formation of nuclei will be greater and the deposit will become more fine-grained. If the current density exceeds the limiting value for the given electrolyte, hydrogen will be evolved.

The experimental procedure used to grow metal leaves from electrolyte solution two dimensionally is shown in Fig. 13.19 and described in the following paragraph.

A petri dish of diameter 20 cm and depth 10 cm is filled with an aqueous electrolyte solution. The platinum cathode is then set at the centre of the petri dish so that the flat tip is placed just on the interface. The electro-deposition is initiated by applying a DC voltage between the cathode and the anode. The anode is immersed in the solution a little below the interface. The metal leaf grows two dimensionally at the interface. The temperature of the system is maintained constant. Depending on the experimental

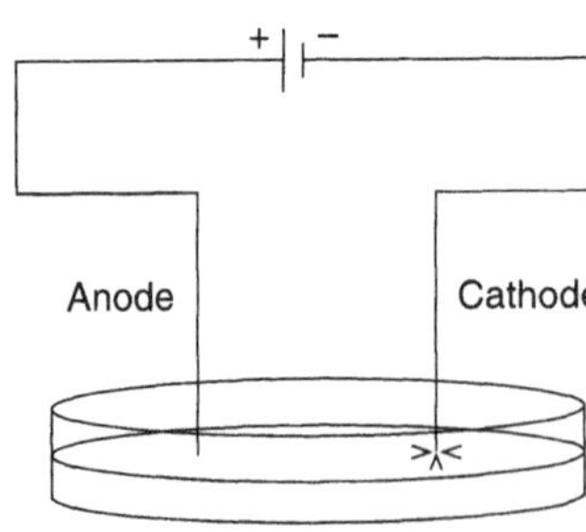

Figure 13.19. Experimental setup for the development of fractal geometry during electro-deposition.

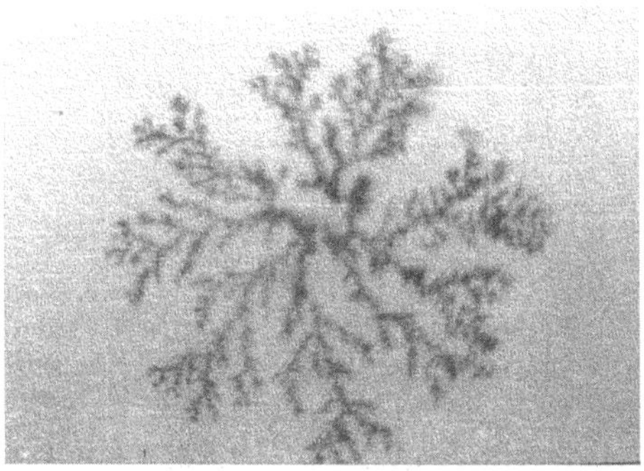

Figure 13.20. Electro-deposited aggregate obtained from aqueous solution of $Pb(NO_3)_2$–$Zn(CH_3COO)_2$ [46].

conditions, the electro-deposits exhibit a variety of structures such as dense radial, needle-like and branched aggregate pattern.

Experimental studies of electrochemical deposition of Pb, Zn and the binary system containing Pb and Zn with two parallel electrodes and a circular anode–point cathode are reported by Das *et al.* [46]. Microscopic view of the electrochemically deposited aggregate obtained from $Pb(NO_3)_2$–$Zn(CH_3COO)_2$ binary system mixed in 1:1 proportion (v/v) as developed by Das *et al.* [46] by employing vertical point anode and point cathode is shown in Fig. 13.20.

Growth kinetics data obey the simple empirical equation $d = ct^m$ for point cathode and circular anode system; d is the radius of the circular envelope grown in time t, and m and log c are the slope and intercept, respectively. Scanned pictures of electro-deposits are characterized in terms of fractal dimension by box counting method. Cathode potential changes with time were also monitored during electro-deposition and dissolution processes. Next amplitude plots indicated that the oscillations were periodic in the binary system while for pure Pb and Zn, it was like random noise. Das *et al.* [47] developed patterns from pure copper sulphate and zinc sulphate solutions as shown in Fig. 13.21

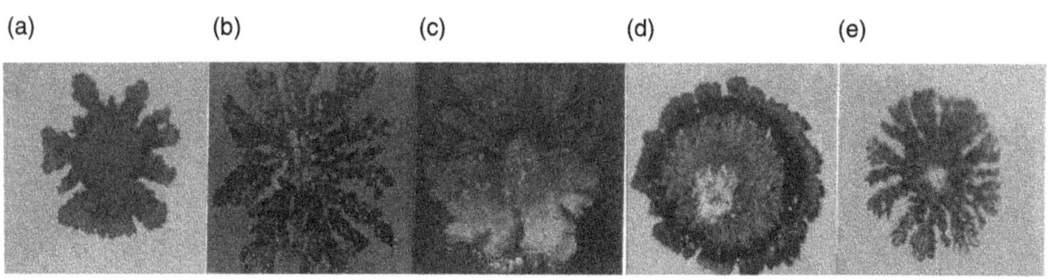

Figure 13.21. Microphotographs of electro-deposited aggregates obtained from (a) $CuSO_4$ (0.1 M), (b)–(d) $CuSO_4$–$ZnSO_4$ binary systems with $[CuSO_4] = 0.066$ M, $[ZnSO_4] = 0.033$ M; $[CuSO_4] = 0.05$ M, $[ZnSO_4] = 0.05$ M; and $[CuSO_4] = 0.033$ M, $[ZnSO_4] = 0.066$ M, (e) $ZnSO_4$ (0.1 M) [47]. See also Colour Plate Section.

plates (a) and (e), respectively. Electro-deposition shown in plate (c) is found to be little characteristic in the sense that the visual observations indicated yellow colour resembling β-brass.

Growth kinetics of the electro-deposits were studied by noting the radius of the circular envelope emerging from the tip of point cathode (in case of parallel electrode setup) as a function of time using a travelling microscope. New results on the non-equilibrium growth of silver have been reported by Das *et al.* [48] in a batch reactor using cells with different electrodic arrangements. Scanned picture of the aggregate was analysed and fractal dimension calculated by box counting method. On addition of copper nitrate, a transition from dendritic to a net-like morphology was observed. Cathode potential changes with time have been monitored in both, batch and flow reactors. Oscillations of small amplitude were observed.

Schilardi *et al.* [49] have developed the 2D copper branched aggregates under different experimental conditions. They found that the electro-deposits mass (M) and the radial of gyration of the object (R) fits a relationship

$$M \propto RD_M$$

with a mass fractal dimension $D_M = 1.67 \pm 0.04$, whereas R and electro-deposition time (t) fulfil the proportionality

$$R \propto t$$

Investigation of the fractal characteristics of the Zn electro-deposition morphology at low applied potential and intermediate concentration was reported by Trigueros *et al.* [50]. They observed that all the mass distribution measure of morphologies displayed a self-similar fractal nature from microns to a few millimetres. Result for the more active growth rate showed a multi-fractal behaviour. Cell dimensions also have a clear influence on pattern morphologies [51].

13.2.6. Growth kinetic studies

To study the growth kinetics during the electro-deposition of metals an experimental setup (Fig. 13.22) was employed by Rastogi *et al.* [52]. It consisted of two identical parallel electrodes at a distance of 5 cm and covered with a thin layer of aqueous electrolyte containing the metal ion. A constant potential with DC supply was applied. The distance travelled by the electro-deposited aggregates emerging from the cathode may be noted with the help of a travelling microscope at different time intervals.

A systematic experimental study on Zinc electro-deposition in an electrochemical cell with parallel electrodes has been carried out by Trigueros *et al.* [53]. The growth velocity of Zn metal leaves was found to be independent of the concentration of

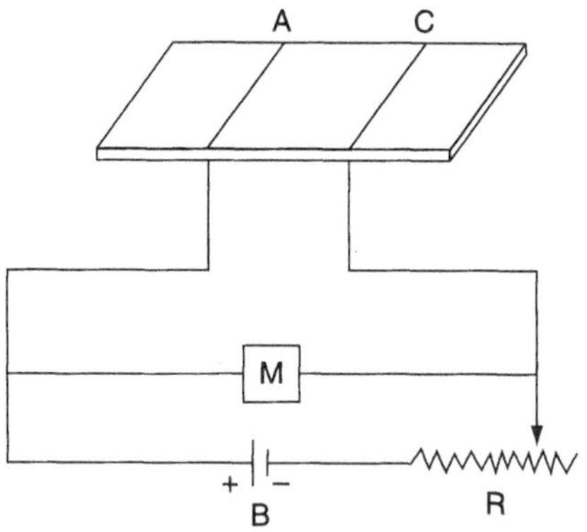

Figure 13.22. Experimental setup for electro-deposition of metals with two parallel electrodes A and C 5 cm apart, B = DC battery, R = rheostat and M = digital multimeter [52].

Zn^{2+} ion from 0.003 to 0.008 mol l^{-1} and from 0.010 to 0.018 mol l^{-1} for a fixed potential of 15 V. The local structure suddenly changed from fractal to dendrite with backbones [54]. A model for the motion of the fluid flow, the electric field and the concentration map around the branches during electro-deposition of metal clusters was proposed by Fleury *et al.* [55] which accounts for the observed distribution of ions in the solution around the metal tips. Tomas *et al.* [56] have extensively studied the aggregation process under a fractal convective flow. On account of the instability after some induction period, different types of structures appear at the cathode. From the electrochemical point of view, the recording of the potential of the cathode is a classical way to get qualitative information about the local kinetics. Therefore potential changes with time were also noted during the process of electrochemical deposition of lead. Dependence of oscillatory characteristics and growth morphologies during electrochemical deposition on the current density, concentration of the electrolyte and presence of colloidal materials such as agar-agar and PVA have been studied by Rastogi *et al.* [52].

The experimental setup of a batch reactor employed by Rastogi *et al.* [52] to monitor potential changes at the cathode during electrochemical deposition of lead metal at an air-water interface is shown in Fig. 13.23. In this experiment the micro-slide is put in a dish containing an aqueous solution of lead acetate in such a manner that a small volume of the solution is just above the slide. Two platinum electrodes P_1 and P_2 are inserted in the solution. The anode P_1 is inserted below the surface of the solution while the lower end of the cathode is put just at the surface of the solution. Potential changes during electrochemical deposition of lead metal were monitored with the help of platinum electrode P_2 coupled with a calomel electrode C.

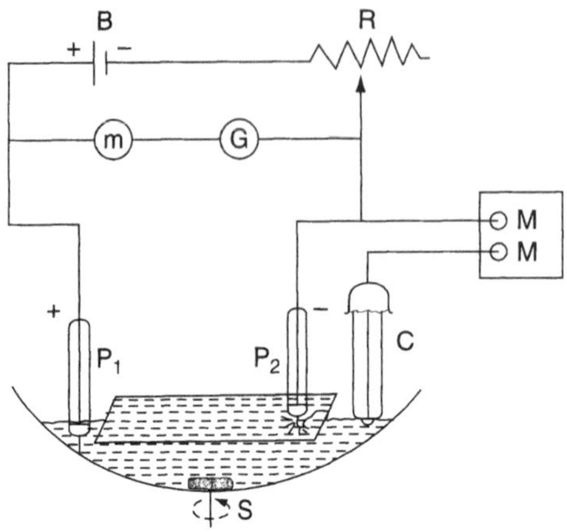

Figure 13.23. Experimental setup to monitor potential changes during electrochemical deposition of metals in a batch reactor [52].

13.2.7. Influence of an inert electrolyte

The influence of sodium sulphate, an inert electrolyte, on electro-deposition of copper has been analysed by Trigueros *et al.* [45]. They have observed the formation of finger-like deposits at large applied potential and inert electrolyte concentration. This new morphology was formed as a result of concurrent electrode reactions taking place together with copper. Electro-deposition growth velocities were found to be linear with respect to inert electrolyte concentration mainly at low values except when open fractal structures are formed. It was observed that the growth of metal electro-deposits without the supporting electrolyte was governed by the speed at which the anions are withdrawn from the deposits. The theoretical models were proposed for the study of electrode processes under diffusion control on rough surfaces and have presented a model which concentrates on the main features of the electrochemical aspects of growth mechanism, incorporating electric drift, diffusion and electroconvective motion.

13.2.8. Flow reactor experiments

In case of batch reactor experiment, described earlier, the concentration of the solution decreases continuously as the structure grows. To maintain the constant electrolyte concentration, lead acetate solution was influxed continuously throughout the experiment using a continuously stirred tank reactor as shown in Fig. 13.24.

Pattern formation and growth behaviour during electrochemical deposition of lead from its aqueous solution have been studied by Rastogi *et al.* [52] employing the experimental setup shown in Fig. 13.24.

The setup consists of a flat bottom petri dish having an outlet nozzle (O) at the corner. A glass slide AA′ is put in the petri dish containing a solution of metal ion in such a

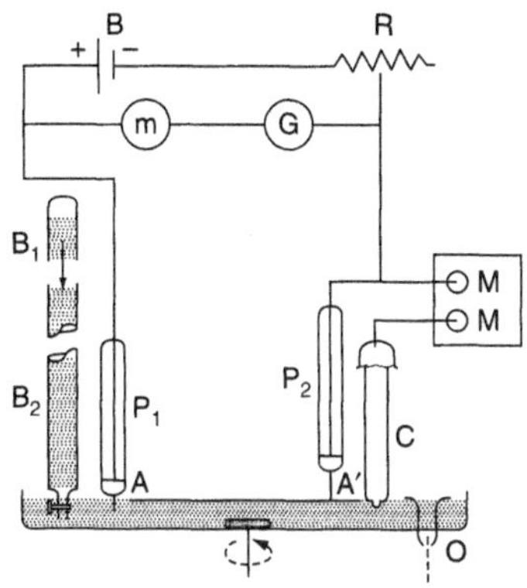

Figure 13.24. Experimental setup to monitor potential changes during electrochemical deposition of metals in a flow reactor [52].

manner that only a small volume of the solution (thickness ~ 1 mm) is just above the slide and is supported by two glass pieces. The concentration and level of the solution are maintained constant with the help of a reservoir (B_1) connected with a burette (B_2). Electrolyte solution was influxed at a flow rate of $0.25\,\mathrm{ml\,min^{-1}}$. Remaining part of the experimental setup was the same as described in the case of batch reactor. Such an arrangement permitted maintenance of constant concentration around the electrodes.

Batch and flow reactor experiments were compared. In case of CSTR, fluctuations were periodic whereas in a batch reactor oscillations were more like random noise when the current was 8.0 mA. It has been found that polymer influenced the oscillatory and growth behaviour during electrochemical deposition of lead. Results also indicated a transition from dendritic to DLA/fractal-type structure on the addition of PVA in the solution.

Suter and Wong [57] have studied the electrical signals due to the deposition of zinc ion in a cell with zinc sulphate electrolyte and observed interesting patterns ranging from random fractals to orderly dendrites. An interesting feature of the investigation is the oscillation in current at various concentrations and voltages. Argoul *et al.* performed experiments on two-dimensional zinc electro-deposition and suggested that electro-deposition and diffusion-limited clusters have similar geometrical structures. Argoul *et al.* have presented experimental evidence for spatio-temporal chaos in diffusion-limited growth phenomena. They have reported preliminary results of numerical computation of a DLA cluster.

13.2.9. Fractal-shaped nano-structures

Metal nano-structures are of considerable interest because of their importance in catalysts, photochemistry, sensors, optical electronic and magnetic devices. Nano-structured platinum is of particular interest for many applications. Platinum nano-structures of different sizes and shapes have been reported earlier. However, the synthesis of additional types of nano-structure is technologically important. The synthesis of fractal-shaped platinum nano-structures of controlled size has recently been described by Song and co-workers. The method is based on the seeding and fast autocatalytic growth approach in which an aqueous solution of platinum salt is reduced by ascorbic acid in the presence of surfactant. The two-dimensional nano-sheets are dendritic. The fractal dimension of these nano-sheets was obtained by analysis of the TEM images. The average fractal dimension was found to be 1.73 ± 0.08 when analysed using the box counting method. The transitions between different morphologies were very similar to those obtained in physical or biological systems. Depending on the applied voltage and electrolyte concentration, different transitions can be obtained.

Highly uniform γ-MnO_2 3D urchin-like and sisal-like nano-structures have been successfully prepared by a common hydrothermal method based on the reaction between $MnSO_4$ and $KBrO_4$. Reaction temperature and the addition of the polymer play an important role in influencing the methodologies of the as-obtained products [58].

Synthesis of highly uniform urchin-like 3D nickel sulphide nano-structures by the reaction between nickel acetate and dithiazone in ethylene diamine was reported by Zhang *et al.* [59] The morphology of the product is similar to fractal structure. Cu_2S is an interesting material because of its semiconduction and photovoltaic capabilities. Monoclinic Cu_2S arrays with 10–100 nm diameter were grown on Cu foil substrates with the aid of H_2S gas [60]. They have reported the discovery that the use of highly polymerized Cu-thiolate as a precursor generates uniform Cu_2S nano-wires with 2–6 nm diameters and length of 0.1 to several μm after precipitation and firing. Transmission electron microscopy (TEM) revealed that the product consists of crystalline Cu_2S nano-wires. The viscosity of colloidal thiolate indicated the degree of polymerization of the precursor which governs the morphology of the final nanoproduct in the solventless thermolysis method.

13.2.10. Biological patterns and bacterial growth

There are many biological patterns that appear to be self-similar [2, 3, 61]. The first quantitative analysis of the geometry of blood vessels in the normal human retina using fractal concepts was presented by Family *et al.* [62]. It has been concluded that the network of blood vessels in the human retina is fractal. The retina's blood vessels have a similar pattern and same fractal dimension as two-dimensional DLA [63, 64]. Biological growth almost invariably leads to the formation of complex shapes, forms and patterns [65].

Biological growth is a complex process which is controlled by a variety of physical, chemical and biological mechanisms. One of the important findings in this direction was that certain bacteria at low concentration of nutrients developed fractal patterns similar to DLA clusters [66, 67]. Physiologists quantified observations of the branching patterns. They observed that the lung tree has fractal geometry [9]. Certain neurons have

Figure 13.25. Typical bacterial colonies.

a fractal-like structure. In the human body, fractal-like structure abound in networks of blood vessels, nerves and ducts. Many other organ systems also appear to be fractal.

Recently, considerable efforts have been made to understand the formation of biological patterns using fractal ideas and models. Of particular interest has been the study of simple biological forms such as bacterial colonies similar to diffusion-limited aggregates obtained during non-equilibrium growth process in non-living systems.

Bacteria when grown on a semisolid agar medium forms a characteristic type of colony which differs in size, shape, surface, elevation, internal structure, colour, opacity and consistency. Typical colony morphologies are shown in Fig. 13.25.

Fujikawa and Matsushita [66, 67] suggested that bacterial colonies also grow in accordance with DLA on agar plates. Because average pore size of the gel network is similar to the size of the bacteria, bacterial cells cannot move in the agar medium. Fujikawa and Matsu Shita [66] found that *B. subtilis* grows in a DLA manner on agar plates. The bacteria is motile and grows aerobically, thus it could grow two dimensionally on agar surface. This colony has been found to be highly hydrophobic compared with colonies of the bacteria such as *E. coli*. The fractal dimension was found to be 1.71 which is in excellent agreement with the expected value of DLA model.

Ben Jacob *et al.* [68] have grown bacterial colonies of *B. subtilis* under different growth conditions ranging from a very low level of nutrient to a very rich mixture, forming a soft substrate to a hard substrate. The colonies adopt various shapes as growth conditions are varied. Patterns are compact at high peptone concentration and become more ramified at low peptone concentration. Optical microscopy revealed that the bacteria perform a random walk like movement within a well-defined envelope. The growth of bacterial colonies present an inherent additional level of complexity compared to non-living systems. Ben Jacob *et al.* have concluded the following generic features for the bacterial growth: (i) diffusion of nutrient, (ii) movement of bacteria, (iii) reproduction and sporulation, (iv) local communication.

Matsushita *et al.* [69] have observed the behaviour of two neighbouring colonies and noted that two neighbouring colonies of *B. subtilis* repel each other and never fuse together. They further concluded that these bacterial colonies grow through DLA processes. They have investigated a novel phenomenon in which a new morphology bursts at a localized point along the interface. The new morphology has higher growth velocity and outgrows the original morphology.

The diffusion-controlled growth morphology and kinetics of *E. coli* colonies under different experimental conditions have been studied by Das *et al.* [70]. Fractal dimension

Figure 13.26. *E. coli* developed on agar plate.

of the scanned picture (Fig. 13.26) was calculated and found to be 1.55. A transition from Eden-like to smooth spreading colony patterns was observed. Influence of vitamins, H_2O_2, antibiotics and *B. subtilis* on growth behaviour has been studied. Burst morphologies in *E. coli* and phthalic acid are observed similar to diffused growth in heptacellular carcinoma and black sea fan. Vitamin C and vitamin B complex showed stimulatory effect to some extent while oxy-tetracyclin and H_2O_2 inhibited the growth of the microorganism. *E. coli* and *B. subtilis* grow independently when co-cultivated on the same plate.

Budrene and Berg [71] proposed that a chemoattractant mediates the observed spot formation in *E. coli*. This is easily confirmed by using a model of growth that includes nutrient diffusion and consumption and bacterial growth and bacterial motion. The addition of a diffusing attractant c that is constantly emitted by the bacteria, together with the bacterial motion towards its gradient, leads to the creation of spot where the bacteria continuously emit chemoattractant. *E. coli* might be used as a repellent attractant in the co-operative formation of complex patterning [70].

The dynamics of the growth of three different strains of bacteria, *B. subtilis* and *E. coli*, was studied under different conditions of low as well as rich nutrient concentrations. It was also found that within the statistical fluctuations in the experimental data, the mean radius of the bacterial colonies grown with a power of time and the exponent characterizing this power law growth had an anomalous value. A simple phenomenological approach for explaining the existence of anomalous power law exponents in bacterial growth was discussed. This approach may be useful in determining the key mechanisms, which control the growth and morphology of bacterial colonies.

Morphology and growth kinetics of the bacteria *Klebsiella ozaenae* on agar surface have been investigated by Das and Kumar [72]. Bacterial colonies grow two dimensionally with random branches as shown in Fig. 13.27 which is identical to non-equilibrium growth processes in non-living systems and generate self-organization. The colony pattern is analysed and fractal dimensional is found to be 1.61, which is in excellent agreement with the expected value of the DLA model. Both phenotypic and

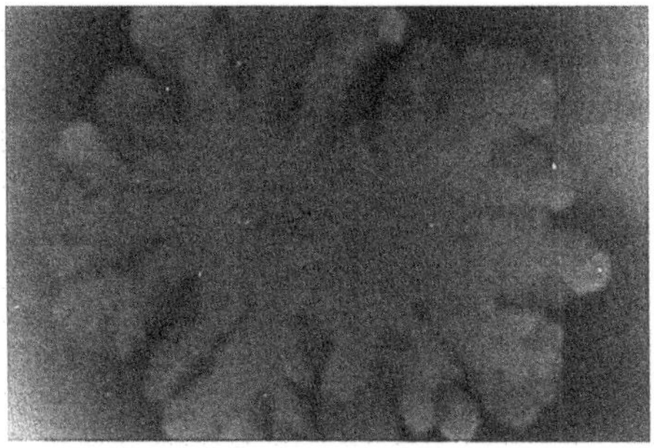

Figure 13.27. *K. ozaenae* developed on agar plate [72].

genotypic adaptations are observed under different experimental conditions. Influence of certain vitamins, disinfectant, antibiotics and the microorganism *B. subtilis* on the growth behaviour of *K. ozaenae* has been investigated.

Kinetics of bacterial growth has been studied in an incubator by measuring the colony diameter as a function of time with a travelling microscope accurate to ± 0.001 cm. The growth kinetics of *E. coli* and *K. ozaenae* on agar surface at various concentrations of peptone and beef extract has been studied. The rate of spreading of the bacterial colonies was not uniform throughout. Initially, the rate was fast, then spread at a relatively slow rate. The rate of multiplication of bacteria is not constant for indefinite period due to exhaustion of nutrient and accumulation of toxic metabolic waste products. The rate was found to increase with increase in peptone level. The process is diffusion controlled as indicated by the equation $d^2 = mt + c$, where m and c are the slope and intercept, respectively. The growth rate of *K. ozaenae* was found to be greater than that of *E. coli* under identical experimental conditions. The microbial growth could be thought of primarily as a consequence of nutrient diffusion and consumption, movement of bacteria and cellular division.

At the cellular level, cell replication is obviously an autocatalytic process represented as

$$\mathrm{X} \longrightarrow 2\mathrm{X}$$

Cellular division is responsible for instability causing a morphological transition. In the present case, Das *et al.* [70] observed that instability increased with decrease in peptone level and hardness.

13.2.11. Burst morphology

Although bacteria do not act on agar-agar, its effect on colony morphology has been studied by Kumar [73]. An interesting observation of the present investigation is the appearance of burst at low agar concentration (Fig. 13.28).

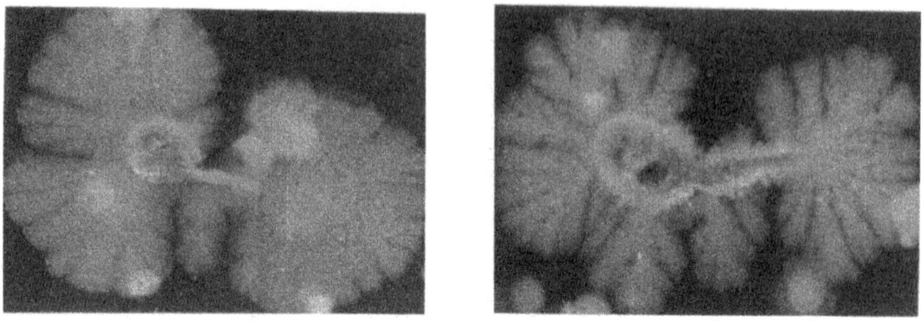

Figure 13.28. Burst morphologies of *E. coli* and *K. ozaenae* colonies at low agar concentration [73]. See also Colour Plate Section.

Colonies of *P. vulgaris* [74] were observed to possess concentric rings segmentation and planet-type structures. The colony morphology differed on different media and was also affected by methylene blue and bromocresol purple.

13.2.12. Chiral patterns

Chiral morphology is an interesting morphology observed during the growth of bacterial colonies on soft agar, solidification of certain salts, crystallization from solution and in other living systems. Scientists are continuously searching for the origin of homo-chirality. It plays an important role in the evolution of living systems.

Macroscopic zonation patterns in cultures of filamentous fungi have been known for a long time. Filamentous fungi and actinomycetes form two groups of industrially important microorganism. Nielsen [75] has proposed the following scheme for the development of hyphal element and pellets from a spore:

$$\text{Spore} \rightarrow \text{Germ tube} \rightarrow \begin{array}{c}\text{Densely branched}\\ \text{hyphal element}\end{array} \rightarrow \begin{array}{c}\text{Agglomerate of}\\ \text{hyphal element}\end{array} \rightarrow \text{Pellet}$$

13.3. Concluding remarks

There is order in disorder in space as we have seen in the case of chemical waves, which is controlled by diffusion. Corresponding to the case of deterministic chaos we have the phenomena of complexity in crystal growth, polymer growth, electro-deposition and bacterial growth governed by DLA and fractal geometry. The concept of fractals can be utilized for investigating social dynamics of the past and present on the land space from the angle of archaeology. In addition it can also be useful in the design of buildings and town planning [11]. In free verse, the concept of "fractal poetry" has also been invoked [76]. It just confirms the concept that there is order in disorder as proposed by Prigogine.

References

1. J.M. Garcia-Ruiz, E. Louis, P. Meakin and L.M. Sander (eds), Growth Patterns in Physical Sciences and Biology, NATO ASI Series, Series B, Physics, Vol. 304, Plenum Press, New York, 1993.
2. B. Mandelbrot, Fractals: Form, Chance and Dimension, Freeman, San Francisco, 1977.
3. B. Mandelbrot, The Fractals Geometry of Nature, Freeman, San Francisco, 1982.
4. J. Nittmann and H.E. Stanley, Nature, 321 (1988) 663.
5. F.W. Billmeyer, Jr, Text Book of Polymer Science, Interscience Publishers, New York, 1962.
6. T. Vicsek, Phys. Rev. Lett., 53(29) (1984) 2281.
7. L.M. Sander, Nature, 322 (1986) 789.
8. F. Sagues, L. Lopez-Tomas, J. March, R. Reigada, P.B. Trigueros, E. Vilaseca, J. Claret and F. Mas, Int. J. Quant. Chem. 52 (1994) 375.
9. A.L. Goldberger, D.R. Rigney and B.J. West, Sci. Am., February, 36 (1990) 35–41.
10. M. Montiel, A.S. Aguado and E.D. Zaluska, Chaos Solitons Fractals, 7(8) (1996) 1187.
11. K. Ravilious, New Scientist, 10 January 2004, p. 43.
12. D.G. Grier, D.A. Kessier and L.M. Sander, Phys. Rev. Lett., 59 (1987) 2315.
13. F. Sagues and J.M. Costa, J. Chem. Educ., 66 (1989) 503.
14. P. Berge, Y. Pomeau and C. Vidal, Order Within Chaos: Towards a Deterministic Approach to Turbulence, Herman and Wiley, New York, 1986.
15. H. Haken, Synergetics, Third Edition, Springer, Berlin, 1983.
16. T.A. Witten and L.M. Sander, Phys. Rev. Lett., 47 (1981) 1400.
17. S. Kumar, Y. Singh and Y.P. Joshi, J. Phys. A Math. Gen., 23 (1990) 2987.
18. T.A. Witten and L.M. Sander, Phys. Rev. B, 27 (1983) 5686.
19. M. von Smoluchoski, Phys. Z., 17 (1916) 585.
20. K. Falkoner, Fractal Geometry, Wiley, New York, 1992.
21. A. Amman, L. Cederbaum and W. Gaus (eds), Fractals, Quasi-crystals, Chaos, Knots and Algebraic Quantum Mechanics, Kluwer Academic, Dordrecht, 1988.
22. R.P. Rastogi, I. Das, K. Jaiswal and S. Chand, Indian J. Chem., 32A (1993) 749.
23. R.P. Rastogi, I. Das, A. Pushkarna, K. Jaiswal, A. Sharma and S. Chand, J. Chem. Educ. (USA), 69 (1992) A47.
24. R.P. Rastogi, I. Das and A. Sharma, J. Chem. Educ. (USA), 71(8) (1994) 694–696.
25. I. Das, A. Sharma, A. Kumar and R.S. Lall, J. Cryst. Growth, 171 (1997) 543–547.
26. M. Daoud and J.E. Martin, in The Fractal Approach to Heterogeneous Chemistry, Surfaced, Colloids, Polymers, D. Avnir (ed.), Wiley, New York, 1989.
27. M. Matsushita, in The Fractal Approach to Heterogeneous Chemistry, Surfaced, Colloids, Polymers, D. Avnir (ed.), Wiley, New York, 1989.
28. J.V. Maher and H. Zhao, in Growth patterns in Physical Sciences and Biology, J.M. Garcia-Ruiz, E. Louis, P. Meakin and L.M. Sander (eds), NATO ASI Series, Series B, Physics, Vol. 304, Plenum Press, New York, 1993.
29. I. Das and S.K. Gupta, Indian J. Chem., 44 (2005) 1355–1358.
30. B.J. McCoy, J. Colloid Interface Sci., 216 (1999) 235.
31. A.M. Chen, L.M. Santha Kumaran, S.K. Nair, P.S. Amenta, T. Thomas, H. He and T.J. Thomas, Nanotechnology, 17 (2006) 5449.
32. I. Das, S.K. Gupta and R.S. Lall, Indian J. Chem. Technol., 12 (2005) 198.

33. V. Horvath, T. Vicsec and J. Kertesz, Phys. Rev. A, 35(5) (1987) 2353.
34. V. Fleury, J.N. Chazalviel and M. Rosso, Phys. Rev. Lett., 68 (1992) 2492.
35. P.P. Trigueros, J. Claret, F. Mas and F. Sagues, J. Electroanal. Chem., 312 (1991) 219.
36. G. Daccord and R. Lenormand, Nature, 325 (1987) 41.
37. M. Kolb and H.J. Herrmann, Phys. Rev. Lett., 59(4) (1987) 454.
38. D. Avnir, D. Farin and P. Pfeifer, Nature (London), 308 (1984) 261.
39. D. Avnir, D. Farin and P. Pfeifer, J. Colloid Interface Sci. 103 (1985) 112.
40. D. Farin, A. Volpert and D. Avnir, J. Am. Chem. Soc., 107 (1985) 3368, 5319.
41. P. Argyvakes, in Fractals, Chaos, Knots and Algebraic Quantum Mechanics, A. Amman, L. Cederbaum and W. Gans (eds), 1988 pp. 53–64.
42. R.P. Rastogi, I. Das and A. Pushkarna, Chem. Phys. Lett., 186 (1991) 1.
43. I. Das, P. Singh, N.R. Agrawal and R.P. Rastogi, J. Colloid Interface Sci. 192 (1997) 420.
44. L.M. Sander, W. Guthingerin and G. Daugelmayar (eds), The Physics of Structure Formation, Springer, Berlin, 1987, p. 257.
45. P.P. Trigueros, F. Sagues and J. Claret, Phys. Rev. E, 49 (1994) 4328.
46. I. Das and S.S. Mishra, Indian J. Chem., 39A (2000) 1005.
47. I. Das, S.S. Mishra, R.P. Rastogi, A. Sharma and S. Shukla, J. Indian Chem. Soc., 78 (2001) 460.
48. I. Das, N. Singh, N. Varshney and A. Kumar, Indian J. Chem. 36A (1997) 920.
49. P.L. Shilardi, S.L. Marchiano, R.C. Salvarezza and A.J. Arvia, Chaos Solitons Fractals, 6 (1995) 525.
50. P.P. Trigueros, J. March, J. Claret, F. Mas and F. Sagues, Fractals, 1(3) (1993) 439.
51. J. Claret, F. Mas and F. Sagues, J. Electroanal. Chem., 328 (1992) 165.
52. R.P. Rastogi, I. Das, A. Pushkarna and S. Chand, J. Phys. Chem., 97 (1993) 4871.
53. P.P. Trigueros, J. Claret, F. Mas and F. Sagues, J. Electroanal. Chem., 312 (1991) 219.
54. Y. Sawada and H. Hyosu, Physica D, 38 (1989) 299.
55. V. Fleury, J.N. Chazalviel and M. Rosso, Phys. Rev. Lett., 68(16) (1992) 2492.
56. L.P. Tomas, J. Claret, F. Mas and F. Sagues, Phys. Rev. B, 46(18) (1992) 11495.
57. R.M. Suter and P.-Z. Wong, Phys. Rev. B., 39(7) (1989) 4536–4540.
58. C.Z. Wu, Y. Xie, D. Wong, J. Yang and T. Li, J. Phys. Chem. B, 107 (2003) 13583.
59. W. Zhang, L. Xu, K. Tang, F. Li and Y. Qian, Eur. J. Inorg. Chem., (2005) 653–656.
60. L. Chen, Y.-B. Chen and L.M. Wu, JACS Commun., 126 (2004) 16334–16335.
61. P. Meakin, J. Theor. Biol., 118 (1986) 101.
62. F. Family, B.R. Masters and D.E. Platt, Physica D, 38 (1989) 98–103.
63. T.A. Witten and L.M. Sander, Phys. Rev. Lett., 47 (1981) 1400
64. P. Meakin, in Phase Transitions and Critical Phenomena, C. Domb and J.L. Lebowitz (eds), Academic Press, New York, 1988.
65. H. Meinhardt, Models of Biological Pattern Formation, Academic Press, New York, 1982.
66. H. Fujikawa and M. Matsushita, J. Phys. Soc. Jpn, 58 (1989) 3875.
67. M. Matsushita and H. Fujikawa, Physica A, 168 (1990) 498.
68. E. Ben Jacob, O. Schochet, A. Tenenbaum, I. Cohen, A. Czirok and T. Vicsek, Nature, 388 (1994) 46.
69. M. Matsushita, M. Ohgiwari and T. Matsuyama, in Growth Patterns in Physical Sciences and Biology, J.M. Garcia-Ruiz, E. Louis, P. Meakin and L.M. Sander (eds), Plenum Press, New York, 1993.
70. I. Das, A. Kumar and U.K. Singh, Indian J. Chem., 36A (1997) 1018–1022.

71. E.O. Budrene and H.C. Berg, Nature, 349 (1991) 630–633.
72. I. Das and A. Kumar, Indian J. Chem., 36A (1997) 197–200.
73. A. Kumar, Dynamic instability and pattern formation in chemical and biological systems, Ph.D. Thesis, Gorakhpur University, Gorakhpur (UP), India, 1996.
74. I. Das, A. Sharma and U.K. Singh, Indian J. Exp. Biol., 29 (1991) 1109.
75. J. Nielsen, Trends Biotechnol., 14 (1996) 438.
76. A. Fulton, Interdiscipl. Sci. Rev., 30 (2005) 323.
77. I. Das, A. Kumar and Namita R. Agarwal, Indian J. Chem., 38A (1999) 301.
78. J.M. Gomez-Rodrignez, A.M. Baro, L. Vazquez, R.C. Salvarezza, J.M. Vara and A.J. Varai, J. Phys. Chem., 96 (1992) 347.

Part Four

NON-EQUILIBRIUM PHENOMENA IN NATURE AND SOCIETY

Chapter 14

SOCIAL DYNAMICS, ECONOMICS AND FINANCE

14.1. Complexity in real systems

Real systems display different types of complexity such as temporal complexity, spatial complexity or both. Some of the typical examples are as follows:

Type of complexity	System
Temporal complexity	Prices and stocks
Spatial complexity	DNA molecule
Both temporal and spatial complexity	Human brain

Some other examples of complex and chaotic behavior are the following:

(a) Sudden changes in history such as revolutions
(b) Unexpected crash of stock market
(c) Self-organization and generation of patterns in biological systems
(d) Working of brain and nervous system.

There is little doubt that Economics and Finance give us example of chaos and unpredictable behaviour in a technical sense. In the context of present euphoria about globalization, one can expect undesirable wild oscillations in Economy as stated by David Ruelle [1].

14.2. Methodology and strategy for study of complex systems

There is need for integrated multi-disciplinary and comprehensive approach for study of real systems with combined efforts (Synergetics). Hence, it is desirable for experts in different disciplines to interact with each other in order to be acquainted with the thought processes and strategies adopted with each other.

Traditional scientific models involve linear systems since these are easier to solve mathematically. More realistic models as we have seen in earlier chapters involve non-linear systems which require computational techniques in addition.

Strategy of study for such systems should involve following sequence of steps:

(a) Identification of sub-systems
(b) Analysis of mutual interactions
(c) Identification of variables and processes controlling the various sub-systems
(d) Conceptual modelling
(e) Mathematical modelling
(f) Testing of hypothesis by qualitative tests and statistical analysis of data. In the case of biological system, direct tests can be performed but not in Social Dynamics and Finance. In the latter case one has to depend on analysis of earlier statistical data.

14.2.1. Models

The word model signifies "a structure for ordering thought". Model may be of several kinds.

(a) Physical or pictorial representation of a situation, e.g. traffic problem.
(b) Diagrams, e.g. demand versus supply.
(c) Conceptual statement about the relationships among variables: when these are known, the general and impressive way as in Economics, the models can be expressed in diagrammatic form.
(d) Mathematical models: it is possible when the relationships are known with greater precision involving mathematical functions.

14.2.2. Identification of cause and effect

For the analytical study of real systems, it is first of all necessary to identify causes and effects. In the case of electro-osmosis in Chapter 4, we have seen that the number of causes and effects are more than two. In this context, it is therefore desirable to take note of the following:

(a) Some effects can arise from plurality of causes. For example, coupling of flows occurs in electro-kinetic phenomena (Chapter 4).
(b) There is a time lag between cause and effect. It may be noted that attainment of steady electro-osmotic pressure takes some time after generating the fixed potential difference across the membrane (Chapter 4). Similar behaviour is observed when temperature difference is maintained across the membrane, and the thermo-osmotic pressure difference is observed (Chapter 3).

(c) Cause occurs before the effect. The cause contains information about the effect that is unique and in no other variables [2]. The other variables only disturb the effect one way or the other.
(d) For instance, in the case of chemical reactions, the positive and negative feedbacks are the cause of temporal or spatio-temporal oscillations, but the temperature and initial concentration of the reactants can also perturb the oscillatory pattern [3].
(e) Effect itself can act as a cause as it happens in the case of autocatalytic reactions or in growth of population (predator–prey reaction) and corruption in the society.

$$\mathrm{A+X \rightleftharpoons 2X}$$

(f) Cause produces an effect. Another cause may produce a similar effect in the reverse direction. When the two effects are balanced, a steady state can be obtained. If the magnitude of either cause changes under specific circumstances, the steady state would be perturbed and can bifurcate to another steady state or an oscillatory state.
(g) In the same system, it is possible to have two or more cause–effect relationships, of the above type. This may act as control mechanisms of oscillatory phenomena, as it happens in the case of dual control mechanism in B–Z oscillator having xylose + oxalic acid as organic substrate as discussed in Chapter 9.
(h) Nature and modification of cause–effect relationship causes bifurcation from one steady state to other, controlled by a bifurcation parameter as discussed in Chapter 6.
(i) There can be relevant causes and irrelevant causes. In thermodynamics, it is easy to distinguish between the two when question of coupling of scalar and vectorial forces comes up. This is not possible according to Curie's principle.
(j) Effect of a particular cause can be influenced by other factors. For example, thermal conduction in gases in initiated by molecular diffusion, but the same can be suppressed by bulk diffusion and eddy diffusion when turbulence occurs.
(k) Interchangeability of cause and effect is also possible. For instance, heat flow can generate temperature difference and mass flow can generate concentration difference.

Causal sequence can be much more complicated such as in chemical reactions involving series of reactions.

Another interesting phenomena occurs in population dynamics and autocatalytic reaction, where effect itself acts as a cause.

14.2.3. Steady state and stability

In real systems, one comes across steady states and metastable states. Since coupling of fluxes and forces goes on uninterrupted, equilibrium state is hardly attained. It is

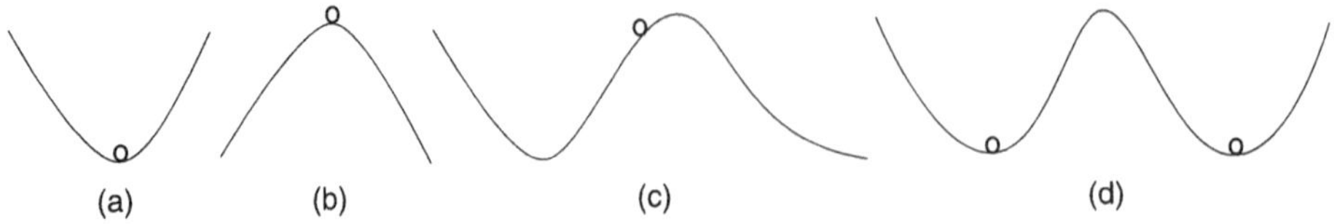

Figure 14.1. Stable, unstable and metastable states.

another matter that steady state may persist for a long time. If we consider the position of a ball on different types of surfaces as indicated below, we can have a glimpse of stable, unstable and metastable states (Fig. 14.1).

A supersaturated solution, a rain-bearing cloud or a supercooled liquid provides examples of metastable equilibrium.

In the context of stability of steady states, the following aspects deserve special attention;

Approach to steady state.
Influence of perturbation on stability.
Adjustment from one steady state to another.

In Chapters 3 and 4, it has been pointed out that on application of force in the form of temperature difference, potential difference or pressure difference, the development of steady thermo-osmotic pressure, electro-osmotic pressure or streaming potential takes some time. Similar situation occurs when these forces are withdrawn, resulting in the decay of steady state. Build-up and decay in the case of electro-kinetic phenomena have been found to be exponential (Section 4.6). Detailed analysis of the relaxation phenomena and co-relations between the relaxation time and membrane composition have been reported.

External factors such as (i) a new force or (ii) a new factor, which may influence the correlation coefficient, can disturb the stability. Furthermore, if new factors appear which disturb the forces themselves, they can also disturb the stability.

Steady state adjustment after disturbance is quite important in social sciences. Disturbances of even minor magnitude in temperature, climate and social tensions can easily shift the system to a new steady state. However, disturbances of major magnitude may take a longer route following a complex mechanism to reach a new steady state.

During health recovery, the relaxation time for the attainment of the next steady state is important. Reorganization of lifestyle is a prerequisite. If a person can analyse his/her own physiology, the results can be quite rewarding. However, age is an important factor, which influences the recovery time involved in approach to steady state.

14.3. Analytical studies of real systems

For biological systems, the desired experimental studies can be performed. These can be strengthened by investigating physico-chemical system, which can serve as a model. However, this is not possible with socio-political and socio-economic systems where similar types of experiments cannot be designed. In the case of social sciences, one has to depend on the following sequence of activities.

14.3.1. Acquisition of statistical data

Sometimes, it may be available or one has to plan for obtaining necessary data [4]. It is sometimes necessary to have cross-sectional data on individual units at a point of time (e.g. data on consumer's income and expenditure on food for a set of families). Another type of cross-sectional data which is often required is related to cross-section of states or of regions. Furthermore, time-series data related to observation over a period of time are also needed.

14.3.2. Identification of variables

These are needed to characterize temporal and spatio-temporal behaviours of the system. For this, it is desirable to distinguish between dependent and independent variables. Furthermore, target variables which play a key role have to identified along with non-target variables which play a secondary role [5]. To illustrate the point, let us examine the dependence of quantity Q_Z, demanded of particular commodity Z which would depend on P_Z, price of commodity Z, P_O price of other commodities, Y income of the consumer and T the suitable measure of the taste. If we take a broader view, Q_Z would also depend on the following: If

$$Q_Z = f(Y_{t-1}, Y_{t-2}) = \text{level of income in previous periods}$$
$$G = \text{taxation and credit policy of the goof.}$$
$$Y_d = \text{distribution of income}$$

we can ultimately express Q_Z as a function of these variables, i.e.

$$Q_Z = f(P_Z, P_O, Y, Y_d, Y_{t-1}, Y_{t-2}, G, Y_d) \tag{14.1}$$

It has to be noted that all the variables are not important. In social and economics sciences, many of such variables are called random variables and expressed as U_e. Taking a simple example,

$$Y_t = b_o + b_1 X_i + U_i \tag{14.2}$$

where X_i is the important variable and b_o and b_1 are constants. Variation in Y_t can be expressed as a function of systematic variation and random variation.

14.3.3. Correlation amongst variables

Correlation may be defined as the degree of relationship existing between two or more variables. Correlation may be linear or non-linear.

Two variables may have a positive correlation, which may tend to increase or decrease in the same direction. They have a negative correlation if they tend to increase or decrease in the reverse direction. There is a third possibility that they may be uncorrelated. Correlation is often expressed in Econometrics in terms of correlation coefficient, which is a measure of the degree of covariability of two variables X and Y. The values that the correlation coefficient may assume are from -1 to $+1$.

14.3.4. Multi-colinearity

Multi-colinearity is important in a situation when the variables are subject to two or more relations. Such a situation arises when there is a strong inter-relationship amongst the independent variables, and it becomes difficult to dissect their separate effects on the dependent variables.

14.3.5. Auto-correlation

In time-series data, the successive residuals tend to be highly correlated and this correlation in known as series correlation or auto-correlation. Auto-correlation is the correlation of the series; not with some other series but with past values of the same series. Thus, auto-correlation refers to the relationship, not between two (or more) different variables, but between the successive value of the same variable.

14.3.6. Probability distribution and random variables

If X is a random variable, the expected value $E(x)$ is given by

$$E(x) = \sum_{i=1}^{n} X_i P_i$$

where the probability distribution is given by

X_1	X_2		X_n
P_1	P_2		P_n

where the probability of a particular value of X is P_i, where i can have values from 1 to n. In the case of random variables, one has to take into account considerations of probability.

Note: In Economics, each variable is influenced by a large number of factors. As an example, the consumption pattern of a family is determined by (i) family income, (ii) prices, (iii) composition of the family in terms of sex and age, (iv) past levels of the family income, (v) taste, (vi) religion, (vii) social and educational status and (viii) wealth, etc.

14.4. Quantification of relationship between cause and effect

14.4.1. Linear steady state

From the angle of predictability, formulation of quantitative relationship is the primary task. At this stage, formalism developed in non-equilibrium. Thermodynamics can serve as a good guide. If cause (forces) and (fluxes) effects can be identified, linear relations can be postulated in the following form for two coupled processes:

$$J_1 = L_{11} X_1 + L_{12} X_2 \tag{14.3a}$$

$$J_2 = L_{21} X_1 + L_{22} X_1 \tag{14.3b}$$

where J_1 and J_2 are the two fluxes; X_1 and X_2 are the two forces. L_{ik} $(i, k = 1, 2)$ are the correlation coefficients. In the case of real systems, situation like tug-of-war exists and these forces are responsible for generating effect and countereffect. When these are balances, steady state is attained. These can be of two types:

$$\left(\frac{X_1}{X_2}\right)_{J_1=0} = -\frac{L_{21}}{L_{22}} \tag{14.4}$$

$$\left(\frac{X_2}{X_1}\right)_{I=0} = -\frac{L_{21}}{L_{22}} \tag{14.5}$$

In Social Dynamics and Finance and even in the case of living systems, steady states can easily be perturbed by random variables or by appearance of new variables. Whereas in physico-chemical systems, the correlation coefficients can be experimentally determined or estimated theoretically; this is not possible in the case of other systems. Only some guess can be made from satisfied data.

Steady-state adjustment after disturbance is quite important in social sciences. Disturbances of even minor magnitude in temperature, climate and social tension can easily shift the system to a new steady state. However, disturbance of major magnitude may take a longer route involving complex mechanism to reach a new steady state.

For self-adjustment during health recovery, understanding of temporal approach to steady state can be highly beneficial.

14.4.2. Non-linear state

The relation between fluxes and forces can be non-linear and even then steady state can be attained under suitable circumstances as discussed in Chapter 7. The relation between fluxes and forces is given by

$$J_1 = L_{11} X_1 + L_{12} X_2 + 1/2 L_{111} X_1^2 + L_{121} X_1 X_2 + 1/2 L_{122} X_2^2 + \cdots \tag{14.6}$$

In the steady state,

$$\left(\frac{X_1}{X_2}\right)_{J_1=0} = \text{constant} \tag{14.7}$$

but the constant would be a function of X_1 and X_2.

In Economics, time derivative of price and production can be considered as flows, whereas demand and supply can be considered as forces. Money flow and profit can also be placed in the first category. In Finance, the above type of interaction is a distinct possibility under specific circumstances. Even in other types of real systems, we can have the above type of complex interaction between fluxes and forces.

Analogous to various non-equilibrium phenomena discussed in various chapters, we can have similar situations in the real systems, such as

Non-linear steady state: bistability in Economics and Physiology
Oscillations: spatio-temporal oscillatory features in population
Chaos and highly complex time series in Sociology and Economics
Fractal growth of cities
Self-organization in Social Dynamics and Physiology

14.4.3. Multiple steady state

As discussed in Chapter 8, multiple steady states can also be observed in real systems. This would happen when a cubic equation governs the relation between two variables. Normally in chemical reactions, such a situation is obtained when the reaction network involves autocatalytic and inhibitory steps. Such a situation is possible in socio-economic and socio-political systems.

14.4.4. Bifurcation and oscillatory phenomena

We have seen in Chapter 8 that a system can move from one steady state to another as a certain variable, which is called bifurcation parameter; it is allowed to vary [6]. This particular stage is called bifurcation point. There can be successive bifurcations. Such bifurcations can easily occur in socio-economic system.

Further bifurcation from steady state to oscillatory behaviour can also occur depending on the relationship between variables as we have seen in the earlier chapters. What is needed is tug-of-war type of situation to generate periodicity.

14.4.5. Spatio-temporal oscillations

These are quite relevant in socio-economic systems. Variation of population in a particular region with time and variation of population density from point to point have their own importance. The first one may be called extensive variable and the second one may be called intensive variable. Spatio-temporal behaviour of population can be studied by the proper choice of variables. Similar phenomena in physico-chemical systems are easy to analyse since the important variables are diffusion coefficients of particular species and those associated with autocatalytic reaction.

14.4.6. Time series and chaos

Most time series exhibit some regular behaviour, but it is useful to decompose them into following components:

Trend T (also known as trend-cycle component)
Seasonal variation S
Irregular or random variation I.

Normally time series are appropriate for the estimation of economic relationships. However, on many occasions, there are problems associated with inter-correlation of the explanatory variables which tend to change contemporaneously with time.

Different economic time series exhibit different trends. Some grow or decline in a linear manner. Some curvilinear trends can be expressed by fitting a polynomial into the series. In some cases such as in TV industry or of any new product, one may get a logistic curve as indicated in Fig. 14.2.

In the case of complex periodic and chaotic time series, role of seasonal variation and random variation can be quite important.

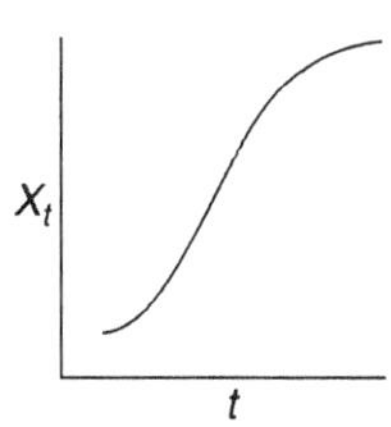

$$X_t = \frac{c}{1 + e^{-(d+\beta)t}}$$

Figure 14.2. Typical logistic curve: x, variable; t, time; c, α and β are constants.

14.4.7. Demography and spatial patterns

Concepts developed in Chapter 13 on fractals are relevant from the viewpoint of population dynamics, demography, growth and planning of cities. Corresponding to diffusion-limited aggregation in crystal growth, we have progressive urbanization stimulated by aggregation due to opportunities of employment, business and facilities. These concepts have been utilized in the study of archaeological excavations.

14.4.8. Self-organization and self-adjustment

We have seen in Part One that when entropy increase in a system is compensated by entropy export to the surrounding, different types of order or organization can result. Hence, order can emerge out of chaos, provided necessary conditions are satisfied. It should be noted that increase of entropy is associated with the increase of disorder in the system. Such a situation occurs in physico-chemical systems, several examples of which have been cited in previous chapters.

Order emerges out of disorder through entropy export to the surroundings and a dissipative structure emerges in open systems which correspond to real systems which we come across in real life. Such entropy export to the surroundings can easily be understood from a few examples from daily life. When one is disturbed or destabilized by anger, the system is normalized by the expression of anger in some form. Similarly, when one is shocked by the death of a near relative, tears or words of consolation through mutual interaction from well-wishers help to stabilize the situation.

In real systems, self-organization occurs in course of time due to interaction of forces, counter-forces and environment. Evolution of living system, human society, historical evolution have passed through different stages via bifurcation [6] since counter-force to combat forces creating disorder emerges in course of time. Similar situations occur in case of fluctuations in stock market, price and investment.

In social dynamics, according to Prigogine, "At points far from equilibrium, the deterministic equation suffice, whereas near the bifurcation point, the stochastic elements become essential". The essential difference between cultural development and cultural growth arises on account of the fact that the former corresponds to instabilities in which stochastic elements play a basic role, while the latter corresponds to deterministic developments.

Processes of change in contrast to biology differ according to non-determinism, and the degree to which they are sensitive to external conditions. It has been pointed out that a broad conceptual framework can be developed on this basis, which can give useful insight into biology, culture and cognition.

In behavioural psychology, adjustment is a process involving both mental and behavioural responses by which an individual strives to cope with inner needs, tensions, frustrations and tries to bring harmony between the inner demands and those imposed by the world in which he lives.

Adjustment can be conceived both as a process and as a state. This distinction is appropriate as it provides insight into dynamics of adjustment phenomena.

Adjustment as a process is the way in which an organism achieves satisfaction of its needs, thereby reducing tensions, particularly under those circumstances, when its customary ways of meeting needs are blocked or proved inadequate.

Adjustment conceived as a steady state, refers to the degree to which the organism is in a state when opposing forces are balanced not only within itself but also in relation to its interaction with the environment.

Adjustment of an individual can be viewed from either personal or social perspective, depending on whether one is assessed by himself or societal norms. Every person has a sense of well-being, regardless of the judgement of the external world. This sense of well-being is the basis of personal adjustment. On the other hand, every society has its own criteria against which adjustment of the individual in the society is evaluated.

As the external environment changes due to age and drastic changes in the environment itself, individual has to change his behavioural patterns. Bistability as discussed in Chapter 8 is expected to be a common phenomena under such circumstances.

In connection with health problems, self-assessment is also essential for positive self-organization, keeping in view the balance of forces which are relevant. Typical examples are physiotherapy, naturopathy and lifestyle adjustment (diet, physical exercise, control of anxiety and anger in the context of heart problems).

14.5. Sociology (social sciences)

14.5.1. Introduction

Human society is composed of sub-systems consisting of its individual members. This system is open. There is continuous interval interaction if a nature and through ideas between the members of the society. In addition, there is interaction with the external environment and the technological surrounding. Members of the society individually adopt different states of behaviour. For deeper study of behavioural changes, it can be assumed as a first step that the causal changes in the attitude of its members, global change of behaviour, can be obtained by introducing appropriate growth variables, for instance in terms of average attitude of groups of the members of the society. Comprehending the variables in different situations and assessing their relative variables are not easy tasks. One has only to depend on statistical data. There is another problem, in contrast to Economics, these are difficult to quantify. Estimation of correlation coefficient is almost impossible.

Social structure and stratification is governed by (i) prestige and esteem, (ii) male and female, (iii) age groups, (iv) caste and class and (v) occupation, status and similar factors.

The evolutionary path of human society is a resultant of two types of forces: the force of fusion and the force of fission. The factors responsible for the first type of forces are tribalism, language, group loyalty, common tastes and values. But wars, conquest, passions and slavery are responsible for creating forces of fission.

14.5.2. Development of human society (biological evolution)

All animals and plants tend to live in groups tied by bonds of mating and by some force of biological cohesion. In higher animals, these bonds are further strengthened by a psychological and social sense of kinship. The group such as biological cohesion has been defined as a race. Development of sub-systems can be understood by the following example. The principal races in the world are Caucasoid, Mongoloid, Aryans and Negroid. In early stage, these races were distributed in different regions of Asia. Originally they lived in Caucasus region of Asia [7]. They migrated in several directions in search of food and shelter and to escape natural disasters such as floods and famines. In course of time, differences in languages, traditions and cultures cropped up in the developing tribal belts and sub-systems grew up.

In examining the problems related to social dynamics, the main difficulty is to identify the relevant variables governing the cause–effect processes involved and their relative importance. In some cases such as problems related to vehicular traffic, this is relatively simple [8].

In order to illustrate how the concepts developed in earlier chapters can be utilized for studying social dynamics, we will comment on problems related to public opinion, urbanization and population dynamics in the next sections.

14.5.3. Public opinion

Actions of social groups are of co-operative nature, but the actions of individuals are determined by complex factors. Social dynamics involves many sub-systems, on account of which the following sequence of steps to be undertaken is necessary, viz.

Identification of sub-systems.
Analysis of the individual systems and their interaction with the surroundings along with cause–effect relationship.
Description of the statistical behaviour using macroscopic variables.

14.5.4. Urbanization of society

Local population increases or decreases according to employment and business opportunities which in turn depends on the market available for goods and services. This acts as a non-linear feedback mechanism in the growth of population density. More complex situation occurs when one is interested in deeper understanding of social and demographic phenomena relative to racial and ethnic population.

The growth of cities can be well understood in terms of fractal growth and diffusion-limited aggregation outlined in earlier chapter.

14.5.5. Ecology and population dynamics

Statistical analysis is the main tool for investigating this area. Statistical study of population (extensive) as well as population density (intensive parameter) is of considerable interest. Former yields information about temporal variation but the latter can yield spatio-temporal variation as well.

The variation of population with population density can be easily graphically represented [9, 10]. Demonstration of the existence of spatial self-organization patterns in population requires sampling over long periods of time at a range of sites. Vole cause severe economic damage and are therefore extensively monitored providing a source of the required data. By identifying spatial autocorrelation in vole populations and annually moving the epicentre of vole plagues as a function of damage extent, it has been inferred that travelling wave pulses in the dynamics of vole population exist [11].

14.5.6. Modelling

Unrestrained growth is postulated in Malthusian theory [12] according to which the linear growth function rises forever upwards. In nature, restrained growth is observed. Accordingly, a modified equation

$$X(t+\tau) = r\,X(1-X) \tag{14.8}$$

has been proposed. In the above equation, X denotes population after a time τ and r is a constant that can be set higher or lower. The new term $(1-X)$ keeps the growth within bounds since as X rises, $(1-X)$ falls. Similar type of equation is used in Economics for logistic map.

Population dynamics involves autocatalysis similar to that discussed in chapter on chemical oscillations. Lotka-Volterra model was proposed involving this concept which involves following steps:

$$\begin{array}{r|l} \mathrm{A} \rightleftharpoons \mathrm{X} & \mathrm{A}+\mathrm{X} \longrightarrow 2\mathrm{X} \\ \mathrm{X}+\mathrm{Y} \rightleftharpoons 2\mathrm{Y} & \mathrm{X}+\mathrm{Y} \longrightarrow 2\mathrm{Y} \\ \mathrm{Y} \rightleftharpoons \mathrm{E} & \mathrm{Y} \longrightarrow \mathrm{E} \end{array}$$

The animal X eats grass A and multiplies. X is the prey and is eaten up by predator species Y and multiplies which also after some time dies. Thus, in the process, autocatalytic reactions play a prominent role. Mathematical modelling yields periodic oscillation [13] in population of X and Y. Computer simulation studies with a similar type of reactor have been reported in Chapter 9.

14.5.7. Bifurcation and self-organization

Social dynamics is getting seriously affected as the human society or sub-systems are getting more and more open although cyclic changes as shown earlier occur in history involving series of bifurcations from one state to another.

The historical process has shifted people from life in basic kinship (Family → Tribe → Regions → State → Nation → Group of nations → Globalization) with low-energy technologies to the complex structures and high-energy processes in which people now manage their existence. The accelerated historical growth and development process in the last century have also created social rifts and polarization resulting in highly differentiated rate of economic development.

Globalization due to advances in technology and communication and mutual interaction between nations, ethnic groups, different strata in the society is a part of evolutionary non-equilibrium process. However, bifurcation from one state to another and self-organization as observed in physico-chemical systems discussed in previous chapters do occur in social systems as well. Earlier, decolonization was one such process which brought a change in outlook in the whole world and eliminated barriers between different countries

The bifurcation process becomes clear when we consider the sequence of changes after the fifteenth century.

Domination
Church clergy, Feudalism
↓
Renaissance
(Traditional way of thinking was criticized and
importance of rational thinking was stressed)
↓
Rise of Nation States (16th century)
Growth of trade and national rivalry (commercial)
↓
Rise of colonization
↓
French Revolution (1789)
↓
Industrial Revolution
↓
Rise of socialist ideas
Communism and Russian Revolution
↓

World War I

↓

World War II

↓

Transformation in Soviet Union

↓

Globalization

Thus, the concept of bifurcation has to be kept in mind to understand the changing world [7].

Forces and counter-forces are all the time in operation in social dynamics and evolution of human society. Autocatalysis (positive feedback) and inhibitory process are simultaneously occurring. But if one is stronger than the other, the change occurs in the same direction. In developing countries, the sequential changes are happening as follows (some stages involve autocatalysis): technology development → market forces → mad race for power and money → social imbalance → corruption → poor governance.

This can only be reversed by good governance stress on moral and ethical values and spirit of sacrifice.

14.5.8. Self-organization in other areas

In psychology, self-regulation is defined by processes that enable an individual to guide his or her goal-directed activities overtime and across changing circumstances including the modification of thought, effect and behaviour [14]. The process of self-adjustment is similar to self-organization. Bistability is also observed in animal and human behaviour as indicated in chapter on multistability (Chapter 8).

Naturopathy is a device for self-adjustment. Even for the living state, along with medication, if some alternative steps are adopted for self-adjustment, the results are sometimes more rewarding.

Human society is a multi-component system, composed of units, i.e. its individual members. The system is open in view of (i) interval interaction of material nature, (ii) interaction through ideas between the members of the society and (iii) interaction with the external environment and technological surrounding [15] (Table 14.1).

Table 14.1. Character of physico-chemical and social systems.

Interactive system	Type of components
Physico-chemical systems	Atoms, molecules nano-particles, colloids, supramolecules
Social systems	Individual, family tribes, castes, races
Interaction material	Villages, cities, states, continent global

In social systems, interaction is related to nature and ideas with members of the society along with the external environment and technological surroundings.

14.5.9. Forces of social change

Some of the important forces of social change are the following:

Rapid cultural growth
Extensive population movement
Urbanization
Mechanization and industrialization
Secularization, or the decline of religions control; decline of ethical values
Globalization

Proper identification of forces is necessary for planning since it is extremely necessary to target the disease and the symptoms.

Many of the results of rapid change which developed nations have become accustomed to are now affecting the developing nations of the world. The stress of rapid cultural change is reflected in

Juvenile delinquency
Severe stresses on the family
Increase in crime as change is speeded up and old social controls are discarded.

In the developing nations, the seat of unrest is usually the large city with its supported rural people who are facing a new life, new ways of government and new moral standards.

The important visible areas of changes in the social setup are as follows:

Changes in the social population
Changes in the social structure
Changes in group interaction patterns.

When large number of people frequently change their place of residence, marital status or place of employment, the society is undergoing a period of rapid change.

14.5.10. Social policies and social planning

To set goals by some future ideal rather than by past accomplishment is probably new in human experience, the product of a technological age.

The stage for planning is set in a dynamic society. A rapidly changing environment can provide a rich source of support for innovation, inventions and constructive planning.

Today "think tank" groups – bodies of scholars and specialists from various fields unconnected with the government – spend thousands of hours on university campuses and in research foundations speculating about the future of the nation and the world.

All plans and programmes are likely to have multiple effects. Planners and policymakers try to evaluate the overall effect of their decisions. If the result is primarily good, then some of the less desirable consequences are tolerated.

Often, the choices are difficult, especially in the areas of the country where the consequences of decisions have far-reaching effect.

14.6. Economics

14.6.1. Introduction

Earlier Economists used traditional concepts in investigating how well economy can adopt to the changes in the market, but at the present stage, economic systems are highly complex. Unpredictability, irreversibility and self-organization are new types of phenomena, which are receiving current attention. Recent developments in experimental and theoretical studies of complex phenomena in physico-chemical systems supported by corresponding developments in mathematics and statistics can serve as a model for complex phenomena in real systems. Under such circumstances, the variables involved and their relative importance have to be assessed. The sub-systems and the nature of their interaction have to be properly identified, which helps in developing relevant hypothesis. In this respect, economic science comes very close to physical sciences.

There is a serious limitation at this stage when the question of its testing comes up. In contrast to physical sciences, it is not possible to design experiments for testing the hypothesis directly. Mathematics and statistics provide the basis for such tests. In such cases, an economist has to depend on (i) statistical data on particular facets of economic behaviour and (ii) statistical records of economic events.

In the next section, it is intended to discuss some typical phenomena in economics in the light of concepts and methodology discussed in the previous chapters.

14.6.2. Equilibrium and non-equilibrium steady states

The concept of equilibrium is quite commonly used in macro- and micro-economics. In fact, in natural systems, one can only expect unstable dynamic equilibrium. It may be noted that one can have stable non-equilibrium steady states. Approach to the so-called equilibrium or to such steady states would depend on a number of variables, which have to be carefully identified. Cause → Effect relationship leads to the concept of Force → Flux discussed in Part One in this book which is quite relevant for real systems. In the steady state, balance of forces occurs leading to balancing of effects. Typical example in economics is demand–supply relationship in the context of time variations in price and production.

14.6.3. Linear relation between fluxes and forces

As a first approximation, linear relation between fluxes and forces can be assumed as in Part One, although quantitative assessment of the phenomenological coefficient would not be easy. Thus for demand–supply phenomenon, one can postulate linear relation between price variation with time dP/dt and excess demand $D = D_t - D_0$ as follows when supply is maintained at steady value S_0

$$dP/dt = L_{11}\ D \tag{14.9}$$

where D_t is the excess demand at time t and D_0 is the corresponding value in the steady state and L_{11} is the correlation coefficient.

In the same manner, one can express the relation between price variation and excess supply $S = S_t - S_0$ where S_t denotes the supply at time t and S_0 is the supply in the steady state as follows, when $D_t = D_0$

$$dP/dt = L_{12}\ D \tag{14.10}$$

Prices fall as supply increases.

14.6.4. Coupling of fluxes and steady state

Under normal conditions, both demand and supply affect the price variation, and coupling of fluxes occurs. The resulting situation can be represented in terms of linear phenomenological equation as discussed in Part One as follows:

$$dP/dt = L_{11}\ D - L_{12}\ S \tag{14.11}$$

A steady state is reached when the fluxes are balanced, i.e. $dP/dt = 0$, Eq. (14.11) yields the relation

$$(D/S) = L_{12}/L_{11} \tag{14.12}$$

In the same manner, rate of production (dP_r/dt) would depend on demand as well as on stock. It increases when demand increases but it decreases with the increase in stock (St). This can also be represented by linear phenomenological equation of the type

$$\frac{dP_r}{dt} = \dot{P}_r = a\ D - b\ \text{St} \tag{14.13}$$

where a and b are constants where $S_t = (St)_t - (St)_0$ and $(St)_t$ denotes the supply at time t and $(St)_0$ denotes the supply in the steady state. Here again, when $dP_r/dt = 0$ in the steady state, we have the relation

$$(\text{St}/D) = 0 = b/a \tag{14.14}$$

Linear phenomenological equations such as Eqs. (14.11) and (14.13) can have more variables in complex situations. In such situations, variables like temperature, region, daytime, technological developments and infrastructure have to be taken into account along with their relative influence, extent of auto-correlation and multi-colinearity.

14.6.5. Non-linear steady states and bistability

Just as the cases discussed in Chapter 7, we can have non-linear relations instead of eq. (14.13) involving higher powers of D and S, although the theoretical interpretation would be quite complex. In the non-linear regime, one can have the phenomenon of bistability which is discussed below.

A system is bistable if, for a given value of perturbation in a variable x, its response to another variable y may take two stable values under steady-state conditions. In case of electro-kinetic phenomenon, current provides the perturbation, while the corresponding potential difference yields the response (Chapter 8).

In economics, a typical example of bistability is obtained, when two types of situations come up, viz.

(a) high production, low price
(b) low production, high price

In this case, a mechanism can be postulated which yields a non-linear relation between production and price similar to that discussed in Chapter 8. Rate of production would be proportional to price; however, a saturation point would be reached when the deviation from normal production becomes too large. Under such circumstances,

$$\mathrm{d}P_\mathrm{r}/\mathrm{d}t = \alpha\ P - \gamma\ P^3 \tag{14.15}$$

Since $\mathrm{d}P_\mathrm{r}/\mathrm{d}t$ would also be proportional to P_r, the combined relation would be as follows:

$$\mathrm{d}P_\mathrm{r}/\mathrm{d}t = \alpha\ P + \beta\ P_\mathrm{r} - \gamma\ (P_\mathrm{r})^3 \tag{14.16}$$

where α, β and γ are constants. For steady state, $\mathrm{d}P_\mathrm{r}/\mathrm{d}t = 0$ so that

$$\alpha\ P + \beta\ P_\mathrm{r} - \gamma\ (P_\mathrm{r})^3 = 0 \tag{14.17}$$

is a cubic equation in two variables, which under specific conditions can yield three real roots [11].

Similar type of bistability occurs in a financial situation which involves

(a) Large deviation from annual production with smaller investment
(b) Small deviation from annual production with large investment

Using similar arguments [16], the following cubic equation is obtained:

$$dX/dt = I + RX - CX^3 \tag{14.18}$$

where X is the deviation from the mean position X_0 of the annual production and R and C are proportionality constants. I is annual additional investment. Thus, bistability in such situation is easily predicted following the procedure discussed in Chapter 8.

14.6.6. Periodicity (a simple model)

For demand, supply and price fluctuations in the market, a simple model can be formulated as follows.

If we consider only relationship of demand and price, we can represent the situation by the following set of equations, when S is fixed:

$$dP/dt = k_1 D \tag{14.19}$$

since price would rise when demand exceeds. Furthermore, since D decreases when price increases, we have

$$dD/dt = -k_2 P \tag{14.20}$$

On combining Eqs. (14.19) and (14.20), we get

$$d^2P/dt^2 = -k_1 k_2 P \tag{14.21}$$

where k_1 and k_2 are constants. Eq. (14.21) yields the characteristics of a simple harmonic motion, which predicts synchronous oscillations in D and P as represented graphically in Fig. 14.3.

In a similar manner one can demonstrate seasonal variation in price and supply when demand is fixed.

In the next section, an alternative model for damped and explosive oscillations has been discussed.

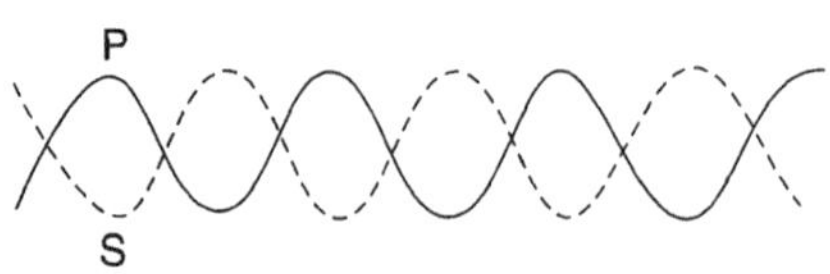

Figure 14.3. Synchronous oscillations in P and D.

14.6.7. Damped and explosive oscillations

In some cases, a delay mechanism may operate which may generate oscillations. A typical example is a situation in economics, when there is immediate response to change in price on demand but a delayed response to the supply side. For such a situation, a simple model based on delay equation can be formulated to simulate damped oscillations and spiralling rise in prices [17].

$$D_t = f(P_t) \tag{14.22}$$

$$S_t = f(P_{t-\tau}) \tag{14.23}$$

where D_t represents the quantity demanded and S_t denotes the quantity supplied at time t. Both are functions of price. However, the quantity demanded is determined by the current price at time t, while the quantity supplied depends on price prevailing at time $(t-\tau)$, where τ is the time interval. Price at time t is determined by market clearance when $D = S$, as the ideal case for equilibrium; one gets assuming linear relations

$$D_t = a + bP_t \tag{14.24}$$

$$S_t = c + d\,P_{t-\tau} \tag{14.25}$$

When the market is normal, the equilibrium price is $\overline{P}$. When this is not so, the difference between the actual price and equilibrium price is $p_t = (P_t - \overline{P})$.

Let Q_t denote the quantity at time t and $\overline{Q}$ denote the quantity at equilibrium. Then, $q_t = Q_t - Q_{t-\tau}$ is the difference in quantity (Stock). Noting that for any situation

$$Q_t = D = S \tag{14.26}$$

it follows that

$$Q_t = a + bP_t = c + dP_{t-\tau} \tag{14.27}$$

Hence,

$$Q_t - \overline{Q}_t = b(P_t - \overline{P}) = d(P_t - \overline{P}) \tag{14.28}$$

So that, we have the relation

$$Pt = (d/b)\,P_{t-\tau} \tag{14.29}$$

If $(d/b) < 1$, damped oscillations would result. On the other hand, when $(d/b) > 1$, spirralling or explosive oscillations would result. It may be noted that (d/b) is the ratio of the gradients of supply and demand as indicated by Eqs. (14.24) and (14.25).

14.6.8. Complex time series

Time series in real systems is usually complex. However, most time series exhibit some regular pattern of movement, and it is common practice to decompose them into following components:

Time T
Seasonal variation S
Irregular or Random variable I

For decomposing, following assumptions are usually made:

Observed series is the product of different components, i.e.

$$X = (T)\ (S)\ (I) \tag{14.30}$$

Observed series X is the sum of different components

$$X = T + S + I \tag{14.31}$$

The trend refers to long-term movement in the series.

On theoretical grounds, time-series data are more important for the assessment of economic growth. However, there are a number of problems with time series in connection with their decomposition. The most important is the problem of inter-correlation of the variables, which tends to change in a synchronous manner with time. In the case of oscillatory reactions discussed in Chapter 9, we have highly complicated reaction network involving numerous reacting species, but since behaviour of chemical species is much better known, through computer simulation, a viable mechanism can be postulated.

References

1. J. Gleick, Chaos, Vintage, Random House Group Limited, London, 1998.
2. C.W. Granger, Am. Econ. Rev., 94 (2004) 421.
3. R.P. Rasotgi, Causality Principle and Non-equilibrium Thermodynamics, J. Sci. Ind. Res., 40 (1981) 565–575, Sec p. 567.
4. A. Kontsoyiannis, Theory of Econometrics, Second edition, Macmillan, London, 1986.
5. G.S. Maddala, Econometrics, Mc Graw Hill Book Company, Singapore, 1987.
6. Ervin Laszlo, The Age of Bifurcation – Understanding the changing world, Gordon and Breach, Philadelphia, 1991, p. 9.
7. R.K. Mishra (ed.), Self-organizng Systems, Selected papers, Eastern Media Series Limited, New Delhi, 1983.

8. P.H. Landis, Sociology, Ginn Co. Laxington, MA, 1975.
9. I. Prigogine and R. Hermans, Kinetic Theory of Vehicular Traffic, New York, 1971.
10. J.D., Reynolds and R.P. Frackceton, Science, 309 (2005) 567.
11. E. Ranta and V. Kaitala, Nature, 390 (1997) 456.
12. P.H. Landis, Sociology, Ginn Co. Laxington, MA, 1975, p. 330.
13. G. Nicolis and I. Prigogine, Self-organization in Non-equilibrium systems, John Wiley and Sons, New York, 1977, p. 160.
14. C.L. Parath and T.S. Bateman, J. Appl. Psychol., 91 (2006) 185.
15. P.H. Landis, Sociology, Ginn Co. Laxington, MA, 1975, p. 211
16. H. Haken, Synergetics, Springer, Berlin, Third edition, 1983.
17. L.W.T. Stafford, Mathematics for Economists, English Language Society and Macdonald and Evans LTD, London, 1975, Chapter XX.

Chapter 15

LIVING SYSTEMS

15.1. Introduction

Living systems are characterized by a high structural order and ordered behaviour in space and time. There appears to be a contradiction between the increasing complexity in the course of biological development and the second law of thermodynamics. However, it may be noted that living system is an open system whose entropy may increase or decrease.

15.1.1. Characteristic features of living systems

Living systems are constantly exchanging matter and energy with the environment. Low entropy substances are imported in the organism whereas high-energy substances are exported as waste. For illustration, it may be noted that in metabolism, two opposing activities, i.e. catabolism (constant expenditure of energy by the protoplasm) and anabolism (intake of energy by the cell for restoration purposes) are involved.

Living state is a highly complex one, but there are a large number of states, all related and accessible to the organism. The bifurcation from one state to another can occur in response to any "stimulus" in the form of nerve impulses, hormonal transmitters released into the blood, chemical changes in the environment etc. The processes in the organism are balanced around steady states near equilibrium. The organism is capable of shifting from state to state with a minimum of alteration in its basic composition and organization.

Biological systems dominated by a point attractor may be disturbed by the action of a drug but they come to their original state as soon as the drug is eliminated or excreted from the body, or they reach a new steady state when the drug concentration is kept constant.

As the frontiers of knowledge are advancing, in the characterization of such systems on a molecular level, as well as in the development of appropriate physical theory, biologists and physical scientists are coming together to evolve a common language for use in investigating the complexity in the living systems.

15.1.2. Complexity in living systems

Living systems involve quite a complicated set-up, such as involving

(a) multiple sub-systems, which are open as well as closed and which interact with other sub-systems depending on the circumstances;
(b) multi-components; and
(c) multi-processes.

As an example, we may consider interconnections of cells, tissues and body organs in the human body as indicated in Fig. 15.1.

The thermodynamic equilibrium is disturbed because of the changing character of membrane transport involving ion-channels, ion-pumps and constant flux of lipids in cellular transport.

15.1.3. Special chemical and dynamical features

It should be noted that

(a) The dynamical variables in chemistry and in a number of biological problems are the concentration of the reactants.
(b) In a chemical–biological system, the chemical processes are associated involving diffusion processes, with transport of matter.
(c) The special feature of a biological system is compartmentalization i.e. the division of the systems into compartment separated by membranes. The system is heterogeneous not only chemically but spatially as well.
(d) In living systems, the non-linear chemical reactions are coupled amongst themselves. But, they may also be coupled with transport of matter and with mechanical and electrical processes as well.
(e) In many chemical–biochemical processes, we have to deal with small number of molecules. The concentration concept itself has limited applicability in these cases, and the probabilities of the various states of molecules have to be introduced as dynamical variables.
(f) The mathematical models of biological processes are often very complicated containing many variables and describe a multistage behaviour.

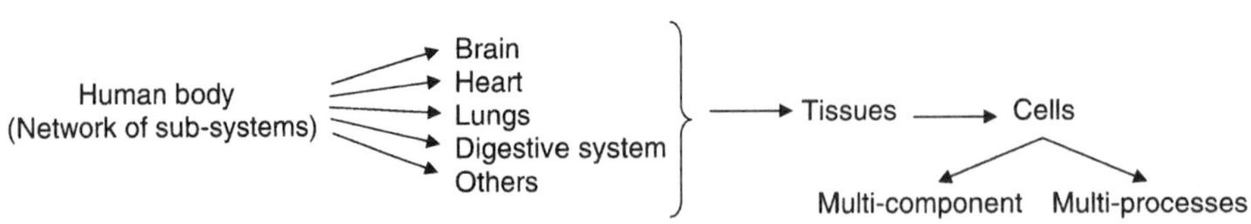

Figure 15.1. Interconnection of cells tissues and body organs.

15.1.4. Thermodynamic aspects of living systems

We know that $dS = d_e S + d_i S$ and hence for the organism, the $d_e S$ term must be negative to compensate for the positive $d_i S$. The organism discards matter with greater entropy content than the matter it ingests, thereby losing entropy to compensate for the entropy produced in internal irreversible processes.

The entropy production of a living organism is regulated by its metabolism. If it refers to chemical reactions only, it can be assessed using the following arguments. Noting that $d_i S/dt = AJ/T$, where A = affinity and $J = d\xi/dt$ = chemical reaction rate. Here ξ is the degree of advancement of reaction. A is given by

$$A = (\delta Q/\delta \xi)_{T,P} + T(\delta S/\delta \xi)_{T,P} \tag{15.1}$$

It is sometimes possible to neglect the entropy production term in the above equation because of its minor contribution. In such cases, the entropy production due to chemical change becomes simply proportional to the heat of reaction [1], so that

$$d_i S/dt = -(1/T)(\delta Q/\delta T)_{T,P} \tag{15.2}$$

In this approximation a coupled reaction ($AJ < 0$) is equivalent to an endothermic reaction.

15.1.5. Coupling of chemical reactions

Normally, the individual reaction rates of different steps are independent and the Onsager reciprocity relations are obeyed in a trivial way [2]. However, when the concentration of species other than the intermediates are kept constant and equal to their equilibrium values, coupling can occur. For such a case, the Onsager relations are found to be non-trivial but still satisfied. [3].

In case of cyclic reactions, the number of independent reactions is one less than the total number of reactions and hence coupling can occur. A few examples of such coupled reactions are the triangular isomerization reactions of cymene [4] and Δ^α – pentenoic acid [5]. However, such reactions are not common in Chemistry. Nevertheless, such coupling is supposed to be common in biological reactions.

Biological reactions such as enzyme reactions involve a reaction network having a series of consecutive reactions. It is often argued which unit, in such set of reactions, is responsible for coupling. In a respiratory cycle, such a unit can be identified [6, 7].

Coupling of chemical reactions in an open system makes possible the occurrence of endergonic reactions forbidden in closed systems, because in these reactions the free energy is increased.

Coupling of endergonic processes to the hydrolysis of ATP is of general importance in Biology. Coupling makes it possible for ATP to perform the universal role as a source of free energy required for endergonic processes to occur. If cells and organisms were isolated systems, ATP would not be able to play its role.

A typical example is the following coupled reaction involved in glucose metabolism:

$$\text{Glucose} + H_3PO_4 \rightarrow \text{G-6} - \text{P} + H_2O \qquad (\Delta G^0 = 12.5\,\text{k J/mole})$$
$$\text{ATP} + H_2O \rightarrow \text{ADP} + H_3PO_4 \qquad (\Delta G^0 = -34.1\,\text{kJ/mole})$$

So that the coupled reaction is,

$$\text{Glucose} + \text{ATP} \rightarrow \text{G-6} - \text{P} + \text{ADP}$$

with $\Delta G^0 = -34.1$ kJ/mole and the reaction can proceed spontaneously since net free energy change is negative.

The net coupled oxidative phosphorylation reaction

$$\text{NADH} + H^+ + 3P_1 + 3\text{ADP} + 0.5O_2 \rightarrow \text{NAD}^+ + H_2O + 3\text{ATP}$$

is the sum total of an exergonic reaction

$$\text{NADH} + H^+ + 0.5O_2 \rightarrow \text{NAD}^+ + H_2O(\Delta G = +221\,\text{kJ/mole})$$

and an endergonic reaction

$$3\text{ADP} + 3\,P_1 \rightarrow 3\text{ADP} + 3\,H_2\,O(\Delta G = -92.0\,\text{kJ/mole})$$

15.1.6. Thermodynamics of chemi-osmotic coupling

Oxidative phosphorylation in mitochondria is coupled to ion-transport. This is a system of coupled processes and not an individual chemical reaction. It provides an example of coupling of scalar and vectorial process.

Using the linear thermodynamics of irreversible processes, one can write the phenomenological relations as follows,

$$J_O = L_{11}\, A_O/T + L_{12}\, A_p/T \tag{15.3}$$

$$J_p = L_{21}\, A_O/T + L_{22}\, A_p/T \tag{15.4}$$

where, A_O = affinity of the oxidation reaction;
A_p = affinity of phosphorylation reaction;
J_O = rate of oxidation reaction; and
J_p = rate of phosphorylation reaction.

We can define the extent of coupling q by the relation

$$q = L_{12}/(L_{22}/L_{11})^{1/2} \tag{15.5}$$

When proton movement is taken into account, coupling of three fluxes has to be considered, the entropy production σ would be given by [6]

$$T\,\sigma = J_p\,A_p + J_O\,A_O + J_H\,\Delta\mu_H \tag{15.6}$$

where $\Delta\mu_H = \mu_H$ (entry into the membrane) – μ_H (exit from the membrane), μ_H denotes the chemical potential of H^+.

The above equation is based on Mitchell Chemi–osmotic hypothesis [8] which consider the following:

(a) Phosphorylation occurs only in membrane that contains closed vesicles.
(b) The membrane separates the interior of the vessel from its exterior. It has a low permeability to protons.
(c) It is assumed that electron transfer between the electron-transfer chain is accompanied by proton movement across the membrane.
(d) Carriers are arranged asymmetrically relative to two sides of the membrane.
(e) Primary force is the protomotive force $\Delta\mu_H$. It determines the synthesis of ATP with the aid of a membrane-bound enzyme-ATP synthetase. This enzyme is responsible for creating a channel for the transport of protons through the membrane.

15.2. Transport through biomembranes

15.2.1. Natural membranes

Soil is an example of natural membrane. A soil may be treated as a bundle of capillaries; however, the pores of the soil are not of uniform dimensions and are connected with one another in various ways through bulges and constrictions [9].

Biological membranes are amongst the other natural membranes which are of great interest. Plasma membrane directs the traffic influx and efflux of the cell [10]. The simplest model of the bilayer lipid membrane is illustrated in Fig. 15.2.

Biological membrane is a dynamically organized system, which consists of mainly lipids and proteins. In many cases, membranes are heterogeneous. They contain whole families of phospholipids and lipids. Phospholipids in membranes are glycerol derivatives that have two esterified fatty acids plus a charged side chain joined by a phosphate ester linkage. Phospholipid molecules characteristically contain hydrophobic (water insoluble) and hydrophylic (water soluble) domains. The hydrophobic ends consist of fatty acid chains. The hydrophylic ends are capable of carrying an ionic (electric) charge which acts as polar ends. The polar groups of the phospholipids are oriented towards

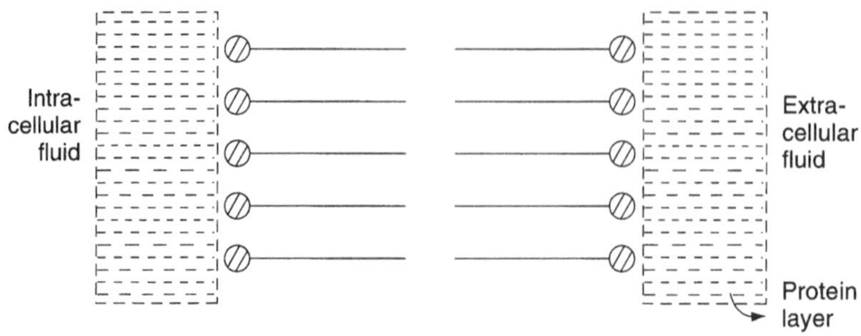

Figure 15.2. A simple model of bilayer lipid membrane.

the protein layers that cover the membrane. Recent studies of a high-resolution structure of a membrane protein reveal intimate contact with the lipid membranes [11].

Biological membranes are supra-molecular systems whose extent in two dimensions considerably exceeds their thickness, which is of the order of 10 nm. The membranes are not passive, semi-permeable shells. They play important roles in the cell. Mitochondrial membranes are much thinner than most cell membranes; their thickness is of the order of 5 nm.

The difference in the concentration of ions between extracellular and intracellular fluid generates a potential difference on the two sides of the membrane, which is called membrane potential. In most of the cases, membrane potential is found to be developed on account of difference in the KCl concentration. Membrane potential for a few typical cases are recorded in Table 15.1.

Biological membranes play a key role in living systems [12–15]. Functions of some typical membranes are indicated in Table 15.2.

Table 15.1. Membrane potentials in biological systems.

System KCl inside/KCl outside		Membrane potential (mv)
Squid nerve axon	19	50–60
Cuttlefish axon	21	62
Frog muscle cell	48	88
Human muscle cell	50	85–100

Table 15.2. Important functions of some biological membranes.

Membrane	Function
Nerve axon	Conduction of nerve impulse
Thylakoid	Conversion of light into energy
Visual receptor	Conversion of light into chemical energy
Mitochondria	Oxidative phosphorylation
Chloroplast	Photosynthetic phosphorylation
Ribosomes	Protein synthesis

15.2.2. Diffusion/permeation mechanism

The following factors play an important role in determining the mechanism of permeation [16]:

(a) Molecular traffic across the membrane moves in both the directions.
(b) There are two classes of solute
  (1) lipid soluble (non-polar); and
  (2) water soluble (polar).
(c) The permeability of lipid-soluble solutes is related to their solubility without any relation to their molecular size.
(d) Water-soluble solutes move through aqueous channels or pores.
(e) Nano pores limit the size of the permeating molecules since the pores are 3–4 A^0 in diameter.
(f) Generally, particles with a molecular weight greater than 200 are excluded from entering the cell by passive diffusion.

The transport phenomena through membranes depends on

(a) the nature of the membrane i.e. size and shape of the capillaries;
(b) size, shape and chemical structure of the permeant molecules; and
(c) nature of the interface.

The following different types of diffusion mechanisms are possible:

(a) *Passive diffusion*: It involves the redistribution of materials by random movement of the particles due to their thermal energy from a higher to lower concentration or electro-chemical gradient.
(b) *Facilitated diffusion*: The solute combines reversibly with a specific molecule in the membrane and the carrier solute complex oscillates between the outer and inner surface of the membrane releasing and building up on either site. The following reaction takes place inside the membrane:

$$\mathrm{C+S \rightleftharpoons CS}$$

where S denotes transported substance, C denotes Carrier and CS denotes complex. The movement is initiated by thermal energy. Membrane proteins that act as specific receptors for a variety of ions, amino acids and sugars are classified as carriers.

(c) *Active transport*: Active transport occurs when solutes are transported against a chemical, electrical or electro-chemical gradient. The cell provides the energy to perform this work by metabolism.

(d) *Group translocation*: The solute is modified chemically. A covalent change is exerted on the molecule, so that the reaction itself results in the passage of the molecule through the diffusion barrier.

It is important to note that active transport is a carrier-mediated transfer from lower to higher electro-chemical potential with the expenditure of metabolic energy, while facilitated diffusion is a carrier-mediated transfer from higher to lower electro-chemical potential without energy expenditure.

15.2.3. Action potential

The role of membrane transport and action potential is very important in the development of self-excited potential (as discussed in Chapter 11) and nerve transmission. For nerve excitation, action potentials play the key role. In the resting state, the polarized or resting muscle cell having a negative surface charge can be represented as in Fig. 15.3(a). When the negative ions migrate to the outer surface of the cell, it is activated or depolarized as indicated in Fig. 15.3(b) and the process is termed "depolarization."

If a stimulus travels through these resting (polarized) cells, the initially activated or depolarized cells will have a negative surface charge, while those not yet activated will have positive surface charge and a current will flow. The rise and fall of potential will follow the pattern shown in Fig. 15.4.

When the cell is stimulated or injured, the negative ions migrate to the surface. There is a relationship between biopotentials and muscle stimulation. When the nerve fibres in the central nervous system are excited, the nerve action potential ensues. This action

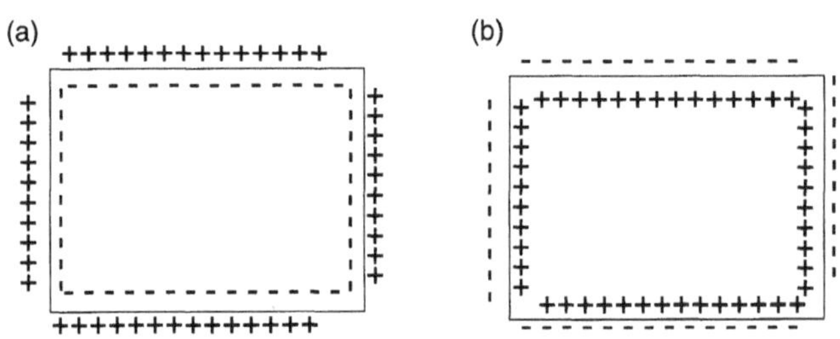

Figure 15.3. (a) Polarised or resting cell; (b) depolarized or activated muscle cell.

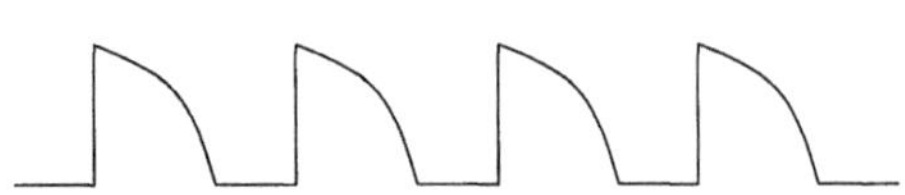

Figure 15.4. Development of action potential.

potential is propagated to the nerve terminal where intercellular transmission across synapse (neuro-muscular junction) occurs.

Transport of ions across the membranes of the cell and organelles is a prerequisite for many of life's processes. Transport often involves very precise selectivity for specific ions. Recently atomic-resolution structures have been determined for channels and pumps that are selective for sodium, potassium, calcium and chloride, for the most abundant ions in biology. These structures provide the desired understanding of the principle of selective ion-transport in terms of the architecture and detailed chemistry of conduction pathways [35].

15.3. Biological rhythm

15.3.1. Biological clocks

Living systems display an innate rhythm at many different levels in adult organisms [17]. Some typical oscillators are mentioned in Table 15.3.

The concept of biological clocks has developed on the basis of day and night circadian periodicity in some characteristic feature of life processes. Biological oscillations having a period close to 24 h, such as change of sleep and awakening, are designated as circadian rhythms. The physiological factor responsible for circadian rhythm lies in auto-oscillatory processes. Their periods are practically independent of temperature and environmental factors. Some typical examples of such oscillatory biochemical reactions are discussed in following sub-sections.

15.3.2. Biological oscillators

Repetitive patterns in biological behaviour are well known. Almost all "organic" processes show certain regularities in a wide variety of species including unicellates, plants, invertebrates, fish-fowl, mammal and man. The circadian rhythm is based on the evolutionary adaptation of the organism to day–night cycle. The time period of biological clocks vary from a few seconds to a few years. Such rhythms have been compared with other naturally occurring time periods in Table 15.4.

Table 15.3. Typical Biological Oscillators and their functions.

Tissue or cell-type	Nature of oscillations	Proposed function
β-cells	Bursts of action potentials	Release of insulin
Anterier pituitary	Action potentials	Release of hormone
Smooth muscle	Slow wave potential changes	Pacemaker activity for Myogenic rhythm
Sino-atrial node	Action potentials	Cardiac contraction

Table 15.4. Time period in different systems.

System	Time period
Atoms/molecules	10^{-15}–10^{-12} (s)
Cells	10^{-2}–10^{2} (s)
Complex organisms	Minutes/years
Population/eco-system	Years/decades
Society/cultures	Hundreds thousands of years
Geological evolution	Hundred thousand–million years
Stars	Billions of years

Recently, Schreiber and Ross have attempted to categorize the following oscillatory enzymatic reactions [18] but in each type of model an autocatalytic and inhibitory step is envisaged.

(a) glycolytic oscillator;
(b) peroxidase–oxidase reaction;
(c) oscillation of cyclic AMP in slime mold cells;
(d) enzymatic pH oscillator; and
(e) calcium spiking in cytosol.

In the following sections, mechanisms of the first two oscillators will be discussed.

15.3.3. Glycolytic oscillator

Glycolytic oscillations are the best-known examples of metabolic oscillations. Glycolysis occurs in yeast cells, cell-free extracts, beef-heart extracts, rat skeletal–muscle extracts and tumour cells. Several reviews on the subject have appeared in the literature [18–22].

Glycolysis involves the multi-step breakdown of glucose into two molecules of pyruvic acid and reduction of NAD to NADH as indicated below

$$C_6H_{12}O_6 + 4NAD + 2ADP + 2P_i \rightarrow 2CH_3CO \cdot COOH + 4NADH + 2ATP$$

Pyruvic acid reacts with NADH in two ways, viz.

$$2H_3\ CCOCOOH + 4NADH \rightarrow 2\ CH_3\ CH\ (OH)\ COOH + 4NAD$$

$$2H_3\ CCOCOOH + 4NAD \rightarrow 2\ C_6H_5\ OH + 2\ CO_2 + 4NAD$$

A simplified scheme of glycolysis has also been formulated [25].

Control mechanism

Phosphofructokinase (F-6-P) catalyzes the quasi-irreversible reaction,

$$ATP + F6P \dashrightarrow ADP + FDP$$

F-6-P plays a key role in glycolysis as an allosteric enzyme whose rate of action is enhanced by its product ADP (autocatalysis) and is inhibited by its substrate ATP. The positive feedback of ADP is represented by a more geberal model

$$\dashrightarrow S \dashrightarrow E \dashrightarrow P \dashrightarrow$$

where S = ATP; E = phosphofructokinase; P = ADP

The enzyme E exists in a number of conformations some of which bind preferentially with ATP. It is activated by AMP also.

The oscillatory characteristics of glycolysis are as follows:

(a) For certain rates of glycolytic substrate injection, oscillations in concentrations of all the metabolites in the chain are observed. The metabolites may vary in a range 10^{-5}–10^{-6} M.
(b) Time period is of the order of minute and depends on temperature.
(c) All glycolytic intermediates oscillate with the same period but with different phases.
(d) The shape, amplitude and time period of oscillation depends on the rate of entry of the substrate.
(e) External disturbances can influence the limit cycle oscillations [40].

The response of allosteric enzyme oscillator to a periodic and to a stochastic input of substrate has been investigated both experimentally [23, 24] and by numerical simulation. Experimentally, it is observed that pulses of ATP or ADP phases shift the oscillations where no such effect is observed with F6P and fructose-1,6-diphophate (FDP). This indicates the important role of adenylate control in the mechanism of glycolytic oscillation [21].

15.3.4. Peroxidase–oxidase (PO) reaction

Considerable attention has been paid to PO reaction, since it has become a model system for the study of complex biochemical dyamics. The peroxide–NADH oscillator has been traditionally studied as an open system of four chemical species combined in a semi-batch reactor. The well-mixed aqueous solution initially consists of native horseradish peroxidase (Per^{3+}), methylene blue (MB^+) and 2-4 dichlorophenol (DCP) to which NADH and oxygen are added at a constant rate.

The PO reaction involves the oxidation of organic electron donors by molecular oxygen catalyzed by horseradish peroxidase (HRP) which also uses peroxide to oxidize substrates in vivo [25].

$$2\ NADH + O_2 + 2\ H^+ \dashrightarrow (HRP)\ 2\ NAD^+ + 2\ H_2O$$

The above reaction is the dominant reaction.

For theoretical/computational studies, the following types of models have been employed:

(a) simple models (typically four variables from existing biological models which have yielded oscillatory behaviour experimentally observed) [26]; and
(b) detailed models based on rate constants of specific reaction steps [27–29].

Three simple models M(I), M(II) and M(III) have been proposed [30] for the PO reaction. The model M(I) is called DOP model, M(II) is a modified DOP model and M(III) is another model. Models (IV) and (V) have also been proposed [31].

Reaction steps	Models in which the step is considered
(1) $A+B+X \rightarrow 2X$	MI, MII, MIV, MV
(2) $B+X \rightarrow 2X$	MIII,
(3) $2X \rightarrow 2Y$	MI, MIII
(4) $X+Y \rightarrow 2Y$	MII, MIV, MV
(5) $A+B+Y \rightarrow 2X$	All five models
(6) $X \rightarrow P$	All five models
(7) $Y \rightarrow Q$	All five models
(8) $X_o \rightarrow X$	MI, MII, MIII
(9) $Y_o \rightarrow Y$	MI, MII, MIV, MV
(10) $A_0 \dashrightarrow A$ (reversible)	All five models
(11) $B_0 \dashrightarrow B$	All five models
(12) $W+B \rightarrow Z$	MIV
(13) $Z+A \rightarrow W$	MIV

where $A = O_2$; $B = NADH$; $X = NAD^*$; $Y = Fe_p^{3+} - O_2^{-*}$ complex; $W = MB^+$;

$Z = MBH$, P and Q are some products. A_0 and B_0 denote the pools of O_2 and NADH, respectively. Model M(V) is more straightforward while the modified M(IV) model takes into account the role of methylene blue. The two models are more reasonable in view of the suggestions of Yamazaki and Yokota [32], who postulated three steps consisting of (1) formation of an inactive product of the enzyme (compound) formed between an active intermediate oxygen radical ion and ferriperoxidase; (2) formation

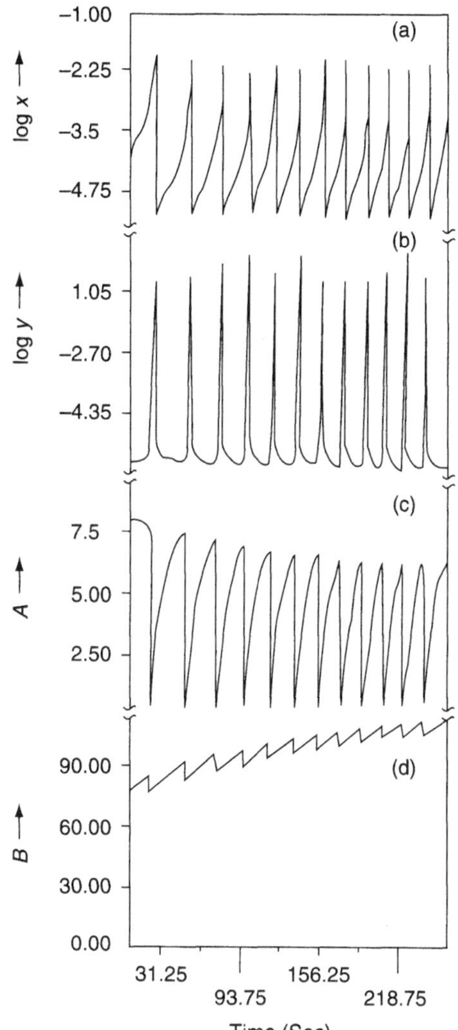

Figure 15.5. (Model IV); (a) Time series in X; (b) Time series in Y; (c) Time series in A (d) Time series in B.

of NAD* by oxidation of NADH; and (3) decomposition of the complex. Computer simulation results for the time series of different variables are recorded in Fig. 15.5 for model IV and Fig. 15.6 for model V using the specific rate constants [31].

It is obvious that several reaction species oscillate during the reaction.

NADPH also undergoes similar oxidation in the presence of enzyme lactoperoxidase [33]. Under a wide range of experimental conditions, the concentration of the

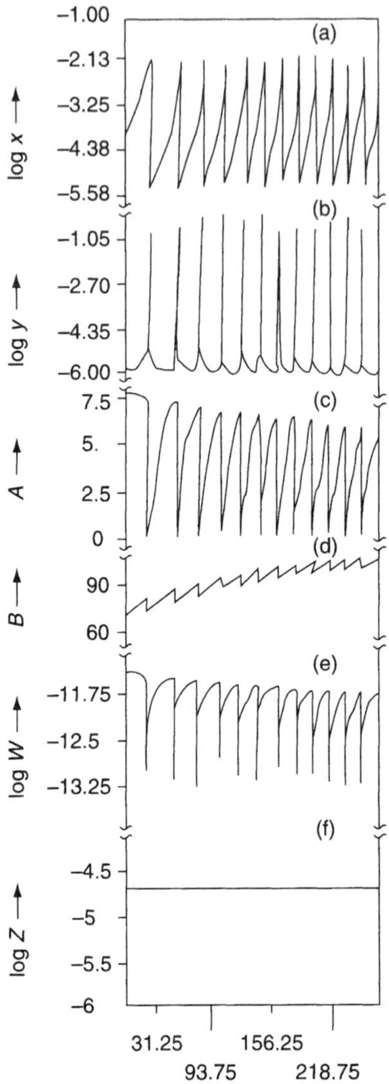

Figure 15.6. (Model V); (a) Time series in X; (b) Time series in Y; (c) Time series in A; (d) Time series in B; (e) Time series in W; (f) Time series in Z.

reactants (Oxygen and NADH) as well as some oxygen intermediates have been found to oscillate with periods ranging from several minutes to one hour, depending on experimental conditions.

Damped oscillatory behaviour in preparations made directly from horseradish root extract have also been reported [34].

The reaction exhibits exotic non-equilibrium phenomena such as:

(a) bistability between two oxygen steady states;
(b) bistability between a steady state and an oscillatory state;
(c) periodicity and intermittency; and
(d) chaos.

15.3.5. Self-excited oscillations: application to physiology

Self-excited oscillations are extremely important from the viewpoint of Electro-physiology and nerve transmission. As an example, sense of taste, smell and injury are related to such self-excited oscillations [36]. Some of the examples of self-excitation which we have come across in previous chapters are

(a) streaming potential (Chapter 4) generated by the flow of fluid through membrane;
(b) sedimentation potential (Chapter 5) generated by the sedimentation of particles by gravity;
(c) Precipitation and dissolution potential (Chapter 6) generated by the precipitation and dissolution of particular electrolyte;
(d) electric potential generated by the density difference between two fluids (Chapter 7);
(e) potential generated at a liquid–liquid interface under specific conditions (Chapter 11);
(f) potential generated at liquid–vapour interface under specific conditions (Chapter 11); and
(g) membrane potential.

The above potentials have various applications in electro-physiology as pointed out in Chapter 11 on dynamic instability at interfaces.

In the context of sensing mechanism of smell involving liquid–vapour interface, [37] the concept of density oscillator has relevance for the sensing mechanism of taste as shown by Srivastava and co-workers [38–39]. The following facts are relevant in this context:

(a) Insoluble substances have no taste.
(b) The sour taste is caused by acids.
(c) The salty taste is caused by ionizable salts. Cations of the salt are mainly responsible for the salty taste.
(d) The sweet taste in most cases is caused by organic chemicals
(e) The bitter taste is caused by (i) long-chain organic chemicals containing nitrogen; and ii) alkaloids.

Taste is sensed through taste buds numbering 10 000 in human beings. The taste bud is composed of ~40 modified epithelial cells. Each taste cell has microville or taste hairs of length ~2–3 × 10–3 m and width ~1–2 × 10–4 m, which are projected in taste pore.

Density difference of saliva and denser solution of taste material can generate electric potential oscillations, but for sensing the specific taste, the role of zeta potential (See Chapter 11) may be more important which would depend on the structure of taste cell. Perhaps, the taste bud having requisite zeta potential would correspond to a particular taste.

Precipitation potentials discussed in Chapter 6 are relevant in the context of injury potential [40]. It has been pointed out that precipitation of the electrolyte occurs at the negative electrode. In human body, the injured part of the protoplasmic structure is electrically negative with respect to the uninjured part. The sensation of injury seems to be associated with the clotting process, which involves the conversion of a soluble plasma protein, fibrinogen into an insoluble protein fibroin. When a blood vessel is damaged, the extrinsic clotting mechanism operates and fibrils of fibrin adhere to the injured vessel to form a clot.

15.3.6. Chaos and fractals in physiology

Blood vessels of the heart exhibit fractal-like branching. The large vessels branch into smaller vessels, which in turn branch into smaller and smaller capillaries. Ultimately, they become so narrow that blood cells are forced to flow through a single file. The fractal character of bacterial growth has already been pointed out in Chapter 12 on fractals.

Chaotic features are quite common in human physiology. Interval between heartbeats varies chaotically. Complex oscillatory features can be easily observed in ECG and EEG. It has been pointed out that chaos in body signals health and periodic behaviour can foreshadow disease [41].

Non-linear dynamics in sudden cardiac death syndrome: Heart rate oscillations and bifurcations [42, 43] is of great relevance, although sustained temporal oscillations are observed in glycolysis and circadian rhythm.

It has been suggested that chaotic systems might be controllable with small perturbations. Techniques based on such approach have been successfully applied to mechanical systems, electrical circuits, lasers and chemical reactions.

15.4. Concluding remarks

In view of the complexity in biological systems, different theoretical techniques have to be applied for different non-equilibrium regions. The range of applicability of linear thermodynamics of irreversible processes is quite limited on account of very thin membranes, nano-pores and large size of bio-molecules. However, the physical concepts of fluxes and forces are still useful for the non-linear region for comprehending the basic features of non-equilibrium phenomena.

Fluxes can be non-linear as pointed out in Chapter 7. Beyond non-linear steady-state range, non-linear dynamics as well as non-linear kinetics can be applied profitably for investigating the far from equilibrium phenomena, using the concepts based on the corresponding studies of physico-chemical phenomena as discussed in Parts Two to Four.

References

1. I. Prigogine, Thermodynamics of Irreversible Processes, John Wiley, New York, 1961, p. 27.
2. R.P. Rastogi, R.C. Srivastava and K. Singh, Trans. Faraday Soc., 61 (1965) 854.
3. R.P. Rastogi and R. Sabd, J. Non-equilib. Thermodyn., 6 (1981) 207–216.
4. R.H. Allen, T. Alfrey, L.D. Yats, J. Amer. Chem. Soc., 81 (1959) 42.
5. D.J.G. Ives, R.H. Kerlogue, J. Chem. Soc., 33 (1940) 1362.
6. B.S. Meyer, D.B. Anderson, Plant Physiology, D.Van Nostrand Corp., New York, 1963 p. 424.
7. A.G. Marshall, Biophysical Chemistry, John Wiley and Sons, New York, 1978.
8. M.V. Volkenstein, Biophysics, Mir Publishers, Moscow, 1982.
9. R.P. Rastogi, J. Sci. Ind. Res., 35 (1976) 64.
10. N. Lakshminarayaniah, Transport Phenomena through Membranes, Academic Press New York, 1969.
11. A.G. Lee, Nature, 438 (2005) 569.
12. H.T. Tien, Bilayer Lipid Membranes, Marcel Dekker, New York, 1974.
13. P.S. Nobel, Introduction to Biophysical Plant Physiology, Freeman International Edition, Tokyo, 1974.
14. M.K. Jain, The Bimolecular Lipid Membranes, Van Nostrand Company, New York, 1972.
15. A. Vander, J.H. Sherman and D.S.Luciano, Human Physiology, Tata-McGraw Hill Book Company, New Delhi, 1975.
16. B.A. Scholtelius and D.D. Scholtelius, Text Book of Physiology, Seventeenth Edition, The C.V. Mosby Company, 1973.
17. A.R. Peacocke, An introduction to the Physical Chemistry of Biological Organization, Clarendon Press, Oxford, 1983, p. 181.
18. I. Schreiber and J. Ross, J. Phys. Chem., 100 (1996) 8556.
19. B. Hess and A. Boiteux, Ann. Rev. Biochem., 40 (1971) 237.
20. A. Goldbeter and R. Caplan, Ann. Rev. Biophys. Bioeng., 5 (976) 449.
21. G. Nicolis and I. Prigogine, Self-Organization in Non-Equilibrium Systems, John Wiley and Sons, 1977, Ch. 14.
22. B. Chance, E. K. Pye, A. K. Ghosh and B. Hess (eds), Biological and Biochemical Oscillators, Academic Press, New York, 1973, Ch. III.
23. M.V. Volkenstein, Biophysics, Mir Publishers, Moscow 1983, Ch. 16.
24. A. Boiteux, A. Goldbeter and B. Hess, Proc. Nat. Acad. Sci., 72 (1975) 3829.
25. A. Goldbeter and G. Nicolis, Progr. Theor. Biol. Academic. NY, 4 (1976) 65.

26. R. Larter, J. Phys. Chem. B, 107 (2003) 415–429.
27. L.F. Olsen, Phys. Lett. A, 94 (1983) 454.
28. B.D. Aguda and R. J. Larter, Amer. Chem Soc., 112 (1990) 2167; 113 (1991) 7913.
29. D.L. Olsen, C.P. Wilksen and A. J. Scheeline, Am. Chem. Soc., 117 (1995) 2.
30. T.V. Bronnikova, V.R. Fed'kina, W.M. Schaffer and L.F. Olsen, J. Phys. Chem., 99 (1995) 9309.
31. L.F. Olsen, Kinetics of Physico-Chemical oscillations, Discussion Meeting Pentsche Busen-geseltscaft fur Physikalische Chemie Aachen, Germany, Preprints, III (1979), 704.
32. R.P. Rastogi, G. Srinivas and M. Das, Indian J. Chem., 33A (1994) 750.
33. I. Yamazaki and K. Yokota, in Biological and Biochemical Oscillators, B. Chance, P. E. Kendall, A.K. Ghosh and B. Hess (eds), Academic Press, New York, 1973, pp. 109–114.
34. S. Nakamura, K. Yokata and I. Yamazaki, Nature, 222 (1969) 794.
35. A.C. Moller, M.J.B. Hauser and F. Olsen, Biophys. Chem., 72 (1998) 63.
36. E. Gonaux and M. Redreck, Science, 310 (2005) 1461.
37. R.C. Srivastava and R.P. Rastogi, Transport Mediated by Electrified Interfaces, Elsevier Amsterdam, 2003.
38. R.C. Srivastava, A.K. Das, S. Upadhyay and V.Agarwala, Indian J. Biochem. Biophy. 33 (1996) 1195.
39. R.C. Srivastava, A.K. Das, S. Upadhyay and V. Agarwala, Indian J. Chem., 33A (1994) 978.
40. U. Roy, S.K. Saha, C.R. Krishnapriya, V. Jayshree and R.C. Srivastava, Instru. Sci. Technol., 31 (2003) 425.
41. A.C. Guyton, Text Book Medical Physiology, Saunders, Philadelphia, 1981, p. 230.
42. A.L. Goldberger, D.R. Rigney and B.J. West, Chaos and Fractals in Physiology, Scientific American, February 1990, p. 36.
43. A.L. Goldberger, D.R. Rigney, J. Mietu, E.M. Antman and S. Greenwald, Experentia, 44 (1988) 983–987.

EPILOGUE

1. Introduction

One of the strengths of thermodynamics is that it does not depend on any model. Philosophy of science involved in non-equilibrium thermodynamics coupled with appropriate experiments under controlled conditions can easily be used for understanding the behaviour of natural systems like living state, socio-political and socio-economic systems. Whereas in case of natural systems, variables cannot be controlled and one has to depend on proper analysis of statistical data; in laboratory experiments particularly in physico-chemical systems, analogous experiments corresponding to specific conditions can be designed. By Mathematical modeling and mathematical interpretation, reasonable predictability is assured in case of real systems based on proper mathematical modelling of analogous situation in model experiments.

In the present book, an attempt has been made to combine theory and experiments to interpret a variety of situations in the entire spectrum of non-equilibrium region.

In the natural systems which are open and in non-equilibrium, flux force or cause–effect relationship controls the entire scenario. Flows and counter flows in the opposite directions dominate the scene. Three cases can arise:

(a) there is no counter flow
(b) there is one set of flow and counter flow
(c) network of flows and counter flows autocatalytic (positive feedback) and inhibitory reaction (negative feedback).

In the case of decomposition there is no counter flow. Case (b) corresponds to the transport phenomena discussed in Chapters 3–6. In more complex systems exhibiting complex phenomena in the non-linear range, there can be network of flows and counter flows including autocatalytic, cross-catalytic and inhibitory reactions as it happens in the case of complex chemical reactions.

It may be noted that, in the case of chemical equilibria, both forward reaction and back reaction simultaneously operate at microscopic level, but not at macroscopic level. On the other hand, steady state is established both at the microscopic level and at the macroscopic level.

In the earlier stages, detailed and comprehensive theoretical and corresponding experimental studies for testing the non-equilibrium thermodynamics formalism have been undertaken for simple phenomena involving coupling of two fluxes and two corresponding forces for the sake of simplicity in the region close to equilibrium which generated steady state, which clearly bring forth the philosophical foundations of the new science of non-equilibrium. These phenomena involve coupling of forces such as (i) forces like temperature difference (or gradient) and pressure difference or concentration difference (or gradient) (ii) forces like potential difference (or gradient) or pressure difference (or gradient) (iii) gravitational force and potential difference (Chapter 6). Case (i) has been investigated for a single system (Chapter 5) as well as for a system containing two subsystems (Chapter 3). The studies confirmed the validity of (i) linear phenomenological relations, (ii) Onsager reciprocity relation and (iii) steady-state relations in the linear range for a wide range in case of a number of phenomena.

Let us try to examine these phenomena from the angle of causality principle. Taking the example of thermo-osmosis (Chapter 3), temperature difference is the starting cause, the effect of which is thermo-osmotic fluid flow, which in turn generates another cause, viz. pressure difference under specific circumstances (e.g. experimental set-up), the effect of which is hydrodynamic fluid flow in the reverse direction. Normally, both these causes and effects operate simultaneously. However, when two opposing flows are balanced, a steady state is reached. Similar type of situation occurs in other steady-state phenomena discussed in Chapters 4–6 including mechano-caloric effect.

Subsequent efforts have been made to extend the domain of validity of non-equilibrium thermodynamics by incorporating newer concepts of extended irreversible thermodynamics (EIT) and non-equilibrium molecular dynamics (NEMD). However, these are limited applicability as discussed in Appendices I–III of Part One.

2. Beyond linear region

Experimental studies on steady states in the non-linear region, discussed in Chapter 7, demonstrate that fluxes can be represented as power series in fluxes and forces. Onsager reciprocity relation between first-order coefficients is found to be satisfied. Second-order coefficients are found to be functions of forces which influence the membrane characteristics in case of electro-osmotic phenomena.

3. Bifurcation phenomena

In the non-linear region, exotic non-equilibrium phenomena such as bifurcation from steady state to bistability and oscillations (in time and space) are observed. New mathematical methodology based on non-linear dynamics has been evolved for investigating such phenomena as discussed in Chapter 8. The essential point to note is that such complex phenomena arise when more than three forces are evolved in addition to the occurrence of multi-process as discussed in Parts Two and Three. Bifurcation process can be understood in mathematical terms, provided bifurcation parameter can be identified using appropriate mechanism. In case of bifurcation from steady state to oscillations in continuously stirred tank reactor, flow rate of the reactant in the reaction chamber is the bifurcation parameter. In case of both bistability and oscillations in chemically reacting systems, the essential requirement is tug-of-war between autocatalytic reaction step and an inhibitory reaction step.

4. Temporal oscillations

Chemical reaction network is a typical example of complexity, where the reactants can interact in a variety of ways depending on the nature of interaction (chemical as well as non-chemical). Oscillatory reactions involve a number of steps, including positive and negative feedbacks. The complexity leads to periodic as well as aperiodic oscillations (multi-periodic, bursting/intermittency sequential oscillations separated by a time pause, relaxation and chaotic oscillations). The mechanism is usually determined by non-linear kinetics and computer modelling. Once the reaction mechanism has been postulated, the non-linear time-dependent kinetic equation can be formulated in terms of concentrations of different reactants, which would yield a multi-variable equation. Delay differential equations are sometimes used to characterize oscillatory behaviour as in economics (Chapter 14).

In most of the cases, one comes across single control mechanism involving one set of autocatalytic and inhibitory reaction. However, in some cases, dual control mechanism is operative as discussed in Chapter 9.

In quite a few cases, chaos can be predicted based on appropriate mathematical model (Lorenz or Rosselor) (Chapter 12). The strange attractor obtained in such cases signifies "Deterministic Chaos". Such models exhibit cross-catalysis and inhibition. Experimentally observed deterministic chaos can be easily ascertained using definite criteria, although prediction of complex time order is quite difficult in many cases. In terms of causality principle, the complexity in the system arises due to complex interdependence of a number of causes and effects.

Quite complicated electric potential oscillations are observed in several phenomena involving (i) liquid–liquid, (ii) solid–liquid, (iii) solid–liquid and liquid–liquid and (iv) vapour–liquid interface (Chapter 11). Modelling has often been accomplished using Van der pol equation, a typical non-linear equation.

5. Spatio-temporal oscillations (oscillations in time as well as space)

Such oscillations are observed in the form of chemical waves or stationary structures (Turing type) in an oscillatory system (Chapter 10). The relative magnitudes of diffusion coefficients of activator and inhibitor are found to be controlling factors in several models. In case of B–Z oscillator, spatio-temporal oscillations are found to be governed by a typical non-linear reaction-diffusion equation. Appropriate experiments yield stationary structures of varying types depending on experimental conditions (Chapter 10). Satisfactory mathematical modelling has been attempted.

Diffusion plays an important role in the development of cities and their fractal nature. Similarly, diffusion of culture controls many socio-political and socio-economic changes.

6. Fractal growth

Fractal-type growth is very common in natural systems such as clouds, coast lines and demography of cities. This is also observable in living systems such as blood vessels and bacterial colonies. Experimentally, it is observable in the physico-chemical systems during crystal growth or electro-deposition. There is order in such cases despite apparent disorder. This has only been possible by recent development of fractal geometry. The structures can be characterized in terms of fractal dimensions (Chapter 13). Mathematical modelling using stochastic considerations (eg. Diffusion limited aggregation model) has been developed, which provides deep insight into growth mechanism.

7. Real systems

Phenomena analogous to physico-chemical phenomena are easily observed in social systems and biological systems as discussed in Chapters 14 and 15. Importance of above studies in the above context is obvious. As a sub-system gets more and more open, new types of self-organized (self-regulated) structures emerge. Furthermore, it has to be noted that philosophical basis of non-equilibrium science can serve as a bridge between basis sciences and social sciences.

Successive bifurcation leading to self-organization yielding a variety of structures is the order of the day, thereby the real systems are generating different ranges of complexity in different spheres of human activity. Forces generated by observable and unobservable factors lead to different types of flows where coupling occurs in multivariable system involving multi-processes. The real challenge is the identification if these forces and how semi-quantitatively these interact. Unbiased assessment of causes (variables) is quite important from the viewpoint of decision-making, administrative efficiency and leadership qualities.

8. Conclusion

Knowledge in the above areas is advancing despite the handicaps. This century is going to be the Century of Science of Complexity. There is continuous tug-of-war between animal instinct and human instinct. Popper's remark "We are not to abandon the search for universal laws and for a coherent theoretical system, nor give up our attempts to explain causality of events we can ascribe" is quite relevant in this context.

Appendix I

DOMAIN OF VALIDITY OF GIBBS EQUATION

1. Introduction

The validity of Gibbs equation is extended by assuming its validity for small volume element by replacing extensive variables as follows:

S by $s(x, t)$ = entropy per unit volume V i.e. $S = sV$
U by $u(x, t)$ = energy per unit volume i.e. $U = uV$
Nk by $nk(x, t)$ = mole number of reactant k per unit volume i.e. $Nk = nkV$

It may be noted that entropy s is not a function of gradients in this formalism [1].
Gradients are taken into consideration in extended irreversible thermodynamics as discussed in Appendix II.

2. Justification of assumption of local equilibrium

From statistical mechanics, it follows that temperature is well defined when the velocity distribution is Maxwellian. Systems for which this condition is fulfilled are complex reactions where the rate of elastic collisions is larger than the rate of reactive collisions. This is generally true for reactions in not too rarefied media and for many biological and transport processes. It may be noted that molecular collisions are responsible for attainment of Maxwellian distribution. Normally, significant deviations from the Maxwellian distribution are observed only under extreme conditions. Distribution is perturbed when physical processes are very rapid. Thus for a gas, local equilibrium assumption would not be valid when the relative variation of temperature is no longer small within a length equal to mean free path.

Fluctuations are caused by the motion of molecules. Molecular motion is responsible for dissipative processes such as diffusion, heat transport and chemical reactions. Fluctuations in the number of particles in a definite volume play a key role in maintaining equilibrium. Fluctuation in the number of particles N^- in a volume ΔV is given by $\delta N^-/N^-$ For a gas under normal conditions, this is of the order 4×10^{-7} when $\Delta V = (1\,\mu\text{m})^3$. For liquids and solids, the same value of fluctuation will correspond to even smaller volume. Consequently, the number density with position on the micrometer scale would be very nearly uniform for most macroscopic systems.

The above conditions, however, are not fulfilled by other processes such as shock waves, highly rarefied gases. Many biological systems have gradients of density, energy density etc. over a distance of 10 Å (such as material near or in a membrane or a cell nucleus) which would not have true local equilibrium.

3. Results of experimental studies

When the gradients are small, theoretical results based on linear thermodynamics of irreversible processes agree well with experimental results. However, when the gradients of intensive variables are large, there are deviations as would be clear from the discussion of experimental studies in Chapters 3–5. The exact domain of applicability of LET and useful information for non-equilibrium region close to equilibrium can be easily ascertained from experimental data as discussed in these chapters.

In chemical systems where N is of the order of Avogadro number, the fluctuations in energy and density would be negligible. In a dilute solution containing two components, the root mean square fluctuation (proportional to $1/(N^{-1/2})$ where N is the number of molecules or particles in a particular volume or sample) would be inversely proportional to number density of solute molecules [2]. The number of molecules inside the cell is much smaller which influences root mean square fluctuation as indicated in Table A1 [3].

Table A1. Root mean square fluctuation in concentration in intracellular fluids.

Species	Approximate no. in cell	rms fluctuation in concentration
Inorganic ions	2.5×10^8	6.4×10^{-5}
Carbohydrates	2.0×10^8	7.0×10^{-5}
Proteins	10^6	10^{-3}
Small molecules	1.5×10^7	2.5×10^{-4}
DNA	4	0.44

4. Recent developments

Recently, a non-equilibrium statistical thermodynamic theory based on stochastic kinetics has been formulated which has been applied to isothermal non-equilibrium steady state for biological systems [4]. Rate equations in terms of the probabilities of enzyme concentration are used instead of concentration. Expressions for the Gibbs free energy and entropy for the isothermal system are obtained in terms of dynamic cyclic reaction.

References

1. D. Kondepudi and I. Prigogine, Modern Thermodynamics: From Heat Engines to Dissipative Structures, John Wiley and Sons, New York, 1998, Ch. 15.
2. N. Davidson, Statistical Mechanics, Mc-Graw Hill, New York, 1962.
3. I.D. Watson, Molecular Biology of Genes, W.A. Benzamine, New York, 1966.
4. H. Qian, J. Phys. Chem. B, 110 (2006) 15063.

Appendix II

EXTENDED IRREVERSIBLE THERMODYNAMICS

Complex systems such as solutions of macromolecules, magnetic hysteresis bodies, visco-elastic fluids, polarizable media require some extra variables in the fundamental equation of Gibbs. Dissipative fluxes (heat, diffusion, viscous pressure tensor and viscous pressure) are included in the Gibbs function in new formalism. In the formalism of extended irreversible thermodynamics (EIT), the dissipative fluxes are the independent variables in addition to classical variables of thermostatics [1].

For simpler phenomena such as thermo-osmosis, electro-kinetic phenomena, thermal diffusion and Dufour effect, the linear thermodynamics of irreversible processes is valid in a wide range as indicated by the experimental results discussed in Chapters 3–5. It may be noted that Onsager relations for thermal diffusion can be proved by EIT [2].

Beyond the linear range, it has been shown by Rastogi and Misra [3] that the non-linear flux equations for mass and heat transport involve non-linear terms containing forces corresponding to the anti-symmetric part of the viscous pressure tensor and the force corresponding to the divergent part of the viscous tensor responsible for bulk viscosity as discussed in Chapter 6. In case of spatially non-uniform systems, a proper expression for Gibbs function should take into account the existence of physical forces as indicated above.

However, there are a host of non-equilibrium situations that demand to look beyond local equilibrium assumption (LEA) and bring them in the thermodynamic fold. The logical solution of this demand is provided by the EIT that is now of about four and half decades standing [4, 5]. The EIT heavily leans on the indicative result offered by the kinetic theory of non-uniform gases. The earliest efforts are those of Grad [6], that for this purpose additional variables are needed. Thus the following functional dependence of an entropy like function, η (which in the limit of LEA reduce to thermodynamic entropy s), is adopted. Namely:

$$\eta = \eta(u, \rho; \chi_1, \chi_2, \chi_3, \ldots, \chi_n) \tag{1}$$

where χ_i's are the required additional thermodynamic independent variables. It is also assumed that η is a continuous and differentiable function. Further, η is required to follow the following balance equation, namely

$$\rho(\mathrm{d}\eta/\mathrm{d}t) + \mathrm{div}\ J_\eta = \sigma_\eta \tag{2}$$

where J_η is the flux density of η and σ_η is its source strength (a scalar quantity). The compatibility requirement is them imposed on Eq. (2) so that when in certain non-equilibrium situations χ_i's become irrelevent it reduces to Clausius–Duhem inequality (for thermodynamic entropy) [7] namely

$$(\mathrm{d}s/\mathrm{d}t) + \mathrm{div}\ J_s = \sigma_s \geq 0 \tag{3}$$

the equation used in classical irreversible thermodynamics.

The most difficult task is to identify the physical quantities that comprise χ_i's. Several researchers [1–6] have considered η itself as non-equilibrium entropy and in the case of a single component fluid, the physical fluxes of heat and dissipative momentum are taken as χ_i's. In this context, Eq. (1) explicitly reads as,

$$\eta = \eta(u, \rho, q, \pi) \tag{4}$$

Two distinct non-equilibrium situations arise in this context i.e., (i) spatially uniform systems in non-equilibrium; and (ii) spatially non-uniform systems. These are discussed below [8]. If the system has the source of irreversibility only in chemical conversions at finite rates, then G function has the following expression, namely

$$G = \Sigma_k \mu_k x_k \tag{5}$$

where μ_k = chemical potential per unit mass of the component k and x_k is its mass fraction.

Let us note that no restriction has been imposed on the extent of irreversibility in chemical conversions at finite rate and also the complexities of reactions. In other words, far away from equilibrium, chemical conversions and complex reactions such as chemical oscillations, chemical chaos, diffusion-controlled reactions having multi-steps etc. are governed by the above equation provided the system is spatially uniform,

In case of spatially non-uniform systems, a proper expression for G should take into account the existence of physical fluxes such as q, π and J_k so that there are additional terms in the expression for G as expressed by the following equation:

$$G = \Sigma_k \mu_k x_k + \mu_q q + \mu_k : \pi + \Sigma_k \mu_k J_k \tag{6}$$

where,

J_k = flux density of component k;
μ_q = chemical potential corresponding to flux q;
μ_k = chemical potential corresponding to flux J_k;
π = dissipative stress tensor; and
μ_π = chemical potential corresponding to dissipation stress tensor.

The last four are the vectorial and tensorial quantities.

Thus, in the formalism of extended irreversible thermodynamics by inclusion of gradients in the basic formalism of linear thermodynamics of irreversible processes, a small correction in the local entropy due to flow appears. With this modification, the new formalism can be applied to the systems such as shock waves, where there are large gradients.

References

1. S. Steniutyez and P. Salmon, Advances in Thermodynamics, Vol. 7, Extended Thermodynmics Systems, Taylor and Francis, New York, 1992.
2. I. Miller and T. Ruggeri, Extended Thermodynamics, Springer-Verlag, Berlin, 1993.
3. R.P. Rastogi and B.P. Misra, J. Phys Chem., 74 (1970) 112.
4. D. Juo, J. Casas-Va'zuez and G. Lebon, Extended Irreversible Thermodynamics, Springer-verlag, Berlin, 1996.
5. D. Juo, J. Casas-Va'zuez and G. Lebon, Rep. Prog. Phys., 62 (1999) 1035–1142.
6. H. Grad, Principles of the Kinetic Theory of Gases, in Handbuch der Physik, S. Fluge (ed.), Band XII, Springer-Verlag, Berlin,1958, pp. 205–294.
7. W. Muschik, Aspects of Non-Equilibrium Thermodynamics, World Scientific, Singapore, 1990.
8. A.A. Bhalekar, Bull. Cal. Math. Soc. 94 (3) (2002) 209–224.

Appendix III

NON-EQUILIBRIUM MOLECULAR DYNAMICS

1. Introduction

Earlier theoretical approaches were concerned with isolated systems that could exchange neither matter nor energy with the surroundings. Theories are developed to organize, describe, and perhaps even predict observations. At the simplest level, theories merely provide relations between observables, such as ideal gas law. At the next level of complexity, we have theories that relate observables to underlying state, e.g. kinetic theory which shows how the observable temperature is related to state through molecular velocities. Such theories at this level can only provide interpretations or explanation of the observables. But if the state itself is unobservable, such theories cannot be used for computing numerical values for the observables [1, 2].

Two strategies have been adopted to counter this constraint. First, higher level theories are developed by reorganizing and reducing detailed information about the state needed to compute the values of the observables. Statistical mechanics follows this course in which observables are not related to underlying state itself, but rather to the probability of the system being in particular states [3–6]. The second alternative usually adopted is molecular dynamics.

The primary aim of molecular dynamics is to numerically solve the N-body problem of classical mechanics. Molecular dynamics methods are used for used for simulating molecular-scale models of matter in order to relate collective dynamics to single-particle dynamics. Typical situations for its application are self-assembly of structures, such as micelles and vesicles.

Molecular dynamics assigns numerical values to states, thereby making states observable, at least for model substances. With numerical values assigned to states, theoretical relations from kinetic theories to compute values by simulation for experimentally accessible observables. Thus, molecular dynamics is more closely related to kinetic theory compared to statistical mechanics. Molecular dynamics, in a sense, is less sophisticated but more direct than statistical mechanics.

2. Molecular scale simulation

This consists of three steps:

(1) construction of a model;

(2) calculation of Molecular trajectories; and
(3) analysis of the trajectories to obtain values of specific properties.

The positions are obtained by numerically solving differential equations. Hence, these positions are connected in time. The positions reveal real dynamics of individual molecules. In other simulation methods, the molecular positions are not temporarily related. In other simulation methods, such as Monte Carlo simulations, the positions are generated stochastically such that a molecular configuration depends only on the previous configuration [7–10].

3. Limitations of molecular dynamics

Molecular dynamics simulations are limited largely by the speed and storage constraints of available computers. Hence, usually simulations are done on systems containing 100–1000 particles although calculations involving 10^6 particles have also been performed [11]. Hence, molecular dynamics approach has the following limitations:

(a) On account of size limitation, simulations are confined to systems of particles that interact with relatively short-range forces, i.e. intermolecular forces should be small when the molecules are separated by a distance equal to half of the smallest overall dimension of the system.
(b) Due to speed limitation, simulations are confined to investigation of relatively short-lived phenomena, those occurring in approximately less than 100–1000 ps. The characteristic relaxation time for the phenomenon under investigation must be small enough, so that one simulation generates several relaxation times.

4. Equilibrium molecular dynamics

Equilibrium molecular dynamics is typically applied to an isolated system containing a fixed number of molecules N in a fixed volume V. Because the system is isolated, the total energy E is also constant. E is the sum of the molecular kinetic and potential energies. Hence, the variable V, N and E determine the thermodynamic state.

5. Non-equilibrium molecular dynamics

In addition to equilibrium molecular dynamics, non-equilibrium methods have also been developed. These methods were originally developed [12–14] as an alternative to equilibrium simulations, for computing transport coefficients. In these methods, an external force is applied to establish the non-equilibrium situation of interest. The system's response to the force is then determined from simulation. Non-equilibrium molecular dynamics has been used to obtain shear viscosity, bulk viscosity, thermal conductivity and diffusion coefficients.

6. Non-equilibrium molecular dynamics simulation (NEMD)

Non-equilibrium molecular dynamics simulations of transport processes can be done in two ways.

The phenomenological coefficients describing the heat transfer and other properties can be expressed in terms of time-correlation functions by using Kubo formalism [15]. The main advantage of this approach is the generality of the resulting formulas, which can be applied in any density regime. However, the calculation of time-correlation function requires simulation times. The computation times require longer time than the characteristic relaxation time of the correlation itself, which makes them computationally expensive.

Using NEMD, it is possible to create stationary non-equilibrium states using temperature gradients produced by placing the material between two heat reservoirs at fixed temperatures. A computational experiment can them be performed to determine the thermal conductivity of the material using Tully's classical trajectory method. The method is computationally less expensive and more accurate than linear response technique (LRT) 4 since it deals with the signal itself instead of its average fluctuations in the equilibrium state.

7. Applications

NEMD has been employed for investigation of transport properties and models for non-equilibrium steady states. It has also been used for obtaining formal solution of Liouville's equation.

We discuss below some recent applications.

Blair *et al.* compared a NEMD simulation of squalene (a low molecular weight fluid with experimental measurements) in the non-linear (Non-Newtonian) regime. Predicted results are in good agreement with experiments [16]. Extensive simulations of transport of superfluid mixture through a carbon nano-pore, in the presence of an external pressure gradient have been reported. A pore-packing phenomenon occurs in which several layers of the fluid fill the pore, if the downstream pressure is large enough [17]. Using Fourier's law, the thermal conductivity of solid material at different temperatures has been computed at different temperatures using NEMD. The method correctly describes the variation of thermal conductivity with temperature [18]. Generalized Langevin-based technique can be used for generating stationary equilibrium thermodynamic states. This ensures the validity of the phenomenological laws, e.g. Fouriier'law (uncoupled flow). The linearity of the temperature gradients obtained indicates that local equilibrium hypothesis holds for the bulk of the material.

Computation of shear viscosity of hard spheres has been attempted using NEMD [11]. Modified non-equilibrium molecular dynamics methods have also been developed for study of fluid flows with energy conservation [12]. NEMD simulations have also been recently performed to compare and contrast the Poiseuille and Electro-osmotic flow situations. Viscosity profiles obtained from the two types of flows are found to be in good mutual agreement at all locations. The simulation results show that both type of flows conform to continuum transport theories except in the first monolayer of the fluid at the pore wall. The simulations further confirm the existence of enhanced transport rates in the first layer of the fluid in both the cases [13, 14].

References

1. S. Chapman and T. Cowling, Mathematical Theory of Non-uniform Gases, Cambridge University Press, 1939 (2nd edition 1952).
2. J.O. Hirschfelder, C.F. Curtiss and R.B. Bird, Molecular Theory of Gases and Liquids, Wiley, New York, 1984.
3. I. Prigogine , Non-Equilibrium Statistical Mechanics, Wiley Interscience, New York, 1963.
4. R. Balescu, Equilibrium and Non-Equilibrium Statistical Mechanics, John Wiley and Sons, New York, 1975.
5. K. Lindenberg and B.J. West, Equilibrium Statistical Mechanics of Open and closed systems, Wiley, 1990.
6. B.C. Eu, Non-Equilibrium Statistical Mechanics, Ensemble Method, Academic Kluwer Publishers, Dordrecht, 1998.
7. W.G. Hoover, Annu. Rev. Phys. Chem., 34 (1983) 103–127.
8. J.M. Haile, Molecular Dynamics Simulation: Elementary Methods, John Wiley and Sons, New York, 1992.
9. J.C. Tully, Acc. Chem. Res. 14 (1981) 188, J. Chem. Phys., 73 (1980) 1975.
10. S. Voltz and G. Chen, Phys. B, 263–264 (1991) 709–712.
11. J.J. Erpenbeck, Non-Equilibrium Molecular Dynamics Calculations of the Shear Viscosity of Hard spheres in Non-polar Fluid behaviour, H.J.M. Hanley, (ed), North-Holland, Amsterdam, 1983.
12. M.E. Tuckerman, C.J. Mundy, S. Balasubramanian and M.L. Klein, J. Chem. Phys., 106 (1997) 5615.
13. A.P. Thompson, J. Chem. Phys., 119 (2003) 7503.
14. Y. Demirel, Transport and Rate Processes in Physical and Biological Systems, Elsevier, Amsterdam, 2002.
15. D.A. McQuarrie, Statistical Mechanics, Harper and Row, New York, 1976.
16. S. Blair, C. McCabe and P.T. Cummings, Phys. Rev. Lett., 88 (2002) 58302.
17. M. Firouzi, T.T. Tsotsis and M. Sahimi, J. Chem. Phys., 119 (2003) 6810.
18. H.J. Castejon, J. Phys. Chem. B, 107 (2003) 826.

INDEX

P

Q

R

S

T

V

W

Y

MIX
Papier | Fördert gute Waldnutzung
FSC® C083411

Kolibri 360 GmbH
Ferdinand-Jühlke-Straße 7
99095 Erfurt, Deutschland
produktsicherheit@kolibri360.de